URBAN FUTURES

Urban Futures brings together commentaries from a wide range of contemporary disciplines and fields relevant to urban culture, form and society. The book concerns cities in the broadest sense, not just as buildings and spaces, but also as processes and events or sites of occupation, in which meanings are constructed in many ways. The contributors draw on their specialist areas of research to inform current debate, but they also speculate as to how cities will be shaped in the twenty-first century.

Not all the contributors share the same viewpoint, but they are all concerned with underlying issues such as social inclusion, cultural expression and the contingency of readings of the city. Each chapter is a free-standing text, but the twelve contributions have been grouped to enable cross-reference and comparison. The book is structured into four sections: framing, moving, practising and shaping. Chapters cover aspects of visual art, architecture, transport, open and public space, ecologies and technologies; and introduce understandings around gender, ethnicity and engagement.

Specific areas of research include homeless people's organisations and restoration ecology in brownfield sites in the USA, post-industrial urban landscapes, post-industrial economics, tourism and cultural planning. The book allows each writer to state their own conclusions, but together they suggest that tomorrow's cities will, while remaining locations of difference and contestation, be rapidly evolving systems in which dwellers assume increasing responsibilities and power.

Malcolm Miles is Reader in Cultural Theory at the University of Plymouth. **Tim Hall** is Senior Lecturer in Human Geography at the University of Gloucestershire.

URBAN FUTURES

Critical commentaries on shaping the city

edited by
Malcolm Miles
and Tim Hall

LONDON AND NEW YORK

First published 2003
by Routledge
11 New Fetter Lane, London EC4P 4EE

Simultaneously published in the USA and Canada
by Routledge
29 West 35th Street, New York, NY 10001

Routledge is an imprint of the Taylor & Francis Group

Typeset in Sabon and Univers by
Florence Production Ltd, Stoodleigh, Devon

Printed and bound in Great Britain by
Bell & Bain Ltd, Glasgow

British Library Cataloguing in Publication Data
A catalogue record for this book is available from the British Library

Library of Congress Cataloging in Publication Data
A catalog record for this book has been requested

ISBN 0–415–26693–9 (hbk)
ISBN 0–415–26694–7 (pbk)

CONTENTS

List of Illustrations vii
Notes on Contributors ix
General Introduction 1

I FRAMING

Introduction 10

1 JANE RENDELL — Where the Thinking Stops, Time Crystallises. . . 13

2 DOROTHY ROWE — Differencing the City: Urban Identities and the Spatial Imagination 27

3 MALCOLM MILES — Strange Days 44

II MOVING

Introduction 62

4 STEVEN MILES — Resistance or Security? Young People and the 'Appropriation' of Urban, Cultural and Consumer Space 65

5 PATRICK LOFTMAN AND BRENDAN NEVIN — Prestige Projects, City Centre Restructuring and Social Exclusion: Taking the Long-Term View 76

6 TIM HALL — *Car*-ceral Cities: Social Geographies of Everyday Urban Mobility 92

III PRACTISING

Introduction 108

7 IAIN BORDEN — What is Radical Architecture? 111

8 PATRICIA PHILLIPS — Public Art: A Renewable Resource 122

9 TIM COLLINS AND REIKO GOTO — Landscape, Ecology, Art and Change 134

IV SHAPING

Introduction 146

10 ANNE R. ROSCHELLE AND TALMADGE WRIGHT — Gentrification and Social Exclusion: Spatial Policing and Homeless Activist Responses in the San Francisco Bay Area 149

11 GRAEME EVANS AND JO FOORD — Shaping the Cultural Landscape: Local Regeneration Effects 167

12 NIGEL CLARK — Turbulent Prospects: Sustaining Urbanism on a Dynamic Planet 182

Conclusion 195

Bibliography 201

Index 223

ILLUSTRATIONS

PLATES

Part I *Frontispiece* Broken columns in front of a converted warehouse, Manchester, UK 9

1.1 Rut Blees Luxemburg, *Caliban Towers I and II* (1998), under a bridge in Shoreditch High Street 15
1.2 Rachel Whiteread, *Holocaust Memorial*, Vienna (1995) 23
2.1 *Islam Now* (1995), by Falkirk Asian Women's Group, Falkirk, Scotland 32
2.2 *Hope* (1995), by Hopscotch Asian Women's Group, Camden, London 34
2.3 *Hemmed in Two* (2000), by Hew Locke 36
2.4 *Hemmed in Two*, a detail, by Hew Locke 37
2.5 *Hemmed in Two*, a detail, by Hew Locke 37
2.6 *Metropolis* (2001), by Lubna Chowdhary 38
2.7 *Metropolis*, a detail, by Lubna Chowdhary 39

Part II *Frontispiece* A loft shop sign, Manchester, UK 61

Part III *Frontispiece* The informal building of pigeon lofts, Skinningrove, the Cleveland coast, UK 107

Part IV *Frontispiece* Reading newspapers, Lisbon, Portugal 145

FIGURE

9.1 Restoration ecology: systems, disciplines, focus and goals 138

TABLES

5.1 Profile of Highbury Initiative Symposia (1988 and 1989) participants 80
5.2 Changes in tenure (number of properties) in Birmingham city centre 1991–7 88
6.1 The relationship between transport and urban form, based on Muller (1995) and Hart (2001) 95

CONTRIBUTORS

Iain Borden is Director of the Bartlett School of Architecture, University College London.

Nigel Clark is a Lecturer in Human Geography at the Open University.

Tim Collins is a Research Fellow at the Studio for Creative Inquiry, School of Fine Arts, Carnegie Mellon University.

Graeme Evans is Director of Research at Central Saint Martin's College of Art and Design.

Jo Foord is Senior Lecturer in Urban Studies at the Department of Social Studies, University of North London.

Reiko Goto is a Research Fellow at the Studio for Creative Inquiry, School of Fine Arts, Carnegie Mellon University.

Tim Hall is Senior Lecturer in the Geography and Environmental Management Research Unit at the University of Gloucestershire.

Patrick Loftman is a Senior Lecturer in the Centre for Public Policy and Urban Change, University of Central England in Birmingham.

Malcolm Miles is Reader in Cultural Theory in the Faculty of Arts and Education, University of Plymouth.

Steven Miles is Principal Lecturer in the Department of Sociology at the University of Plymouth.

Brendan Nevin is based in the Centre for Urban and Regional Studies, University of Birmingham.

Patricia Phillips is Dean of Creative Arts at the State University of New York (SUNY), New Paltz.

Jane Rendell is Lecturer in Architectural Theory at the Bartlett School of Architecture, University College London.

Anne R. Roschelle is Assistant Professor of Sociology at the State University of New York (SUNY), New Paltz.

Dorothy Rowe teaches art history at the University of Surrey, Roehampton Institute.

Talmadge Wright is Associate Professor of Sociology at Loyola University of Chicago.

GENERAL INTRODUCTION

This general introduction attempts to say what kind of book this is. It begins by outlining the book's aims, scope and organisation. Since each of its four parts has its own more detailed introduction only a brief note on the chapters they contain is given here. The major part of this introduction is then taken up in an effort to describe the terrain in which the book is situated.

AIMS

The book has two main aims: first, to bring together perspectives from a range of subjects which share a focus on contemporary urban questions, so that together the twelve chapters take discussion beyond the limitations of any single discipline; and, second, by juxtaposing material which shows differing ways in which cities are shaped, to emphasise that cities are more than mere accumulations of forms and spaces, and that the processes of their future determination are open to change as well as investigation. The relation between reflection, theoretical framing, and the collation of data varies from one chapter to another, but all, it is intended, emphasise that cities are made not given, are products of human thought and action, their futures open to speculation which embodies critical understanding.

Several chapters deal with culture, either in the specialist sense of cultural production – as in art and media – or in the broader sense, established in cultural studies since the 1960s, of expression of a set of values in the acts of everyday life. In that sense, which tends to be all-pervasive, thought and action are moulded by the culture or cultures in which they are situated. But culture, too, is produced through thought and action in a continuous and simultaneous reciprocity of shaping and being shaped. Cities likewise carry the evidence of past conceptualisations of the city while endlessly modifying them in daily use and occupation. Just as the form of a language *is* its mundane usages as well as its rules, and both adapt according to circumstance, so cities correspond to ideas of citizenship and history, to power structures, and to predominant spatial practices. But the point is again that circumstances can be changed, are not simply given, but are a space for intervention.

SCOPE AND ORGANISATION

As a collection of essays, some more theoretical, others more empirical but all contemporary in focus, on how the futures of cities might be shaped in the early twenty-first century, the book makes no claim to forecast social or cultural trends, nor to prescribe what cities should be like. There has been no committee to sift the findings, and neither have the authors been asked to revise their chapters in the light of others in the book. While this precludes the continuity of a more managed script, such as a set of conference papers on a single theme, it allows a diversity of attitude and voice which may be more provocative. Having said that, the scope of the book is to interrogate the conditions and processes of change in contemporary cities in the affluent world of Europe and North America. This could be seen as a limitation when the highest rates of urban growth are in cities such as Lagos, Manila and São Paolo, yet the book's scope is already widely stretched and there is elsewhere a growing literature in liberation ecology and development studies which shows what the affluent world might learn from the non-affluent world. This, of course, does not preclude in due course another book to complement this one, looking further afield. This book, however, does recognise the voices of the marginalised within the affluent world.

The organisation of the book in four parts places beside each other texts which share some point of reference or departure, such as contemplation of an image or a story, or concern with the development of a practice. Beyond that the chapters diverge in material, method and position, as well as voice. The presentation of different kinds of enquiry in proximity, it is hoped, will lead the reader to question her or his assumptions as well as those of the writers.

The book's four parts are headed framing, moving, practising and shaping. This loosely charts a path from reflecting on stories of a city to its everyday mobilities, to its construction, and to questions as to whose city it is, in whose voices its story can be told, and by what paradigm coloured. The three chapters in Part I – framing – hint at this broader organisation: Jane Rendell reflects on an angelic, interstitial space between past and future; Dorothy Rowe looks at the emergence of difference in visual culture; and Malcolm Miles comments on dystopian narratives to suggest alternative trajectories. In Part II – moving – Steven Miles looks at the spatialities of young people's urban consumption, asking if they are rebellious or complicit, or if they negotiate a position between the two; Patrick Loftman and Brendan Nevin deal with models of city centre redevelopment in relation to social exclusion and mobility; and Tim Hall asks how urban futures are moulded by the car, taking mobility as a key factor of urban experience. The practices reviewed in Part III are architecture, by Iain Borden, who sees its future as being outside its conventional boundary; public art, brought by Patricia Phillips into a radical framework in relation to public space; and restoration ecology, adopted by artists Tim Collins and Reiko Goto in an attempt to integrate the agendas of public space and bio-diversity. Part IV – shaping – begins with a chapter by Anne Roschelle and Talmadge Wright on the impact of homeless people's organisations on urban change in San Francisco; this is followed by an inquest by Graeme Evans and Jo Foord on urban regeneration, also based on empirical work; and Nigel Clark asks how sustainability might be advanced by seeing the world not as a site of stable systems, as perhaps was the dream once of planning, but as a constantly evolving and dynamic turbulence. The end of utopia? Perhaps, in an ending of expectations for a static world which retains the perfection of its first design. Finally, a brief conclusion, in that spirit of a messy world, does not resolve the questions raised nor seek to synthesise the arguments, but offers questions for a future dialogue.

CHANGING CITIES, SHIFTING DISCOURSES

This introduction turns now to the terrain in which the book is situated. Without being a history of urbanism, it sets out some points of reference on a map, seeing the map as articulated by questions which influence urban decision-making and are encountered in disciplines such as architecture, planning, geography, sociology, ecology, and cultural studies. For example, there is a divergence between scientific rationalism – as in the Chicago School's work in the 1920s and 1930s, in which cities are described rather than viewed critically and from which planning facilitates a given form of progress – and an equally rational but cultural critique in which all, including the direction and ownership of progress, is open to change through conscious intervention. As stated above, the book tends to the latter position, but is a product of a situation in which both approaches are still evident. Where approaches intersect tends to be in trans-disciplinary discussion. But how effective is that in opening possibilities for new ways of facilitating change?

Although planners, architects, sociologists, geographers and cultural critics, among others, continue to operate within conventional subject boundaries and professional organisations, and thus retain conventional methodologies, they also meet in conferences and trans-disciplinary academic courses in which cities – as social, economic, cultural and political as well as built entities – are the focus. This, together with new models of operation such as action or radical planning has contributed to new understandings of cities as events and processes of becoming, and to new agendas recognising a diverse range of urban publics and their needs. Enquiries into the conditions of urban living, too, recognise a potential to move beyond the kinds of insight available within a single academic or professional field. This extends rather than contradicts the already broad approaches found in urban sociology a century ago, notably in Georg Simmel's analysis of the new conditions of a metropolitan environment. And if the optimism of the modern dream – of engineering a new and better society by design – has gone, its failure adds weight to the view that urban problems are multi-layered and complex, and require to be addressed in ways no less intricate. Drawing on research in architecture, cultural planning, cultural studies, ecology, geography, and sociology, this book is a contribution to a critical, trans-disciplinary field. These disciplines have different methods of enquiry but share a concern today with how cities are or have been shaped. Some are based in direct investigations of cultural expression while others, such as geography and sociology, look to cultural production and reception as indicative of social currents; similarly, cultural planning denotes a culturation of the planning process when cultural activities are seen to contribute to city marketing as well as a politics of identity.

This, however, raises a question as to what constitutes culture. Is it art in the museum, graffiti, drinking tea, the decoration of a room, or the tending of an allotment? Beyond that is another question as to whether urban cultures differ not only from each other but also from the rural or suburban. And then, a question as to how far the urban reaches: the notion of a countryside, for instance, becomes urban when it is seen less as a wild stretch of land outside a city and more as an agricultural production zone in the city's footprint of consumption, or a leisure zone into which urban dwellers make weekend forays. Culture is understood in this book inclusively, as making evident a set of values in everyday acts and occupations of space, through which the urban is characterised, as well as in cultural production in media such as photography and film, much of which tends to be concentrated in urban centres. One problem with such a definition is that it includes everything; but the way in which people act differs, indicating the culture specific to a social group. Culture, then, is a site of identity formation, and within any society will be multiple and mutable. While strata of the high, the low, and the intermediate can be seen in particular media – like classical music, folk music and jazz – conventionally associated with social class though not without connotations of gender, ethnicity and age, such multiple categories may be more evident in studies of the ways in which daily life is conducted. Because those ways are tactical and negotiated, cultural analysis reflects a state of society but equally opens possibilities to foresee future states other than the given.

The idea of culture as a way of life owes much to Raymond Williams. In *Culture* (1981), Williams argues that the sociology of culture in the mid twentieth century achieved a convergence between an idealist position discerning the ethos of a society from its cultural objects, and a materialist position explaining cultural forms in terms of socio-economic organisation. For Williams, the advantage of this is that it leads to the insight that cultural practices are elements in social constitution. As he explains, culture is a signifying system through which a social order is communicated, experienced, investigated and reproduced – an agent of change as well as a way to understand the past (Williams, 1981: 12–13).

Culture, then, as a shaping force, is a recurrent interest in this book. But the book is situated, too, in the field of urbanism and this involves more than cultural analysis. Urbanism is defined by a concern with urban living, and with conceptualisations of the city. It is also largely a response to modern and metropolitan cities. While similar methodologies in sociology or geography can be applied in studies of non-urban situations, early urban sociology, as in the Chicago School, revolves around the question of what is unique in city life. The conditions and scale of cities in the industrialised countries by the beginning of the twentieth century were, of course, new, and followed mass immigration as well as technological innovation. New publics experienced new modes of transport and forms of energy – the tram, the electric light – and new feelings of speed and anonymity. Not surprisingly, much commentary on cities, whether lamenting the conditions of the poor or celebrating city freedoms, emphasises that to live in a large city feels radically unlike living in a town or a village. Whether that is still the case, as cities devolve into urban villages and people in remote rural locations communicate globally through the net, is an open question, but cities such as New York, Paris and Berlin retain a mystique of cosmopolitanism.

The study of urban lives and spaces, as distinct, began in part at least in sociology, but through the twentieth century the field has widened to include the work of geographers, historians, economists, cultural critics, and ecologists as well. Among the texts which establish the field is Georg Simmel's essay of a century ago, 'The metropolis and mental life' (1903), which itself links sociology with psychology (an emerging science at the time). Simmel writes:

> The individual has become a mere cog in an enormous organization of things and powers . . . the metropolis is the genuine arena of this culture which outgrows all personal life. Here in buildings and educational institutions, in the wonders and comforts of space-conquering technology, in the formations of community life, and in the visible institutions of the state, is offered such an overwhelming fullness of crystallized and impersonalized spirit that the personality, so to speak, cannot maintain itself under its impact. . . . This results in the individual's summoning the utmost in uniqueness and particularization.
>
> (Simmel, 1903)

Hence the characterisation of the metropolitan citizen as blasé, inured to so much sensation, retreating into a liberating anonymity.

James Donald, in *Imagining the Modern City* (1999), interprets Simmel's essay as stating ambivalence between defensiveness in face of experiences which threaten to overwhelm, and affirmation of the city as providing 'the texture of modern experience and the fabric of modern liberty' (Donald, 1999: 71). The latter position is taken by Elizabeth Wilson (1991), seeing metropolitan environments as 'spaces for face to face contact of amazing variety and richness' (Wilson, 1991: 158), which are particularly liberating for women. It could be added that the freedom not to know one's neighbours or their business, and be known by them, but to participate in groups of common interest across a city, is liberating for all. The metropolitan then becomes the cosmopolitan, the site of a mixing which kindles freedom from convention and constraint, which allows difference not least because difference is already ubiquitous, becoming a threat only in the suburb. The metropolitan can also, however, be romanticised. Frequently – for Jane Jacobs (1961), for instance – the informal mixing of citizens which makes a democratic city is identified with a public realm of streets and gathering places. These open sites have generally been gendered, men's spaces, and Donald, citing Simmel's 1904 essay on fashion, notes that Modernist architecture – as in the domestic architecture of Adolf Loos – in any case redirects attention away from views outside to the privacy of the house interior, as windows are screened and seating faces inwards (Donald, 1999: 107–9). If life takes place indoors, what price then the notion of the city as site of free exchange?

The shift from the optimism of a modern urban outlook as expressed by Jacobs to a present of security and surveillance against a backdrop of privatisation of space and exclusion of the poor from visibility (Davis, 1990) is exemplified by what is called, particularly in North America, new urbanism. This is the built form of neo-liberalism, epitomised by the Disney Corporation's move into real estate at Celebration, Florida:

> Using the photographs and measurements . . . assembled from thirty Southern towns and cities, the Celebration developers created a contemporary pattern book to provide the town's builders with strict interpretations of the design and planning principles required to create a cohesive community. . . . But the *Celebration pattern book* was fun to read, too, with its pages and pages of line drawings and details.
>
> (Frantz and Collins, 1999: 64)

Fun, perhaps, for some but also nostalgic, and restrictive when the choice of shrubbery and colour of curtains is regulated by a code to which residents are required to sign up. None of Simmel's kaleidoscopic city here, only uniformity and a safeness which denies the threat perceived in difference in a town which is universally sub-urban.

Between Simmel's risky but creative city of montage and Celebration's homogenised façades is the work of the Chicago School. Informed by Simmel, the Chicago sociologists investigated specifically urban forms and types. Perceiving major cities as a new social laboratory, they borrowed terms from the natural sciences. The classic case of this is the concentric-ring diagram proposed by E.W. Burgess (1925) to explain the effects of continuing waves of migration in the production of transitional zones in Chicago. Burgess, making a self-contained image as value-free as modern art, drew his diagram without reference to any features of the built or natural landscape, and used a biological model of growth to naturalise conditions of conflict. He also assumed the centrality of a business district (CBD) based on Chicago's Loop, a description which became prescription as the diagram was widely reproduced as a model for any city. James Duncan argues (1996: 257) that Burgess used natural science to lend his work legitimation. But because biology is beyond history, its use in urban theory makes change explicable but closed to intervention. The extent to which the diagram has been influential is, however, disputed: David Sibley calls it 'a universal statement' obscuring other views (Sibley, 1995: 127) and Duncan suggests that without the diagram the paper would have fallen into obscurity (1996: 256); but Mike Savage and Alan Warde comment that while most writers on urban sociology mention it they do so 'nearly always to show that the subject of interest to them could not be fitted easily into its dimensions' (Savage and Warde, 1993: 16).

In contrast to Burgess' reductive generalisation is the work of the Frankfurt Institute for Social Research (or Frankfurt School, re-established at Columbia University after 1935 as the International Institute for Social Research), since extended by Jürgen Habermas and Susan Buck-Morss. The position of the Frankfurt School is from the outset politicised and haunted by the rise of fascism following the failure of revolution in Germany after 1918. It seeks insights into future possibilities for change through social research and cultural criticism, and begins by questioning the assumptions on which theory is based. Max Horkheimer established a foundation for critical theory in an inaugural lecture as Director of the Institute in 1931, 'The present situation of social philosophy and the tasks of an institute for social research', and in 'Traditional and critical theory' published in the Institute's journal, *Zeitschrift für Sozialforschung* in 1937. Seeing knowledge as socially produced he writes of an 'interpretation of the vicissitudes of human fate – the fate of humans not as mere individuals, however, but as members of a community' to argue that the material of philosophy 'can only be understood in the context of human social life: with the state, law, economy, religion – in short with the entire material and intellectual culture of humanity' (Horkheimer, 1993: 1). The limitations of scientific philosophy are its isolation of the object of study from the world of which it is a part, and its isolation of the disciplines through which the world is known. Social research, in contrast, views social organisation through multi-disciplinary enquiry involving philosophy, sociology, economics, history and psychology (Horkheimer, 1993: 9). Critical theory differs, too, from scientific theory in not manipulating reality in predictable ways but seeking liberation through critical reflection. Interaction between disciplines promotes understanding which, like psychoanalysis, identifies patterns of repetition to offer, in that recognition, release from its burden.

The map of urban theory at the opening of the twenty-first century includes several further strands of thought, from Foucault's interrogation of power to more recent work on globalisation. It is dangerous to try to generalise from this, but perhaps one issue which recurs is that of the balance between regulation and a free market, not only in trade but in planning. Wilson, for instance, opposes the conventional authoritarianism of the planning process but argues there remains a need for regulation in the public interest, calling for a renewal of planning to produce a city for all. Leonie Sandercock, too, argues in *Towards Cosmopolis* (1998a) for a radical revision of planning for multi-ethnic cities. These approaches are an antithesis of Burgess' prescriptive theories, and allied to them are David Byrne's application of complexity theory to urban policy (Byrne, 1997), and Iris Marion Young's refutation of the liberal notion of assimilation in favour of recognition of group difference (Young, 1990). Cities, then, are increasingly perceived as sites of difference, the intricacies of which can no more be understood through monolithic methods of enquiry than their futures can be mapped by diagrams. People do not occupy space according to neat formulations but negotiate their tactics as desire urges and environment allows. Ernesto Laclau (1996) proposes that what is viable for democratic societies today is a negotiated position between freedom and unfreedom, and Wilson charts a not dissimilar path between living in a capitalist society and resisting excess:

> To deny that there are any pleasures to be had in capitalist society is to depart too radically from the perceptions of most of those around us. It is only in recognising and analysing those pleasures and their source that we can develop a view of something that might be better.
>
> (Wilson, 2001: 71)

These approaches share engagement in a situation of shifting contingencies with which it remains possible, is perhaps a responsibility, to work.

The book's contributors were not required to adopt any such position, yet the book is characterised by acknowledgement that urban futures will be achieved in ways which are practical and material as well as reflective and critical. Its juxtaposition of a range of critiques states implicitly that mono-disciplinary approaches are inadequate, a view in which the editors claim no originality: Manfredo Tafuri argues that radical change will not happen within the conventions of architecture:

> It is useless to propose purely architectural alternatives. The search for an alternative within the structures that condition the very character of architectural design is indeed an obvious contradiction of terms.
>
> (Tafuri, 1976: 181)

If so, then urbanism, as a discourse in reciprocal relation to the disciplines and practices which contribute to it, is more than a set of inter-disciplinary conversations. Such conversations are not without interest, and sociologists, geographers and architects have recently found a shared interest in spatial practices, all influenced by Henri Lefebvre (1991b), but if all that is broached is the comparison of methodologies rather than a new criticality then the result is academic tourism. Urbanism, then, needs to be a trans-disciplinary terrain in which the problems of city living are reconsidered. And if urban conditions consist in ecologies of complex relation, these relations are not conflict-free. As Michael Peter Smith writes in *Transnational Urbanism*:

> People's intentions and actions, and the meanings given to them, are derived from human experience. As positioned subjects, occupying multiple social locations, people often experience inner tensions and conflicts in forming their own sense of agency . . .

But also, as he continues:

> Despite the difficulties involved in sorting out all the untidy contingencies involved in moving from acting subjectively to conscious human agency, the quest for an agency-oriented urban research and practice is nonetheless preferable to resolutely maintaining a global gaze which focuses our consciousness disproportionately upon the global economy, reified as a pre-given 'thing' existing outside of thought.
>
> (Smith, 2001: 6)

The implication is that it will be in the actions of dwellers that new cities, the architectures of which are social as much as built, will take shape. It will be untidy, and perhaps demand a new and less neatly packaged, though still critically coherent, literature to support it.

PART I
*F*raming

Broken columns in front of a converted warehouse, Manchester, UK (Photo: Malcolm Miles)

INTRODUCTION

This part includes three critical reflections on images or texts which serve as points of departure for a wider reconsideration of change in cultural and urban histories. Each, then, begins from a study of specific imaginaries as manifest in visual or verbal cultures, in order to investigate questions of temporality and spatiality, of difference, and of narrative, and how these are perceived in ways which themselves convey the tensions and fluidity of urban conditions and psyches.

Jane Rendell cites Walter Benjamin on a painting by Paul Klee, *Angelus Novus* (1928, Israel Museum, Jerusalem): an angel, emblem of thresholds, caught at the threshold of time, merging past and future in the instant. Benjamin writes of this image in his 'Theses on the Philosophy of History', completed in 1940 (Benjamin, 1992: 255–66), the first passage of which offers an image of an automaton playing chess, always winning but its strings in fact pulled by a hidden, hunchback expert player. He likens the automaton to historical materialism, the philosophy of determination by circumstance which Marx criticises as lacking a principle of active intervention to make the dialectic. Part of Benjamin's intention is to reintroduce, in but also beyond dialectical materialism, a sense of immanent (pervasive) rather than imminent redemption: 'only a redeemed mankind receives the fullness of its past' (passage III). This is in the face of the technologically achieved devastation of the 1914–18 war in Europe and the rise of fascism; the state of emergency is 'not the exception but the rule' and the task in hand 'to bring about a real state of emergency' against fascism (passage VIII).

For Rendell, reading Benjamin viewing Klee, the moment at which time is stayed is the point of dialectical contradiction, the argument stilled and paradoxically eased into motion in the tension of its polarities. Rendell sees in Benjamin's work a creative tension between sub-title and content, repositioning the new in the old, for which the emancipatory angel is a metaphorical figure between theory and practice.

In 1922, Benjamin took the name *Angelus Novus* for a proposed journal, though it was seen by the publisher as commercially non-viable and never appeared. And he was forced a few years later to withdraw, or have failed, his *Habilitationsschrift* (the final doctoral thesis required in Germany for an academic career) which the examiners found incomprehensible (Gilloch, 2002: 15). Its subject was German seventeenth-century tragic drama (*Trauerspiel*), in which for Benjamin history is likened to a process of natural decay, represented in this genre as fragmentation, ruination and mortification (Gilloch, 2002: 15). This is close to elements of his (much later) theses on history, which Richard Wolin in *Walter Benjamin: An Aesthetic of Redemption*, singling out the passage on the angel, summarises as depicting 'a vast heap of ruins that grows incessantly higher with the passage of time' yet which is lit from the future by 'a redemptive hope' (Wolin, 1994: 61). Wolin adds that such an understanding can be communicated only in metaphorical terms. Esther Leslie, too, in *Walter Benjamin: Overpowering Conformism*, sees the angel as standing on ruins, 'staring in half-disbelief at the ruins, devastated by the failure to co-operate, made manifest in the sheer destructive capacity of technological progress' (Leslie, 2000: 7). Benjamin's doctoral study was eventually published in 1928 – now translated in English as *The Origin of German Tragic Drama*. In it he refutes criticism as (definitive) interpretation, which assumes a power relation, and instead proposes an immersion in the material on which reflection takes place, a heightening of consciousness in which the truth of the work speaks and the work itself is extended, again through the oblique medium of metaphor. Rendell, adapting categories used by Benjamin, describes her text as angelic in form, situated between subjectivity and objectivity, between critical and creative writing. Its form is, unlike any other essay in this book, that of a creative rather than simply discursive practice, structured as ten meditations on images of emptiness followed by a reiteration of the first heading as coda. Those images – for instances the photographs of Rut Blees Luxemburg and Catherine Yass, or Tacita Dean's work on abandoned (because out-dated) technologies of surveillance – are in keeping with Benjamin's understanding of history as ruin, but also as what we have.

Dorothy Rowe, in another way, uses an intersection of text and visual imagery, from Zadie Smith's novel *White Teeth* (2000) and the work of artists Sonia Boyce, Zarina Bhimji, Sutapa Biswas, and others. Rowe adapts Griselda Pollock's idea of differencing the canon of art history, mapping categories of race and gender as well as class onto contemporary cultural production, specifically in London. Drawing, too, on David Sibley's *Geographies of Exclusion* (1995), Rowe adds that difference is not only a position constructed in marginality but potentially an enabling one as well, a making of visibility, literally and metaphorically, for difference. This would accord with the refusal of assimilation in favour of group identity formation for which Iris Marion Young argues (1990), as referenced by Malcolm Miles in the third chapter in this section. The argument for a dispersal of conventional structures of power through a reconfiguration of the centre–margin model, then, as a multiplicity of sites in contending, contesting and contingent relation but in which the concept of hierarchy is deconstructed through positive recognition of difference, occurs in both cultural and political stories. As Rowe states, the transgression of boundaries is a tangible point of assertiveness.

Rowe cites the exhibition *Shamania: The Mughal Tent* (1991, Victoria and Albert Museum, London) devised by Shireen Akbar as addressing the cultural and educational issues facing British South Asian women and children. This draws in another kind of threshold, that of mass migration, which is sometimes seen as specific to late twentieth-century globalisation. While isolation and loneliness are frequently recurring images in this collective textile work, Rowe notes, too, the type of an imagined homeland as characteristic of a diasporic culture. This is not a nostalgic notion, but problematises questions such as where is home, how the site of domesticity and everyday life is transported yet always somewhere else, something other (not least from the dominant culture). Rowe agrees with Anthony King's (2000) case that urban studies needs a language in which to address otherness, arguing that it is in visual culture's representations of difference that this can be found.

In the third chapter, Malcolm Miles offers a critical reconsideration of the disaster scenario as informing and perhaps coercing urban trajectories. Drawing on Edward Soja's (2000) critique of Mike Davis' well-known writing, he notes the invisibility for Davis of resistance, the absence of voices from margins, and the overwhelming gloom of an analysis positively identifying contradictions in neo-liberalism's happy fiction but subsumed in an apocalyptic myth which has its own adrenalin-producing attractions. There may be a cold war on the streets, but its representation goes uncomfortably close to the genre of the (men's) war comic. And if narratives influence how the plot evolves, and what parts people might play in it, then it seems that the construction of narratives, rather than their pronouncement as interpretations of the world for others, is a necessary site of intervention.

Miles draws on postmodern and feminist geography (Massey, 1994, for instance), as well as radical planning (Sandercock, 1998a) and political science (Young, 1990), to piece together something of an alternative set of positions. Thinking back to Rendell's use of Benjamin, and Benjamin's redemptive history read back, as it were, from the end, the disaster scenario in urban writing, both academic and fictional, posits a bleak absence of any such hope; neither the cold war on the streets nor the threat of toxic conflagration or civil disintegration promises any lamp to illuminate the present's ruin. In contrast, accepting a kind of non-integration in which difference is seen as a category in planning and political life, and through assertion of identity formation, the immanence of Benjamin's critical project might be translated into, in part, a reconfiguration of the idea of a public sphere. That is, into a location not so much of that informal mingling of citizens projected perhaps nostalgically onto spaces such as the commons (in North American history) but more a mental and communicative space in which the subject's self-understanding derives from its perceptions of and by others – at root an Arendtian concept (1958) but one which can be enhanced through postmodern frames of difference, contiguity and recognition of the transformative aspects of everyday lives.

1 JANE RENDELL

Where the Thinking Stops, Time Crystallises . . .

I THE ANGEL OF HISTORY

> A Klee painting named *Angelus Novus* shows an angel looking as though he is about to move away from something he is fixedly contemplating. His eyes are staring, his mouth is open, his wings are spread. This is how one pictures the angel of history. His face is turned toward the past. Where we perceive a chain of events, he sees one single catastrophe which keeps piling wreckage upon wreckage and hurls it in front of his feet. The angel would like to stay, awaken the dead, and make whole what has been smashed. But a storm is blowing from paradise: it has got caught in his wings with such violence that the angel can no longer close them. This storm irresistibly propels him into the future to which his back is turned, while the pile of debris before him grows skyward. This storm is what we call progress.
>
> (Benjamin, 1992: 259)

Starting with Walter Benjamin's comments on Paul Klee's *Angelus Novus*, this chapter looks at the present as a place between past and future, a place where past actions and future intentions meet. *Angelus Novus* shows an angel caught at the threshold of time, allowing the coming together of the past and the future in one instant. Occupying a threshold position in space, time and consciousness: between history and myth, between dream and awakening, between antiquity and modernity, Klee's angel was a key image for Benjamin. The figure of the angel is dialectics at a stand-still – a frozen moment encapsulating dialectical contradiction. At this moment, the present is allowed access to the past. Here the past and the future, the 'has-been' and the 'not-yet', come together in a single configuration creating, for Benjamin, a 'monad'.

In *One-Way Street*, Benjamin created dialectical tensions in his own work; he played on the juxtaposition of subtitle and content in each of his prose pieces, using the subtitles to bring to life hidden meanings in the prose text. For art practice using techniques of juxtaposition may involve using text or titles to displace perceived meanings, or placing an object in a site in order to recontextualise meaning. Here the positioning of something new into an 'old' or existing context may work to displace certain preconceived meanings of the past, creating alternative histories and suggesting new futures.

For me, the angel provides an emancipatory impulse for thinking between places, times, people, things and ideas. Referencing the figure of the angel as a messenger or threshold figure works metaphorically as a device for moving between theory and practice, as well as across disciplinary boundaries between architecture, art, critical theory, geography and philosophy. This chapter is not structured as part of a linear and progressive argument, but rather as a collection of self-contained pieces, with similar themes that overlap and reiterate notions of the angelic as 'the between'. The chapter is angelic in content and form. The style is angelic thinking, a mode which is both subjective and objective, critical and creative. The structure is an angelic topography, an architecture with a complex pattern that can only be compared to the intricate structure of a snowflake not discernible to the naked eye, or a multifaceted crystal that changes in the light, or even to a kaleidoscope that fragments an existing view of the world. Angels are not simply messengers, they are also figures of transformation. In their status as flux, as ever-changing, they challenge traditional modes of representation and offer opportunities to think about space and time differently. In this case, an angelic temporality is one where memory is not nostalgic but imaginative, and which chooses neither to look backwards nor forwards, but to focus on the potential offered by engaging with the present moment.

From Rut Blees Luxemburg, Katherine Yass and Uta Barth, to Tacita Dean, Jane and Louise Wilson and Victor Burgin, contemporary artists working in photography and video seem obsessed with capturing the present as a frozen moment, often in relation to permanently or temporarily abandoned buildings. Images of buildings in the architectural press are rarely cluttered by inhabitants or even traces of inhabitation. This current fascination with places devoid of people is different. In these photographs of architecture, the emptiness is not about keeping the details clean, the gesture is more generous. Here the emptiness is not created in order to view the object better, but in order to provide a place for the viewer to imagine. These images are suggestive. These places have not always been and will not always be empty. Their very emptiness in the present passing moment allows us to project all kinds of alternative scenarios onto them – past and future. Like detectives we search for clues, traces of past occupations; like script writers, we set up props for future activities.

A similar kind of interest can be found in the work of artists intervening in abandoned spaces, such as Ann Hamilton; dealing with material expressions of absence and presence, such as Rachel Whiteread; or exploring decay and transience, such as Anya Gallaccio. Here we have insertions into found spaces, not simply as the presence of a gaze, but as the physical action of placing something material in a site. Whether inside a traditional gallery, or in the urban realm outside the gallery, these artists tend to engage closely with the situations and contexts they find themselves working within. The kinds of objects they 'place' in these spaces do not necessarily operate through dialectical juxtaposition, but through the insertion of moments which allow time to expand. In this way their tactics are analogous to the work of those making photographic images. In both cases, despite the differences in media, the viewer is asked to engage with the object they are presented with in such a way that they have no choice but to become conscious of the passing of time.

II LUSTRE

the quality or condition of shining by reflected light
a sheen or gloss
an iridescent metallic decorative surface on ceramics
the glaze used to produce this
a thin dress fabric with a fine cotton (formerly also silk or mohair) warp and worsted weft
a glossy surface
any fabric with a sheen or gloss
to make attractive be or become lustrous.

III GOLD SHEEN

> . . . like the febrile light with which Blees Luxemburg's photographs are suffused, is a reality rendered artificial in the brief suspension of time.
>
> (Bracewell, 1997: 11)

Caliban Towers I and II is one in a series by Rut Blees Luxemburg, entitled *London: a modern city*. It images two high-rise buildings aspiring to touch the skies. Luxemburg's photographic technique, shooting at night with a long exposure, gives the inanimate architecture a strange luminescence. Empty façades cease to be simply passive; the buildings captured in the frame have a deviant glow. As part of the public art project *Wide*, curated by muf and funded by the London Arts Board and London Borough of

Plate 1.1 Rut Blees Luxemburg, *Caliban Towers I and II* (1998), under a bridge in Shoreditch High Street (part of *Wide*, a project curated by muf, funded by the London Arts Board and London Borough of Hackney) (Photo: Jane Rendell)

Hackney, this photograph was placed under a bridge between Hoxton and Shoreditch, one mile or so down the road from the very housing projects the image depicts. What does this intervention say about time? Within a few months, the photograph was removed by the council – no more than a memory on that particular stretch of street – but for a short while in 1998, this modernist aspiration was juxtaposed to a specific past and future. This particular image seemed to be critiquing, through the title of the work – Caliban is a monster – that the modernist project constitutes a form of monstrosity, a 'place of crisis' rather than the progressive vision the modernist designers might have projected. And indeed, up the Hackney Road, on a sunny Sunday in July, while *Caliban Towers I and II* were resident in south Hoxton, a block of flats just like them was demolished, dust in nine seconds. Along with the commercial billboards, pigeon shit and rough graffiti, the advent of fine art photography finding its way into a grubby bit of Hackney could easily be read as a comment on the future of the area in terms of the gentrification typical of postmodern urbanism rather than the democratic socialism of the modernist high-rise dream.

Following on from *London: a modern city*, Luxemburg's latest work, *Liebeslied* (2000), provides further observations on emptiness, time and nocturnal urban space. These images, also taken at night with a long exposure, saturated in the golden colour of street lamps, have an elegiac quality. They too are concerned with spaces of dereliction, neglect and decay, less with a mourning for the utopian project, more with emotional themes connected with loss and emptiness. In the accompanying catalogue, the philosopher Alexander Garcia Duttman cross-references the images in a text entitled 'My suicides'. I am intrigued by the absence of only one of the photographs, the one that interested me most. No. 13, *Meimsuchung/Affliction* (2000), is strangely omitted from Duttman's text. Why?

On first glance, No. 13 *Meimsuchung/Affliction* looks quite familiar. It is a photograph of a typical terraced house, except that the house itself has been removed. What we are looking at is the interior, the gap this one-time home has left behind. One of the side walls, originally the party wall between this house and the one we can see in the distance, is the main interest of the image. With pitched roof and chimney pot the house creates a child-like silhouette against the dark sky. The brickwork is coloured warm russet, but is in tatters; covering it are shards of wallpaper and crumbling paintwork. The main feature of the photograph, in the remaining wall of what would once have been the living room, is a dark hole – the fireplace. The flue has been removed, the hollow channel it used to fill can be traced up the wall. Flanking the fireplace are archways sunk a few inches deep into the brick face. No longer filled with books or pictures, the decor disintegrates into powder.

Top right we are shown a corner of the well-lit house behind this one. This is unsettling since it suggests that the house is actually much smaller than we imagine it to be from its outline. There is something quite odd about this seemingly ordinary house in terms of size and relation to its neighbours. But what interests me more, is the shiny rectangle that runs across the top third of the wall. I think it is a mirror. In it we can see reflected the house that stands exactly where the viewer is positioned. It is not clear what we are looking at. Is this a staged shot? In which case why? What does it say about this empty house and its neighbours? Almost all we know of the context is contained in the reflection in the mirror.

When I look at the other images comprising *Liebeslied* I note that there is liquid in nearly every shot: small globules on the tarmac, rivulets running down the pavement, trickles following cracks in walls, pools of still, black water, planar surfaces of night river. All this water is thick, dark, sultry, glossy, almost as gorgeous as the limpid images of themselves. They make the abandoned places they depict appear beautiful. As they give themselves up to the light, their lustrous surfaces become mirrors, multiple sites of reflection.

IV BLUE LUSTRE

> I think of the space between positive and negative images as a gap.
>
> (Yass, 2000: 81)

Catherine Yass also works with photography, empty architectural spaces and light. Her technique, developed by accident, is to make prints of a positive and negative image overlaid. The negatives show all the lightest areas in blue, and given that the negative images are slightly smaller than the positive images, a bright 'halo' marks the edges of these patches. The rich blue of her work is striking and has been compared to the use of lapis, a precious stone. Certainly the use of saturated, intense colour, as well as the presence of halos, makes it easy to describe Yass' images as modern icons; though her palette, bruised reds, violet-blue and sickly lemon, is quite a shift from the usual red, blue, gold of religious symbolism. Yass' work is often shown in light-boxes, in dark rooms, shifting the relationship that the usually darker photographic object has to a well-lit white gallery, and also increasing the magical glow of the images. It is also interesting to note that the blue areas do not allow light to penetrate, creating a situation when the images are back-lit where the light can only slip through the gaps around the edges of the blue, further enhancing the halo effect.

Many are of the places Yass photographs are institutional service spaces, for example, *Corridors* (1994), *Stall* (1996), *Stage* (1997), *Cells* (1998), *Baths* (1998), *Toilets* (1999). These kinds of marginal areas are usually ignored, rarely photographed in colour, and seldom explored in relation to their formal properties, in terms of light and shadow. But Yass' images investigate exactly this: the spatial qualities of these liminal zones.

In between making the two images, positive and negative, there is a time lag. It is the space one can occupy between these two moments in time that interests Yass. By playing with light, with positive and negative, Yass teases apart further the gap between the views, between the way we are used to seeing light and its reversal, shadow. The position of the viewer switches constantly, between reading the image as a positive view of light and shadow, as we might see things in 'the real world', to an image that renders the familiar strange. This gap between one view and another operates between the familiar and the strange. Yass' images make ordinary places appear quite odd.

The concern then appears to be less about communicating the 'real' physical dimensions and qualities of the photographed places and more about presenting a 'screen' where viewers can project their own emotions into the two-dimensional picture plane. The technique of superimposition of positive and negative already creates strange collisions of real and imaginary space, of grounded surfaces and slippery patches of floating colour whose material substance cannot be discerned. Texture, with its hidden depths and reflecting planes of luminous smoothness, creates different kinds of territory for the imagination. The fact that the places are always empty, if only for a certain time, enhances this ability of the mind to play with space. The effect of the blue is perhaps the most effective in this context. Confusing any one reading of the 'true' depth of field, the blue areas appear to hang in the air in an indeterminate position between viewer and subject matter. Yass plays with this viewing position, for example in the *Corridor* pieces she says that she intentionally placed the blue closest to the edges of the image so that it gave the viewer something to hold on to – in order to allow them to enter the work.

Like Rachel Whiteread's strategy of making the absent present, Yass' technique of collapsing of two viewing conditions into one another could be described as a temporal tactic. What she has been developing in her work is the creation of a temporal gap between positive and negative images, and the potential of this gap to oscillate the viewers' position. What interests me is how this device operates differently depending on the content and context of the photographic image. *Bankside* (1996), Yass' photographs of the Tate Modern empty, before reconstruction really got going, work best for me. Unable to choose between seeing the building as a composition of materials that have decomposed and been worn away over time and as a series of surfaces, serene, glossy and seemingly indifferent to history, the viewer is left hovering in the present. I find this an intriguing time-frame, or way of looking at an architectural place that is right on the edge of moving from one life into another. But for a brief second time freezes: Yass holds us there by offering ambiguity.

V BLURRED FOCUS

> the 'location' of the work hangs somewhere between the viewer and the wall, in the empty space we are looking through.
>
> (Barth, 2000: 15)

The third photographer I want to bring into this discussion of time and empty spaces is Uta Barth. Like Yass, Barth is also interested in intervening in the viewing relationship that the subject has with the object that they are looking at, and she does this in a number of ways. Sometimes, as in *Ground* (1994–7), she blurs the focus, making it impossible to read anything within the frame but a territory and some objects contained within it. Often these are images of mundane and everyday domestic settings, where we find ourselves peering at what might be, for example, the edge of a curtain, No. 46 (1994); pictures on the wall, No. 42 (1994); or the air-vents along the edge of the floor, No. 96.3 (1996). In numbers 3, 9 and 10 of *Field* (1994–8), the images of ordinary street lights in the city at dusk appear to hang in the sky like huge harvest moons, lending an unearthly quality to the scene. In playing with our desire to see these things clearly and to discern their every detail, Barth brings an intensity to the act of looking.

On other occasions, Barth focuses our attention on previously un-observed margins and gaps between things. It is the edge of her frame, not the centre, that Barth is interested in; and she makes these edges matter. Objects we would expect to find in the middle of an image, we find right at the periphery, just about to slip out of view; while centre stage we are staring seemingly at nothing at all. For . . . *in passing* (1997), Barth made a series of photographs that prioritise the gap between two people by placing this absence in the centre of the image. In *nowhere near* (1999), we view through selected portions of a framed window, patches of the horizon blurred, out of focus, far, far away. In her most recent work, . . . *and of time* (2000), the photographs show light caught on the wall-to-wall carpet and beige painted walls of the most boring rooms. The only things to capture our imagination here are the fleeting shafts of light itself.

But what do these images tells us about time? It is clear from Barth's work that time cannot be understood without space. By holding the object of our gaze away from us, Barth creates a distance between the viewing subject and the viewed object – it is a distance that she controls. By bringing the experience of looking and the desire to see right into focus, Barth positions us very much in the present tense. Like Yass' play with the gaps between different viewing positions, Barth's focus on the absences in normal viewing conditions makes time more tangible: we are made aware of ourselves looking in an ongoing present. As in Luxemburg's *Liebeslied*, Barth's *Ground* and *Field* create a lustrous surface, a glossing of the present moment, that allows the viewer to become absorbed in their own act of looking.

II (LUSTRE AGAIN): THE ACT OF SELF-ABSORPTION

[I feel unsure here of what I have said. Is what I see in these photographs what is 'really' there or am I only looking at what I am looking for? Am I, like the photographers, lost in my own self-absorption? The places here are empty, they allow us to project ourselves onto the image. They are full of time, of the minutes passing by as we try to see, and of time ignored as we drift off into dreams. Are these images full of potential, are they dialectical images where our awareness of the present allows the past to become visible? Or is time held still by our own reflection? The self-absorbed look polishes and shines until we can see our own reflection. The tendency in the views of these contemporary urban photographers, and in my own comments on their work, is to gloss the external world of things and places, creating a lustrous surface. Are we doing this to reassure ourselves? Is the present more comforting than the past? Is self-reflection the creation of a critical distance or is it simply narcissism?

This seems to me akin to nostalgia. Doreen Massey has written of nostalgia as a selfish act, as a way of keeping the past in place according to one person's view of how things happened. Nostalgia holds everyone in check, not allowing them space to move forward. Unlike modernism's wish to move forward, to progress, a postmodern attitude to the past is a nostalgic version of history, whose coordinates are determined by a wistful backwards glance. Looking back from a distance up ahead, things and places seem to gain a glossy sheen. Memories take well to stain-remover and to lustre.]

VI TWO KEYS: POSITIONS OF ENTRY

> *key an instrument fitting into a lock for locking or unlocking it, usually made of metal and with more or less elaborate incisions, etc., to fit the wards of the lock*

key as representing the power of custody, control, admission
key as a symbol of office
a thing that opens up the way to something; especially a sure means to or a desired objective
a place from which the strategic advantages of its position gives control over a territory
a solution or explanation of what is unknown, mysterious or obscure
an alphabetical or other system for the interpretation of a cipher, an allegorical statement
a text explaining the abbreviations or symbols used in a book
the figures in a photograph or picture
a book containing solutions of mathematical or other problems
translations of exercises in a foreign language

VII BETWEEN REAL AND IMAGINED

> The narrative comes from the location, our connection to the space that we're filming in.
> (Wilson and Wilson in Corrin, 1999: 10)

For some time now, Jane and Louise Wilson have been investigating centres of power and coercion and abandoned spaces. In *Stasi City* (1997) and *Gamma* (1999) both these conditions of interest converge. *Stasi City* was the unofficial name coined by a West German journalist for a prison and phone-tapping centre – the headquarters of the DDR's intelligence service, the *Staatsicherheit*, that included a former Stasi prison, Hohenschönhausen. Now vacated, the location appealed to the Wilsons since during the Cold War no one knew the function of these buildings at the end of a suburban street in East Berlin. Likewise, when Greenham Common came up on a list from the Ministry of Defence of buildings that had been decommissioned, the Wilsons, with their recent memories of Greenham, CND, and the women's movement, were curious to find out more. Although subsequently rendered obsolete, in both cases the buildings occupied the very centre of power for a key moment during the cold war. In different ways their architecture enabled scenes of domination to be enacted, psychologically and physically, through real acts of violation and the threat of future abuse.

When shown at a recent exhibition of their work at the Serpentine (2000), both *Stasi City* and *Gamma* comprised installations combining video projections, still photographs and objects. These media work in different ways to represent and communicate spatial experience, through the visual content of the images, the time-based sound track of the video and the physical scale and presence of some of the furniture from the building itself. The Wilsons state that they are interested in 'animating and revealing a specific space', but the viewer cannot access *Stasi City* or Greenham, only Jane and Louise Wilson's construction of these places. This is not to say that the artists are dictating their understanding of these architectures of control; many of the tactics they use generate a self-conscious questioning in the viewer about the power and the occupation of space. For example, the use of double wall projections in the corners of rooms provides simultaneously two different views. The viewer may find themselves at the missile-control panel and alone in an empty decontamination chamber at the same time, leaving the choice of viewing position undefined. Within the gallery installations it is possible to talk about the intersection of multiple spaces: architecture as the subject matter of the places where the film was shot, the architecture created by the installation in the space of the gallery and the place that they intersect in the mind of the viewer.

In *Stasi City*, the images show a series of rather ordinary empty rooms with doors, desks, chairs and phones; but these office interiors and hallways are the interrogation rooms of the cold war. These two double doors are set wide enough apart to conceal a person. With this red button, a nuclear missile is launched. With such traumatic histories, it is enough just to see these places in images to cause fear and apprehension. These seemingly indifferent environments bear the traces of psychic dramas on an enormous scale. In some abandoned hospital theatre, in a scene that appears peaceful, we watch paint crumble from the ceiling, in an office room, tidy chairs and telephones with their lines unplugged give nothing away, only the upturned chair in another empty room indicates the possiblity of some kind of disturbance. The sounds suggest a little more, the cranking of machinery, breathing, footsteps. We start to feel uneasy, we imagine the worst. Are these stage sets constructed by the artists or do these images show rooms untouched since they were evacuated? What happened here? What really happened here, and do we really want to know? These images make reference to a history we would rather had never existed, they challenge us to take a position. The work plays on all kinds of tensions in our desire to reconstruct the past and our refusal to adopt either of the positions the narratives offer, perpetrator or victim.

VIII BETWEEN DREAM AND AWAKENING

> [Tacita Dean's] films are haunted by architectural relics which seem to embody outmoded or bankrupt beliefs, but at their time of execution offered much.
>
> (Wallis, 2001: 13)

Another artist working with video who has been attracted to abandoned places is Tacita Dean. In several of her works, specifically *Delft Hydraulics* (1996) and *Sound Mirrors* (1999), she has been drawn to highly functional architectures whose original reasons for construction are no longer valid or viable. In *Delft Hydraulics*, Dean focused on a machine, located in De Voorst, Holland, designed to measure wave impact in order to study coastal erosion, but which due to the emergence of new technologies became redundant. The story is quite similar for *Sound Mirrors*, huge listening devices that were constructed along the Kent coast between 1938 and 1939 to pick up the sounds of approaching aircraft during World War II. These strange-looking objects were also rendered obsolete as they were unable to discern between different kinds of passing traffic.

Both these architectural spaces have unusual properties, but rather than portray them as 'ahead of their time', as avant-garde and high-tech, Dean's films emphasise their arcane qualities. These dysfunctional objects critique a certain mode of scientific thinking which links rationality, abstraction and timelessness, as well as aspires to a technologically driven and controllable linear future. In their futility, they stand sorry witnesses to the passing nature of time.

Dean's interest in the heroic but doomed endeavour is best understood alongside *Teignmouth Electron* (1999) and (2000), her work around Donald Crowhurst's failed attempt to sail around the world alone, or *Trying to find the Spiral Jetty* (1997), her homage to Robert Smithson, who died while trying to document his work *Spiral Jetty*. Images of Crowhurst's abandoned boat in Cayman Brac in the Caribbean and the Great Salt Lake where we know a piece of land art lies submerged, speak of a certain sympathy for these epic voyagers who had such hopes for the future.

What interests me here is the way in which the making of the work articulates Dean's engagement with objects, places and time. In many of her films, Dean tends to focus on one viewpoint, for a long time, a very long time, long enough for our attention to wander and for random thoughts to enter. This positions the viewer somewhere between focused meditation and random drifting, between the ways in which we are expected to regard a work of art in a gallery, through a selective viewpoint that attempts to edit out all else, and the way we experience buildings in the city, as backdrops against which other thoughts meander in and out. Then, without warning, Dean will switch the viewpoint and offer us another long gaze at the same place or object but from a different angle. I'm not sure what is achieved in this shift of vision. I'm tempted to compare it to the distracted ways in which we view objects in everyday life simultaneous to all kinds of other activities. But since these are constructed in sequence rather than in parallel, this doesn't really work, it feels more like drifting off to sleep, that horrible moment on the train when my head drops forward and I jolt myself awake. This reminds me of Benjamin's comment on the importance of the awakening state as a coming into consciousness.

Dean does a similar kind of thing in *Foley Artist* (1996) and *The Sea, with a Ship; afterwards an Island* (1999), Dean's drawings made from reading the stage directions of the opening of *The Tempest*. In both these works, she investigates the unusual places hidden between real, represented and imagined space. How often when we watch a film, and, aided by the sound track, drift off into another world, do we bring ourselves back into reality and consider the places in which the sounds themselves were made? How often when we see a play performed, actors following stage directions against a backdrop, do we wonder how the author imagined the spaces? This is a different version of Benjamin's account of the work of the historical materialist, the job of revealing the conditions of the making of an event. But unlike the kind of Marxist history that investigates the social conditions of its own making, Dean's work is more akin to Benjamin's own approach. She opens a door for the viewer to enter a previously concealed imaginative realm and from this new position offers a new vista on the past.

IX BETWEEN FRAGMENT AND GESTURE

> The more the film is distanced in memory, the more the binding effect of the narrative is loosened. The sequence breaks apart. The fragments go adrift and enter into new combinations, more or less transitory, in the eddies of memory: memories of other films, and memories of real events.
>
> (Burgin, 2000: 29)

Victor Burgin has investigated the relationship between space and the unconscious, history and psychoanalysis in much of his work, but in his most recent videos he deals explicitly with architecture, narrative and absence, specifically in relation to two architectural monuments of early and late modernism. The Barcelona Pavilion designed by Mies van der Rohe and built for the International Exhibition in 1929, subsequently demolished and reconstructed from photographs in 1986, remains an icon of early modernist ideals of the open plan of free-flowing, homogenous space. Dominique Perrault's Bibliothèque Nationale de France in Paris is exemplary of late modernism, the repetitive grid of no-quality, the anonymous glare of reflective glass, the perfected perspective of rectilinear geometry. These works are incredibly complex in their evolution, and as always with Burgin, an intricate construction of intentional and accidental moves. But in both, Burgin draws connections between the tendency of psychoanalytic theory to universalise and the desire of the historical materialist to talk of the particular in terms of time and space.

In *Elective Affinities* (2000), Burgin's video work juxtaposes views of the interior of the Barcelona Pavilion with footage of the Spanish Civil War. A connection is made between *Dawn* by George Kolbe, the bronze nude female statue in the pool at the Barcelona Pavilion, and a woman soldier in the civil war. Both adopt the same gesture, their forearms are lifted across their faces, to shield themselves, their eyes, from the sun, or as Burgin intimates, from some other form of intrusion, the searching gaze of the camera or gun? What made Burgin choose to connect these two female gestures? This is what fascinates me. If the gesture embodies a desire to protect oneself by refusing to look, does it feel the same for both women? Is it a symbolic act, or is it a signifier capable of communicating a variety of moods and emotions? Here holding up one's arm to shield one's eyes operates to make a comment on the relationship, temporally and spatially, between the universal sense of time and space suggested by dawn, the eternal return of the repeated rising of the sun, and the non-differentiated homogenous spaces so loved by proponents of international modernism, and the specific site of this pavilion in the cultural history of local resistance to fascism in Catalunya.

A similar attempt to juxtapose the universal and the particular operate in another recent video work, *Nietzsche's Paris* (2000). Here Burgin locates himself in Dominique Perrault's Bibliothèque Nationale de France in Paris, a site that resonates in a number of ways with a three-way relationship between Friedrich Nietzsche, Lou André Salomé and Paul Reé. From reading their correspondence Burgin discovered that for a period of time in 1882, these three were intending to live together in Paris, but without warning Salomé and Reé left Leipzig together. Nietzsche thought they had gone to Paris, a city he had never visited, but which Burgin speculates must have held an important place in his imagination given his knowledge and love of French literature and philosophy. Earlier in that year, in August, Salomé and Nietzsche had spent three weeks together in the forest of Tautenberg debating philosophy.

At the centre of Perrault's library, surrounded and cornered by gridded glass towers full of books, is an inaccessible courtyard full of mature and densely planted trees. Burgin's work consists of four panoramas, taken from points on each side of the quadrangle between the corner towers and edited to create one full, but impossible, circular gesture. The forest, a place of enchantment and learning, features in each camera sweep. Is Burgin suggesting that Salomé and the forest can stand in for one another? Both are out of reach and unobtainable. Certainly he brings a specific story that took place in eight months in 1882, in a real space, the forest, and an imagined city, Paris, into close proximity to the present, to the inaccessible forest at the heart of Paris' cultural repository of knowledge.

VI (TWO KEYS AGAIN) WHEN TWO KEYS FIT THE LOCK

[All three of these video artists seem to be working at the place where real and imagined space intersect, and their attempts to articulate particular conditions of this overlap, provide the viewer with an experience in between dream and awakening. The differences between dreaming and waking knowledge interested the surrealists as well as Benjamin and were at the heart of many of the disagreements between them. Where surrealists, like Aragon and Breton, defended the importance of the unconscious in providing a different and potentially emancipatory view of the 'real' world, Benjamin felt they were turning their back on history by mythologising it. Benjamin chose instead to try and develop a methodology where dreaming could be considered in relation to a collective rather than individual or universal unconscious, and where the event of awakening reminded the historian of the importance of recognition for revolutionary action.

But in inserting a key into a lock, the artist does not open the door to the past, but unlocks the door to their own version of history. As Benjamin has said, there is nothing more important to the writing of history than the conditions of the present. This includes the mind of the historian and the artist. The difference is that the historian likes to solve a puzzle, but art ends when the puzzle is solved and the key fits the lock.]

X TAILOR'S CHALK

Tailor's chalk: hard chalk or soapstone used in tailoring, etc., for marking fabric as a guide to fitting

XI AN ONGOING MEMORY

> I felt I was trying to describe a memory, but an ongoing memory.
>
> (Whiteread, 1997: 31)

For almost a decade, from her first public work, *House* (1993), the cast of a Victorian terraced house in east London, to her most recent intervention, an inversion set on top of the empty plinth in Trafalgar Square, Rachel Whiteread

Plate 1.2 Rachel Whiteread, *Holocaust Memorial*, Vienna (1995) (Photo: F. Konecny)

has been working with notions of absence. In plaster and resin, taking hollows, gaps and cavities, sometimes of existing structures, other times of imagined spaces, she makes emptiness manifest. Whiteread's tactic is to make absence present.

For the *Holocaust Memorial*, Jüdenplatz, Vienna (1995), Whiteread's winning competition entry was for a concrete cast of a library, with double doors, a ceiling rose, lined with thousands of books, that would sit at the northern end of the square on the excavations of a thirteenth-century synagogue. The site has a turbulent history; in the fifteenth century Jews committed mass suicide by going down into the crypt and burning themselves alive. Like most public monuments, Whiteread's memorial is a solid and visible historical marker, but its quiet presence does not attempt to confirm the facts, rather it makes their absence tangible. Her rewriting of history is not the insertion of more text, but a marking of the places that history has never allowed to exist: the gaps between the lines, the silences between the words, the stories that get left untold. Although the monument appears permanent, Whiteread hopes that its pale surface will get marked by the passing of time.

In New York, in a recent work for the Public Art Fund, Whiteread has created a transparent resin cast of the inside of a wooden water tower. Once you notice them, water towers form a striking feature of the New York skyline; but quite often, despite their physical presence, as utilitarian objects they remain invisible. Placed in the context of the Soho, Whiteread's *Water Tower* (1998) is a new addition to the roofscape that slips in and out of visibility depending on the weather conditions. As in her other projects, she marks presence in a quiet and almost invisible way, drawing our attention to all the other 'working' water towers in the neighbourhood. In so doing, her *Water Tower* operates as a new kind of ephemeral landmark, and as a device for refunctioning the water tower as more than a utilitarian rooftop object. Whiteread's projects tend to develop out of a close study of the location she is working in. Her interventions often transform the existing meaning of a site, allowing audiences to re-think and re-read the terrain from different perspectives, making manifest histories previously rendered invisible.

XII PRESENT AS PALIMPSEST

> The garden as an idea seems to fit perfectly with all my main preoccupations – the garden as palimpsest perhaps, letting the past show through and seeing what happens in the future.
>
> (Gallaccio, 1999: 60)

Anya Gallaccio is known for working with particular kinds of material – decaying and decomposing substances, such as flowers, fruit and chocolate; ephemeral and unstable elements, such as steam, melting ice and chalk. As many of the materials she uses are natural, they are not stable and are constantly caught in a state of flux. Her work engages with the time of transformation. Sometimes it is the time of decay. There are the 800 red gerbera of *Preserve Beauty* (1991), placed between glass and the window of Karsten Schubert Ltd, London, or the bed of 10,000 slowly rotting tea roses which comprised *Red on Green* (1992), at the ICA, London. Gallaccio shows us the impossibility of trying to hold time still, such a futile act, and yet we are all obsessed by it somehow. She holds our attention as the moments pass, engaging with our sense of smell and touch and altering our perception of time. So engrossed do we become in the aroma of decay, that we hardly have time even to think. In other work, Gallaccio emphasises the other side of the uncontrollability of natural things, the process of growing rather than breakdown. In *Keep off the Grass* (1997), Serpentine Gallery Lawn, London, she sowed common vegetables and flower seeds into the decaying patches left by other sculptures and more recently in *Glaschu* (1999), Lanarkshire House, Glasgow, she planted a green line into the concrete floor. The garden seems to provide a threshold, for Gallaccio, where past and future meet in the present.

In other works Gallaccio captures an ongoing process of change in a more elemental way, but by focusing on the transitional points of material transformation: the melting point of ice or the boiling point of water. For *Prestige* (1991), Gallaccio filled the space of an abandoned tower at Wapping with twenty-one whistling kettles linked to a compressor. For a later installation commissioned by Women's Playhouse Trust also at Wapping, *Surfaces and Intensities* (1996), Gallaccio made a 32.5 ton cube of ice bricks measuring 3x4x3 on the boilerhouse floor. This ephemeral sculpture slowly melted away, aided by the large chunk of rock salt embedded in the centre. Both these pieces of work made reference to the importance of water in the original function of the building, but the tactic was not to insert another object into the place in order to highlight or make visible an aspect of history, but rather to involve the audience in a process that related to the temporal and material properties of the building's previous function. Gallaccio makes large scale sculptural objects, but like Whiteread, these are not objects that can overpower us, rather these are reticent objects which quietly refuse to give themselves away.

This critique of the value the art world assigns to the permanent and ownable object is nowhere more manifest than in *Two Sisters* (1998), a public work Gallaccio made for ArtsTransPennine, a series of arts interventions that ran right across the north of England, from Liverpool to Hull. *Two Sisters* was a 6m tall, 60 ton column of locally quarried chalk, bound with plaster, and placed in Minerva Basin, Hull. The site was chosen by Gallaccio for its shifting nature. The coastline around Hull is being continuously eroded and, according to Gallaccio, parts of it will eventually end up in Holland. Hence the name *Two Sisters*, which refers to a Dutch expression used to describe two architectural features of a similar nature, next to one another. The work disappeared over a five-month period, refusing to remain a stable entity. But the task of making *Two Sisters* was not insignificant – it involved large-scale machinery and a team of helpers, locating the event in the drama involved in producing the work, rather than in the object itself.

XIII STILL LIFE

> [Hamilton has an] inclination to fold history back on itself, treating it as a reversible narrative.
>
> (Hamilton, 1999)

In all of Ann Hamilton's installation work, cloth is central; so too, is historical research and memory. In *Whitecloth* (1999) at the Aldrich Museum of Contemporary Art, Hamilton discovered that before becoming a gallery space, the building had been a general store, a meeting place, a post office and a place of worship, as well as becoming Ridgefield's First Church of Christ, Scientist from 1929 to 1966. Hamilton became interested in the history of the location of New England in general and the importance of religion to cultural life. The central material the audience encountered in a number of forms through the installation was white cloth: as a handkerchief fluttering through the gallery, possibly a virgin's veil, a shroud, an artist's canvas; as a sheet hovering just above a table, reminiscent of a seance scene, the last supper, the lovers' bed. White cloth in its various manifestations referenced specific aspects of the history of the space, as well as contemporary art discourse, but operated most powerfully as a device for exploring the tension between sensuality and purity that Hamilton believed was key to the building's occupation over time.

Hamilton uses erasure as a tactic for articulating her concern with time. Erasure occurs in the physical alterations she makes to the architecture she works within. In *Tropos* (1993) Hamilton replaced the translucent windows with clear glass in order to flood the gallery with light. In *Whitecloth*, she felt that the outside of the gallery, a white weather-boarded façade, belied its interior and her first move was to remove the dry-walls that had been installed to cover all thirteen windows. Erasure also occurs in Hamilton's work through the effacing of text. In *Whitecloth* this was done by creating pieces of writing where, by painting in the gaps between the letters, the space between the words threatened to subsume the script itself and so obscure its meaning. A woman sat quietly at a desk in the centre of the gallery space in *Tropos*; from a distance it looked as if she was writing, but a closer inspection revealed that she was erasing every line of text she read with a burning stylus.

Hamilton's interest in the relation of writing and sewing explores ways in which we make connections between things, words and spaces. Hamilton's installations could be described as 'still lives' inspired by her historical research and made manifest as compositions, consisting of various carefully positioned fabrics, objects, texts and performed actions. The links that can be made between the various different pieces of the still life operate through emotion and the intellect, but engage most effectively when drawing on the ways in which memory blurs the boundaries between present and past. For example, Hamilton has used 'welle' or the 'weeping wall' in a number of installations, but each time it works in a different way depending on the proximity of other objects. In *Whitecloth*, the proximity of a wall slowly dripping with tiny drops of water on to a worker's jacket hung up on the back of a chair allowed the viewer to make any number of associations between labour, sweat and tears, depending on their own personal history. For Hamilton, cloth is the most powerful medium for working with memory, precisely because it operates like skin between inner and outer. Analogous to the wall of a building, cloth covers the body, creating a boundary that separates and connects inside and outside. The environments Hamilton creates do not provide readings of history that 'tell it as it is', but provide sensual surroundings that induce a form of poetic contemplation: a still life.

X (TAILOR'S CHALK AGAIN) OVER WRITING

[Tailor's chalk makes marks for action, like chalk on the blackboard, they are marks that will be rubbed out, that have a certainly short life span. But unlike blackboard chalk, they are marks for another writing that has not yet taken place, for a cutting, a folding, a sewing. The new mark yet to come will, in the act of its own making, erase this one incidentally even before it is intentionally scrubbed out. Our words and gestures make room, prepare the way, for those that follow that will necessarily erase them, before they themselves get erased. History is the story of continual transformation.]

I (THE ANGEL OF HISTORY AGAIN) WHERE THE THINKING STOPS, TIME CRYSTALLISES . . .

[Benjamin's observations on historiography, certainly in his 'Theses on the philosophy of history', are a fascinating insight into his thinking about time in general. While Benjamin warns of the dangers of ignoring the role of the present in our constructions of the past, *Angelus Novus* refers to Benjamin's understanding of the dialectics of time: an image that holds dialectics at a stand still. It is important that we acknowledge the effect of 'now' on our readings of Benjamin himself writing in the 1930s. The methods adopted by historical materialists writing history, now, in the twenty-first century, need to respond to their own specific situation and the aesthetics of the contemporary condition. In art practice today, working dialectically seems to be less about creating tension through additional means, through techniques of juxtaposition or montage, and more about making images, objects, spaces which contain contradictions within them and allow uncertainties to be briefly held in place. In different ways that is what all the artists discussed here do; through devices akin to lustre, keys and chalk, each creates a moment of particular resonance, where we become conscious of the ambiguous presence of the present itself.

I am not sure exactly what this suggests in terms of 'temporal tactics' or ways of addressing time in what we do as practitioners and theorists engaging with urban futures. I have been trying to theorise a position that avoids the pitfalls of the modernist utopian blueprint for the future, but also the regressive post-modern turn, of the nostalgic look backwards at the past. My hope is that by following a deconstructivist move – to focus on the potential of the undecideability of the present as standpoint – it is possible to stop still and contemplate both past and future at once. This act of contemplation is potentially dangerous as it can tend towards navel gazing, but if we remain true to the tenets of critical theory, it is surely possible to be at once self-reflective (to look backwards) and to imagine different futures (to look forwards) – for this is the time where the thinking stops.

Materialistic historiography . . . is based on a constructive principle. Thinking involves not only the flow of thoughts, but their arrest as well. Where thinking suddenly stops in a configuration pregnant with tensions, it gives that configuration a shock, by which it crystallises into a monad.]

2 DOROTHY ROWE

Differencing the City: Urban Identities and the Spatial Imagination

> They open a door and all they've got behind it is a bathroom or a lounge. Just neutral spaces. And not this endless maze of present rooms and past rooms and the things said in them years ago and everybody's old historical shit all over the place
>
> (Zadie Smith, *White Teeth*, 2000: 514)

The frustration that Irie Jones articulates in this public outburst to her parents and their friends whilst travelling by bus across London (from Willesden Lane to Trafalgar Square, from suburban periphery to national centre), seems an appropriate starting point for a chapter that concerns itself with speculations on urban futures and cultural identity. In her irritation, the teenage Irie compares her own family existence (as the mixed-race daughter of a black Jamaican-born, England-raised mother and a white English father, who is also pregnant by one of the identical twin sons of her father's Bengali best friend and war compatriot) with her imagined vision of how she perceives other families in Britain to be. The idea that there are any 'neutral spaces' is of course just that: an imagined concept that deftly serves to foreground Irie's own struggle for a post-colonial subjective identity within contemporary Britain.[1] *White Teeth* is a resolutely urban novel. It both exposes the contradictions inherent in any easy search for a stable cultural identity within a small metropolitan cross-section of society that is constantly in flux, and facilitates explorations, such as this one, into the possibilities for cultural identity and urban futures. Indeed, as Stuart Hall has observed:

> cultural identity is a matter of 'becoming' as well as 'being'. It belongs to the future as much as to the past. It is not something which already exists, transcending time, place, history and culture. Cultural identities come from somewhere, have histories . . . they are subject to the continuous 'play' of history, culture and power.
>
> (Hall, 1990, cited in Hall and Sealy, 2001: 103–4)

In particular, it is the role of the city in the continuous re-invention of cultural identity and the concomitant 'play' of history that is central to this chapter. City identity is something that I have explored within the context of western art historical modernism for the past few years (Rowe in Meskimmon and West, 1995; Rowe, 1995; and Rowe, 2003). Caught up as an Asian woman in the hybridity of my own adoptive Anglo-German upbringing, Imperial and Weimar Berlin have been and still are central to my professional concerns as an art historian. However, it is the recognition that European urban modernity was predicated not only on the power of industry, speed and communication technology 'at home', but precisely on the economic prosperity that such developments could be used to forcibly harness 'abroad', that any consideration of European modernist culture should by default take account of its interdependence on the colonial 'other'. It is the significance of modernism and its implications for the present that has directed me towards a 'different' conceptualisation of the city for the purposes of this chapter.[2] As Homi Bhaba has exorted, 'the Western metropole must confront its post-colonial history, told by its influx of post-war migrants and refugees, as an indigenous or native narrative *internal to its national identity*' (Bhaba, 1994: 6). In the course of future research I shall return to the colonial question in modernist Germany, but for the present I should like to turn to the implications for shaping the city that are to be found in the work of many contemporary black and British-Asian artists working, as Chila Kumari Burman has intimated, 'beyond two cultures'.[3]

My title, *Differencing the City*, takes its cue from recent developments in art history that posit 'difference' as a concept that is mediated by a politics of race, class and gender. Griselda Pollock implies that 'differencing the canon' is not about the replacement of one set of canonical works by another as devised by feminism, but that it is a rather more nuanced activity that continually questions the borders of knowledge, desire and power (Pollock, 1999). To 'difference' something is not only to provide alternative readings from the edge, to acknowledge the power of alterity in racial inequality, social inequity and gender liminality, rather it is to foreground sites of contestation from within. Following recent work by Adrian Kear, I would even suggest that it could be conceptualised as the parasitical activity of Michel Serres' formulation (Serres, 1982). As Kear explains, for Serres the parasite functions as relational catalyst, like a secret agent interrupting order, producing disorder or generating a different order (Kear, 2001: 38). Although Kear uses the analogy for a discussion of the role of public site-specific performance installation in the exposition of embedded social and ideological practices of a particular 'host' site, the implications of such parasitical activity have a peculiar resonance for an analysis of the artwork produced by many black artists working in Britain today, for whom the act of 'interruption' of a dominant paradigm can often play a key role within a performative creative strategy (Kear, 2001).

Further to this, the notion of *Differencing the City* is prompted in response to David Sibley's formulation of difference in relation to the role of the boundary or border. In Sibley's 1995 study, *Geographies of Exclusion*, the concept and the physical fact of the boundary as a prime demarcator of 'difference' is explored. As Sibley notes, transgressing boundaries offers the transgressor moments of both anxiety and exhilaration, whilst in other circumstances boundaries 'provide security and comfort' (Sibley in Miles *et al.*, 2000: 269). However, in Sibley's formulation, the position of alterity as a site for the operations of hegemonic exclusion remains rather bleak:

> The alterity personified in the folk devil is not any kind of difference but the kind of difference which has long-standing association with oppression – racism, homophobia and so on. The moral panic will be accompanied by demands for more control of the threatening minority. . . . In the 1980s . . . different moral panics with slightly different scripts signalled the continuing presence of racism.
>
> (Sibley, 1995, in Miles *et al.*, 2000: 273)

Whilst in no way disputing the general premise on which Sibley categorises alterity, the concept of difference as merely a position of the oppressed leaves very little space for a consideration of alterity as an enabling position and it is, I would like to argue, precisely in the creative work of many black British artists operating within the spaces of the contemporary city that a more energising perspective on 'difference' can be found.[4] As Gary Bridge and Sophie Watson have commented:

> Differences are constructed in, and themselves construct, city life and spaces. They are also constituted spatially, socially and economically sometimes leading to polarization, inequality, zones of exclusion and fragmentation, and at other times constituting sites of power, resistance and the celebration of identity.
>
> (Bridge and Watson, 2000: 251)

It is the centrality of the boundary to metropolitan discourse that is pivotal to conditions of identity location and it is in this nexus between the city and the individual that Sonia Boyce's 1988 mixed media montage *Talking Presence* can be read. Boyce's metropolitan collage on photographic paper fuses the historically and geographically specific with references to an iconography of western modernity that simultaneously disrupts and satirises the exotic black body of western colonial visual convention. History and geography, time and place, collide in this meditation on post-colonial London. The viewer bears witness to a British 'order of things' that is dislodged by the dominant 'presence' of black people. Two black nude bodies, male and female, reclining and sitting respectively (in a subtle gender inversion of the visual rhetoric of western modernity), are placed immediately before the viewer. They are positioned close to the frontal plane and lower edge of the collage, one on each side, framing, presenting and contemplating the scenes in front of them, yet remaining outside them. They look on, situated in what might be termed a 'paravisual' position, both extending Genette's definition of the paratextual as well as alluding to Serres' notion of the parasite as an 'interrupter'.[5] For Genette, the paratext is 'neither on the interior nor on the exterior: it is both; it is on the threshold' (Genette, 1997: xvii, cited in Stewart, 2000: 26). Although, as Stewart points out, Genette's definition of paratexts relates to the written word, to the marginalia of the book form, here I think it has resonances for a consideration of the formal visual positions in which the figures are placed, and in turn the cultural positions that are signified by their visual relationship with this London scene. Their poses recall the languid exoticism of Gauguin's Tahitian nudes, yet Boyce's figures look away from the viewer, highlighting the recurrent trope of bodily objectification central to western visual modernity by their refusal to present themselves as the sole objects for such commodification. Boyce is 'talking' the 'presence' of black British identity via her adroit manipulation of visual signs. The figures may be mute but their presence is not. The paravisual boundary from which they enunciate their position resonates with Homi Bhaba's reformulation of Heidegger's concept that 'the boundary is that from which something begins its presencing' (Heidegger, cited in Bhaba, 1994: 1–5). Catherine King observes that 'although the figures are restful and might appear acquiescent, the title suggests that as representatives of the black community they are no longer mute' (King, 1999: 260).[6] The idea of affirming presence from the boundary also recalls Boyce's interview with Zarina Bhimji, published to accompany Bhimji's 1992 installation *I Will Always Be Here*.[7] The exhibition consisted of sixty glass shoe boxes, each containing different materials that acted as a trigger to 'a particular set of memories', accompanied by large wall-mounted photographs of 'poignant arrangements – flowers, strands of hair, nipples, tissue paper dresses, white chiffon headscarves'. In responding to a general question put by Boyce about the use of language and accessibility, Bhimji remarks that:

> I find I'm entering into the area of things that are not structured any more, boundaries are not distinct. I am experimenting with moving the boundaries. I am trying to talk about things which can be difficult to grasp. It is essential to find the universal language, to go beyond personal references, yet it can only come out of personal experience.
>
> (Bhimji in conversation with Boyce, 1992, *I Will Always Be Here*: 1)

The transgression, transformation and dissolution of boundaries becomes a tangible position from which both Bhimji and Boyce can assert existence and presence in their explorations of broader issues of subjectivity.

The montaged scene that viewers are invited to contemplate from the edge of Boyce's image is a symbolic cultural landscape of the city coupled with an intimation of a distant colonial 'elsewhere'. In the background, icons of the church and the state (signified respectively by St Paul's Cathedral and the Houses of Parliament), and the centrality of river-trade to colonial expansion (Tower Bridge), are juxtaposed with the vernacular architecture of the Victorian tenement and the post-war skyscraper. At the centre, familiarity and estrangement are coupled via the inclusion of a black cab and red bus, indexical signs of the London tourist postcard that implicitly question who the tourists in this city might be.[8] The interrogation of concepts of 'home' and the 'stranger' are at the heart of this image of modern London and they are further underlined by the inclusion of a fragment of an interior wall, on which is hung a framed picture of an ocean-faring vessel. The ship and the collection of shells and other objects in front of it hint at distant countries, different homelands and 'the journeys of the colonized and colonizers' (King, 1999: 260). The historical context of colonialism and Empire is also alluded to via the two fragments of birds, cut from a William Morris textile and wallpaper design (*Strawberry Thief*, 1883). The wallpaper is a reminder of the Victorian drive towards global expansionism and colonisation, whilst the birds are a poignant symbol of migration. The allusion to Morris in this work references Boyce's earlier four-part work on paper, *Lay Back, Keep Quiet and Think About what Made Britain So Great*.[9] The 1986 work 'also contextualizes present-day British life within an historical narrative', depicting symbols of the British colonies in three of the panels, with a self-portrait in the final panel (Beachamp-Byrd, 1997: 23). The background pattern within all four images is derived from designs by Morris & Co. In both works Boyce excavates Morris's position at the heart of British heritage culture and discloses the colonial narratives that underpin such cultural symbols. In *Talking Presence*, the dislocated interior wall, visually sandwiched between the two black bodies at the front and the metropolitan scene behind, is the fulcrum in this story of 'differencing the city'.

The dialectic between strangeness and familiarity, difference and similitude, 'home' and 'elsewhere' within the modern metropolis, is one that informed the conceptual development of the Victoria and Albert Museum's *Shamiana: The Mughal Tent* project in 1991.[10] Devised by Shireen Akbar, this inter-generational national textile project was intended to address some of the cultural and educational issues faced by Britain's population of South Asian women and children, as well as widening access to the collections of a national museum not normally seen by them as part of their cultural framework. Using the Nehru Gallery as their starting point, women's groups from community centres across the country were invited into the museum to study aspects of South Asian histories and cultures in order to formulate ideas for their own textile panel, which would be displayed as part of an overall larger tent, the *Shamiana*.[11] As Akbar explains:

> We chose the idea of the tent for its many connotations. . . . Tents can be romantic and exciting, such as nomadic and circus tents, or they can be physically or psychologically threatening, related to the feelings of being a refugee. All of the participants of the core groups shared the common experience of immigration. For them the symbolism of the tent was powerful and easily understood.
>
> (Akbar, 1999: 15)

As a synchronic symbol of shelter and transience, the tent signifies the uncertainties of travel and migration and a sense of impermanence emphasised by the fact that for many South Asian people who migrated to Britain for economic reasons during the 1950s and 1960s, it was never intended to be a permanent home. Fahmida Shah observes that for some of the Asian women participating in *Shamiana* it was the subsequent 'growing generation and cultural gap with their children which spurred many to address their isolation and loneliness' through the project (Shah, in Akbar, 1999: 25). Akbar explains that the focus was initially on Asian women since they were identified as 'the most isolated of all immigrant groups; they carry much of the responsibility for the informal "cultural" education of their children; and they are skilled in traditional textile work' (Akbar, 1999: 16). The focus on textiles was a deliberate choice of medium since many of the older generation women were already skilled textile practitioners so it provided them with a sense of confidence in their abilities to execute the work and it raised the public profile and status of their craft. As Deborah Swallow indicates, citing Rosika Parker, it is also a medium that has a significant gendered history in terms both of the construction of femininity and, more appropriately in this context, its subversion, 'limited to practising art with needle and thread, women have nevertheless sewn a subversive stitch – managed to make meanings of their own in the very medium intended to inculcate self-effacement' (Parker, 1984, cited by Swallow in Akbar, 1999: 31).[12] Again, the interview between Sonia Boyce and Zarina Bhimji concerning Bhimji's 1992 installation is evocative here. In *I Will Always Be Here*, the assertion of subjectivity implied by the title is coupled with a more exploratory approach to identity. Sonia Boyce points out that she sees 'an interest in textiles running through the work' and Bhimji replies:

> I never thought about that but I suppose it is true. I did make a conscious decision not to work in textiles at college because of the hierarchy – it is so easily dismissed as women's work, not serious art. I moved into photography because of my fascination with light, but textiles do recur as a motif. I do take pleasure in them – the feel of pure silk or cotton, almost like skin.
>
> (Bhimji in conversation with Boyce, 1992: 3–4)

As Swallow goes on to outline (and as Rosika Parker, Lisa Tickner, Jane Beckett and Deborah Cherry have all demonstrated), the significance of textiles to the history of female identity politics should not be underestimated.[13] Fahmida Shah, one of the organisers, notes of the *Shamiana* project that:

> the majority of the panels carry clear messages with mixed emotions from the women about their past and present lives. *Islam Now*, a panel made by women in Falkirk, Scotland, makes a strong statement about what it is like to be an Asian woman living in an alien culture, where the pressures of being a wife and mother can at times almost seem imprisoning. One young woman in the group depicted this by stitching a gilded cage surrounding an eye that sheds a tear.
>
> (Shah, in Akbar, 1999: 26)

The gilded cage, created by Nadira Sadiq and part of the larger overall design of the panel, is a visual metaphor for the 'sense of loneliness of being apart from other family members who are still in the home country' (cited in Akbar, 1999: 110). As Homi Bhaba has noted, the use of 'metaphor . . . transfers the meaning of home and belonging, across the "middle passage", or the central European steppes, across those distances and cultural differences that span the imagined community of the nation-people' (Bhaba, 1990: 291). Although referring to the perspective of the western migrant in exile, Bhaba's observations also have resonances for its reverse, for the eastern migrant in western exile. He observes that 'the nation fills the void left in the uprooting of communities and kin and turns that loss into the language of metaphor'. The idea of an imagined homeland becomes a rich source of metaphoric yearning and is played out here in Sadiq's piece. However, this is not an isolated example, rather it is a familiar trope of diasporic visual culture in Britain and is negotiated in different ways by a variety of individual artists according to inflections of race, class, sexuality, gender, generation and, as Gary Bridge and Sophie Watson have remarked, 'able bodiedness' (Bridge and Watson, 2000: 251).[14]

Meera Chauda's 13x15ft photo-mural entitled *Where is Home?* was produced for a public site in Newham, East London and was commissioned and executed in association with the *Art of Change* in 1995.[15] The mural presents the upper body of a South Asian woman in traditional sari with a plain red bindi on her forehead, filling the frame of a temple-like structure, which also evokes the structure of a mausoleum. She looks out of an open window from the temple onto a scene that the viewer also sees in reflection on the pane of the window. The reflected scene depicts a western-style suburban house. *Where is Home?* remains an unanswered question: eastern temple or suburban house; one could argue that it is both, linked as they are through the visual device of the window and its reflection. Chauda uses the architecture as an index to complex desires.[16] The sense of indeterminacy is central to her practice as an artist and it also evokes Hanif Kureshi's perspective on doubled identity:

Plate 2.1 Islam Now (1995), by Falkirk Asian Women's Group, Falkirk, Scotland (Photo: courtesy of the Trustees of The Victoria and Albert Museum)

> In the mid-sixties . . . we didn't know where we belonged, it was said . . . we were 'Britain's children without a home'. The phrase 'caught between two cultures' was a favourite. . . . Anyway, this view was wrong. . . . It had been easier for us than for our parents. . . . For me and others of my generation, Britain was always where we belonged. . . . Far from being a conflict of cultures, our lives seemed to synthesize disparate elements.
>
> (Kureshi, 1986: 160–1)

The notion of synthesis is one that is also implied by Chila Kumari Burman's location of her subjectivity as being 'beyond two cultures' rather than 'between' them and it picks up on a general paradigm expressed by many of the younger generation of South Asian women who joined the *Shamiana* project as it progressed, offering a slightly different generational perspective from their forebears. Fahmida Shah explains that 'young people explored issues of growing up within two strong cultures and through their panels expressed their aspirations for the future' (Shah, in Akbar, 1999: 29). Indeed one participant categorically affirmed that she had 'actually discovered "me" and what I wanted to become' and she continues by observing that:

> We have not lost our Asian identities, in fact, I would say that we are now closer to it [*sic*] than we ever have been. Those who try can achieve what they wish, as long as we ourselves remember that we should integrate but not assimilate.
>
> (Firth Butt, cited by Shah, in Akbar, 1999: 29)

The aspirational aspect of this statement is one that is frequently repeated in the many group statements that accompanied the presentation of the individual panels to the museum, as well as in the visual themes of the individual panels. *Hope*, produced in 1995 by Hopscotch Asian Women's Group in Camden, north London, is one of the most vivid examples in which the urban context of identity politics is explicitly recognised within a panel produced for *Shamiana*. According to the group statement about the piece:

> We wanted to represent all the groups who use our Centre – Bangladeshi, Indian, Pakistani and Nepalese – and show from a woman's point of view the different lives we lead. The elements of women's lives now, which we have tried to represent, are: the working woman earning a living with the computer and the sewing machine; studying in the library; and enjoying her children. London is depicted by Big Ben, the plane tree and a pigeon; beliefs by the temple and the crescent moon. All is approached through a beautiful archway, representing traditions and roots . . . representing hopes and wishes for the future.
>
> (Tasneem Khan, in Akbar, 1999: 100)

As in Boyce's *Talking Presence*, the legible city of scopic consumerability is delineated by a landmark, a symbol of its authority, in this case Big Ben, yet that legibility is simultaneously re-inscribed by the inflections of multicultural identities operating within the representational spaces of everyday urban life.[17]

The exploration of identity, difference and the city is one that is further pursued in the recent abstract work of artists such as Hew Locke and Lubna Chowdhary, both working three-dimensionally. Hew Locke's *Hemmed in Two* (2000) is described by the artist as 'part boat, part city, part animal, part infestation' (Locke, 2001). A key feature of the work is its allusion to a sense of transience and impermanence, as well as its indeterminacy and deliberate abjection. Made from cardboard, tape and dolls, its organic structure is painted on the outside with acrylic, whilst inside it devours off-cuts and remnants of its larger self, recycling its own waste products. Each time the work is displayed, Locke changes its structure according to the space in which it is sited. The metamorphosis of the work in relation to site is orchestrated to ensure that *Hemmed in Two* does not fit the space in which it is shown. It climbs the walls, sprawls across the floor and attempts to escape from the gallery window. Resonant of Kurt Schwitters' unfinished *Merzbau*, it 'confounds easy categorization' (Locke, 2001). Its visual form suggests a host of possible meanings ranging from narratives of colonial migration to the prospect of homelessness

Plate 2.2 Hope (1995), by Hopscotch Asian Women's Group, Camden, London (Photo: courtesy of the Trustees of The Victoria and Albert Museum)

in 'cardboard city', to the sprawling degradation of the urban slum.

For Locke the work is about the 'beauty of failure', the regenerative aspects of something that does not succeed in the way it was meant to. The use of cardboard is particularly significant in this regard since, as Locke notes, the work 'is made out of a material that is very useful, a very everyday material', which is also 'a perishable material' (Locke, 2001). Cardboard acts as the interface between having a home and not having one for the many urban poor in the contemporary global city but it also fails in this function since it is inadequate to the task.[18] Whilst there is no beauty in this particular evocation, in its role as material for art in *Hemmed in Two*, cardboard is manipulated so that it does not easily accede to a scopic regime of controlled visual consumption. It fails to be functional and to be easily graspable, yet it succeeds as artwork because of this. For Locke, this 'part city . . . part boat . . . part animal, part infestation' is 'something you would like to get rid of at times but you can't quite get rid of it . . . it's just there, you have to deal with it; that's in a sense what it's about' (Locke, 2001). It is parasitical (an infestation), and it is insistent. It speaks its presence through an uncanny relationship with beauty. Externally it is covered with export signs as an allusion to the absurdity of the regular critical reception of Locke's work as 'an exotic export from some other country' whereas, he remarks, 'in actual fact, my work is exported from Kennington' (south London) (Locke, 2001). He continues:

> The boat-like quality in the piece comes out of a long-term interest in issues around migration and things like that. I've been making boats on and off for a number of years and this sort of symbol of migration is difficult to avoid and I'm not trying to avoid it. I should say that I was born here. I grew up in Guyana and I came back to live here and the issue that has faced me, that I mentioned earlier, was that I was always forever seen as an export of another country (my work is always seen this way) rather than what it is, a British product.
>
> (Locke, 2001)

Locke cuts to the heart of some of the key issues surrounding British post-colonial identity politics and bears witness to Homi Bhaba's remarks that:

> It is to the city that the migrants, the minorities, the diasporic come to change the history of the nation. . . . In the West, and increasingly elsewhere, it is the city which provides the space in which emergent identifications and new social movements of the people are played out. It is there that in our time, the perplexity of living is most acutely experienced.
>
> (Bhaba, 1994: 170)

The centrality of the city to the formation of subjective identity is an issue that is also explicitly played out in a group of the clay works by the Tanzanian-born artist, Lubna Chowdhary. Born of Indian parents and educated in Britain, Chowdhary has lived in a number of different cities but currently lives and works in London.[19] The centrality of the metropolis to her practice is explored by fellow artist, Sutapa Biswas, in an essay concerning the work of three contemporary women artists of South Asian descent. Biswas explains that:

> Chowdhary has always been fascinated by technology and she views the city as a rational entity. She sees it as representing a sense of order which is challenging to her on a personal level. . . . It is her obsession with order and her desire to achieve it which guides her process as an artist.
>
> (Biswas, 2001a: 206)

As an inhabitant of the city, Chowdhary's work appears to be concerned with mapping her experience of the organic and the rural, derived from her perceptions of rural India onto her experience of the technological, industrial global city. Biswas contemplates the binaries between order/disorder, urban/rural, industrial/organic that structure Chowdhary's practice and dismantles them in a 'differencing' of perspective and experience of the city that is a common feature of her own practice as an artist. In her own work Biswas constantly questions the borders of cultural construction and

Plate 2.3 Hemmed in Two (2000), by Hew Locke (courtesy of the artist)

definition. Thus, in her 1997 fifteen-minute video projection onto frosted glass, *Untitled (The Trials and Tribulations of Mickey Baker)*, Biswas disrupts a whole range of cultural assumptions concerning love, youth, gender, sexuality and authority. The work involved the filming of a naked male model (hired from an acting agency), deliberately requested by Biswas to be both 'of portly stature' and 'in his mature years'. Mickey Baker was placed statically in front of the living room bay window in Biswas' London home for about an hour whilst she filmed him (Biswas, 2001b). The work drew on her own preoccupation and fascination with the charge and anticipation embodied in Vermeer's *Woman in Blue Reading a Letter* (1662–4) and Hopper's *A Woman in the Sun* (1961). Biswas' work expands the framework between artist and viewer. The theme of the piece concerns love, longing and desire, issues that are normally culturally encoded in glossy depictions of beauty and youth but here they are transposed onto the figure of an ordinary ageing man. As Rohini Malik Okon observes:

> *Untitled* is a work about love; more particularly it is about our longing for love and our vulnerability in seeking it. The figure of Mickey Baker, standing naked in front of a window is exposed and immediately vulnerable to the viewer's gaze.
> (Malik, 2001)

Borrowing familiar tropes from art history, Biswas transforms them in a sympathetic rendering of human desire that places her own (art)historically located position as 'authorial' subject central to the work. Mickey Baker performs the piece through Biswas' choreography.

In relation to her contemplation of Chowdhary's perception of the ordered, industrial western city, Biswas intervenes with her own experience of the metropolis:

Plate 2.4 Hemmed in Two, a detail, by Hew Locke (courtesy of the artist)

Plate 2.5 Hemmed in Two, a detail, by Hew Locke (courtesy of the artist)

Plate 2.6 Metropolis (2001), by Lubna Chowdhary

Plate 2.7 Metropolis, a detail, by Lubna Chowdhary

> As someone who has spent most of her life living in London, my own experience of London as a city is different from Lubna's. To me, compared with Paris, Los Angeles or Frankfurt, which appear architecturally more uniform (at least in their centres), London's urban development and growth is organic. One interesting aspect of London is the way in which its huge population spills over onto practically every aspect of the formal city. In India, most of the industrialised cities similarly experience this spilling over of human existence, albeit more visibly. What is interesting about Lubna's work is how she transfers signs of this spillage from one structure onto another in a harmonised and organic way.
>
> (Biswas, 2001a: 208)

In Chowdhary's series of clay works, *Metropolis* (1993–4), small-scale sculptural forms are grouped together in metonymic relationship with urban form. Each object maps an iconography of industrial modernisation, such as a modernist grid pattern, onto an organically derived form. The use of hand-moulded clay worked on an intimate scale to evoke the concept of a large urban mass is a perfect formal solution to Chowdhary's interest in the dialectic between technology and nature, the urban and the rural. *Metropolis* is a 'poetic' and 'mythic' city that addresses the co-existence of order and disorder, 'legibility' and 'indeterminacy'. It belongs with Michel De Certeau's theorisation of 'another spatiality' in which 'a migrational or metaphorical city . . . slips into the clear text of the planned readable city' (De Certeau, 1984: 93 and 199–203). The 'migrational' city in this instance also references

Chowdhary's interest in the nexus between urban London and rural India.

Finally, then, my consideration of *Differencing the City* leads me to the location of the rural in the construction and differentiation of metropolitan identities and in particular the photographic practice of Ingrid Pollard. Born in Guyana, but educated and now long-resident in Britain, Pollard's hand-tinted photographic work confronts and disrupts constructions of 'English' identity as encapsulated in the myth of the rural idyll. *Pastoral Interlude* (begun in 1986) was an early photographic series that focused on the English countryside as a site of racialised dislocation. The images situate solitary black figures in harmony with the landscape, accompanied by a few lines of 'paratext' that remind the viewer of the oppressive and often violent colonial histories that contributed to 'a lot of what MADE ENGLAND GREAT' (Pollard, 1987 *Pastoral Interlude*, text excerpt). Bringing the series into the present, as well as reminding the viewer of the past, Pollard comments that:

> it's as if the Black experience is only ever lived within an urban environment. I thought I liked the Lake District; where I wandered lonely as a Black face in a sea of white. A visit to the countryside is always accompanied by a feeling of unease; dread.
>
> (Pollard, 1987 *Pastoral Interlude*, text excerpt)

Pollard's work deliberately 'differences the city' by leaving it behind and turning to the country in order to interrogate constructions of British rural identity. Interestingly, in a later work, *Miss Pollard's Party* (1992), this 'interruption' is re-presented in the city via a quintessential urban form of visual display, the billboard poster. Although alienation and dissonance are a key feature of these dislocated experiences, Pollard also reflects on their potential as 'a metaphor, a skeleton on which I explored ideas about place, space and where we all fit in to the world scheme' (Pollard, *Autograph*).[20] The globalising vision for her practice is a profound disruption of the more comfortable cultural categories of 'black' and 'urban', that underpin the continued marginalisation of black experience within cultural constructions of British identity – an identity that draws its strength from the selective amnesia of rural 'tradition' in which 'multiculturalism' can often be experienced as anathema.[21] As a comment on the possibilities for urban futures, it would be encouraging to think that the 'differencing' of the city through transformative visual practices, such as the ones that I have explored here, might go some way towards more inclusive approaches to accommodating the needs of the global communities that inhabit our cities. As Anthony King has suggested:

> If urban studies are to address issues of ethnicity, religion, nationalism, cultural identity, they need a language and a set of concepts to do so. The question is not simply who is writing the city, or even where he or she is coming from. It is rather the positionality, and the theoretical language adopted.
>
> (King, 2000: 267)

Perhaps it is through the visual that such a language may be located?

FURTHER READING

Akbar, Shireen (ed.) (1999) *Shamiana: The Mughal Tent*, London: V&A Publications.

Bhaba, Homi K. (1994) *The Location of Culture*, London: Routledge.

Chanting Heads 2001: A Glimpse of Eleven Visual Artists Working in Britain Today (CD ROM), AAVA, NSEAD and the John Hope Franklin Centre in association with the University of East London, the Clore Duffield Foundation and the Arts Council of England. Essays by Rohini Malik Okon; teacher's notes by Paul Dash.

Doy, Gen (2000) *Black Visual Culture: Modernity and Postmodernity*, London: I.B. Tauris.

Ghosh, Amal and Lamba, Juginder (eds) (2001) *Beyond Frontiers: Contemporary British Art by Artists of South Asian Descent*, London: Saffron Books.

Hall, Stuart and Sealy, Mark (eds) (2001) *Different*, London: Phaidon.

Transforming the Crown: African, Asian and Caribbean Artists in Britain 1966–1996 (1997 exhibition catalogue), New York: Franklin H. Williams Caribbean Cultural Centre/African Diaspora Institute.

NOTES

1 I use the term 'post-colonial' cautiously here since, as Sara Suleri has observed, it too often becomes a convenient abstraction 'into which all historical specificity may be subsumed' (Suleri, 1992: 757). Suleri's critique is particularly resonant in its attempts to articulate the epistemological problems attendant on the articulation of a black feminist discourse within Anglo-American academia.

2 Anthony King makes the interesting point, however, that 'there are any number of reasons to challenge the taken for granted assumption that the "natural" study of contemporary urbanism should properly begin with the so-called "industrialist capitalist, modernist city" in the West'; for further elaboration see King (2000). Nevertheless, for an excellent example of recent art-historical writing on European modernity that problematises the status of feminism and colonialism in a reading of visual culture during the second half of the nineteenth century in Britain, see Cherry (2000). For a recent exploration of the notion of 'difference' in black visual culture, see Hall and Sealy (2001).

3 The term is derived from Burman's 1993 mixed media installation for Birmingham Museum and Art Gallery, *Body Weapons and Wild Women Beyond Two Cultures*. It also became the subtitle for Lynda Nead's 1995 monograph on Burman. From this point onwards, throughout the chapter, I shall frequently use the term 'black' as a generic term with reference to artists of Afro-Caribbean, African and Asian descent for the sake of ease of expression although it remains a deeply ambivalent term. As Mora J. Beauchamp-Byrd has stated:

> Although it is commonly used in England to describe all non-whites, the term 'Black', with its broader association with people of African descent, was also problematic. While some Asian artists avoided aligning themselves with a 'Black' political stance, others embraced it, struggling with their colleagues of African descent to ensure that their works were placed in the public realm. During the mid to late 1980s artists and arts administrators of both African and Asian descent became increasingly incensed by arts funding policies which tended to lump all 'Black' artists and organizations together.
>
> (Beauchamp-Byrd, 1997: 25)

For another useful summary and discussion of the politics of the term 'black' see Doy, 2000: 4–10.

4 In saying this, I do not intend to 'lump all black artists' working in Britain today in the same category. Although there are many areas of common ground, particularly in terms of issues of cultural identity politics, there are also, of course, innumerable differences inflected by variations of ethnicity, gender and sexual orientation, some of which I shall consider as the chapter progresses.

5 I would also suggest that the formal properties of montage are an ideal medium for highlighting the juxtapositions implied by the notion of the 'paravisual' as I conceptualise it here: the idea of adding something on top of or next to something else in order to transform the meanings of both.

6 Whilst the notion that the figures are 'representatives of the black community' is a rather loose generalisation, the thrust of King's overall analysis of the work is generally pertinent (King, 1999: 260).

7 Zarina Bhimji, 1992 exhibition *I Will Always Be Here*, Birmingham, Ikon Gallery, 4 April–9 May 1992. A full transcript of the interview can be found on the artist's file at the African and Asian Visual Artists Archive (AAVAA), University of East London.

8 For a more sustained analysis and deconstruction of the visual role of the postcard in the construction of consumable urban identities, see Miles, 2000: 9–35.

9 Arts Council Collection; charcoal, pastel and watercolour on paper; four parts, each 152.5x65cm. In Araeen, R. (ed.) (1989) *The Other Story*, London: Hayward Gallery, the country in the title of the work is listed as Britain but in Beachamp-Byrd (1997: 23) the country in the title is listed as 'England'. Here I have chosen to use Araeen's earlier listing.

10 For a full catalogue of this project including several essays and colour reproductions of the different tent panels, see Akbar, 1999. The catalogue essays outline the practicalities of initiating such a project to ensure that it was a group-centred initiative in which the women took charge of their own decision-making. They also outline the constraints of language, time and cultural differences that were embedded in the project from the outset and how these were resolved to bring the project to successful fruition from both the museum's and the women's point of view. The catalogue also has full listings of all of the panels and their creators as well as all of the exhibition dates and tour venues that hosted displays of the tent after its completion and display at the V&A in 1997. I should like to thank Fahmida Shah and Hajra Shaikh (V&A Education) for lending me a promotional video (shot by Jim Divers) about the project whilst it was still in process.

11 *Shamiana* is the term for a ceremonial tent of the Mughal emperors (Akbar, 1999: 15).

12 In the epigraph to her catalogue essay 'The panels as works of art' in Akbar, 1999: 31, Deborah Swallow cites a passage from Parker as follows: 'The role of embroidery in the construction of femininity has undoubtedly constricted the development of the art. What women depicted in thread became determined by notions of femininity and the resulting femininity of embroidery defined and constructed its practitioners in its own image. However, the vicious circle has never been complete.'

13 See Cherry and Beckett in Deepwell, K. (ed.) (1998) *Women Artists and Modernism*, Manchester: Manchester University Press; Parker, R. (1984) *The Subversive Stitch: Embroidery and the Making of the Feminine*, London: The Women's Press; and Tickner, L. (1987) *The Spectacle of Woman: Imagery of the Suffrage Campaign 1907–1914*, London: Chatto and Windus.

14 The particular issue of able-bodiedness is one that is addressed by the Pakistani-born artist Samina Rana who died in Britain in 1992, wheelchair-bound after a serious injury as a child. For more information about Rana's work and statements by her about the visual dominance of negative stereotypes of disabled people see Samina Rana 'The Flow of Water', in Gupta, 1993 and Doy, 1999: 144–6.

15 Thanks to David Bailey and AAVAA (African and Asian Visual Artists Archive) at the University of East London for facilitating access to information about this and many other works, January 2002.

16 For more information about Meera Chauda's art, see the press release from *Self Assembly*, a touring group exhibition from the City Gallery, Leicester. It notes of the works on display in the exhibition that 'Meera Chauda questions the authority of the stories we are told in her fantastical hybrid images combining Hindu religious narratives and European children's stories'. The exhibition toured to Rugby Art Gallery and Museum from 19 June to 12 August 2001.

17 I borrow the term 'legible' from De Certeau's discussion of the 'indeterminate', see De Certeau (1984) *The Practice of Everyday Life*, Berkeley: University of California Press, 199–203, whilst the concept of 'representational spaces' is suggested by Henri Lefebvre's delineation of the distinction between ordered 'spaces of representation' and the lived spaces of disorder in the everyday; see Lefebvre, Henri (1974) *The Production of Space* (translated by Donald Nicholson-Smith, 1991), Oxford: Basil Blackwell, 38–79.

18 For a sustained critique of homelessness through recent interventionist art projects see Deutsche, R. (1996) *Evictions: Art and Spatial Politics* Massachusetts: MIT Press; Wallis, B. (ed.) (1984) *Martha Rosler: If You Lived Here*, Seattle: Bay Press; and Wodiczcko, K. (1999) *Critical Vehicles: Writings, Interviews*, Massachusetts: MIT Press, amongst others.

19 For further biographical information see Biswas, Sutapa, 'The Autobahn', in Ghosh and Lamba, 2000: 206.

20 Full details and references can be found at www.autograph.abp.co.uk (Autograph: The Association of Black Photographers), directed by Mark Sealy. Further information can also be found on *Chanting Heads 2001* CD ROM (available from AAVAA, University of East London) and from Autograph (1995) *Ingrid Pollard Monograph* London: The Association of Black Photographers

21 In her essay 'But is it worth taking the risk?' Jacquelin Burgess explores how women of different racial origins negotiate access to green spaces and the anxiety that 'difference' often causes to both self and other in open rural

spaces. Her research was based on interviews with different user and non-user groups, 'women and men; young people and those in later years; women of Asian and Afro-Caribbean communities as well as white women'. One of the Asian Women's Groups (aged between 20 and 50) whom she interviewed discussed their feelings about Bencroft Wood: 'But if there are two Asians, a man and a woman, then you do get these looks from the white community. Whereas, when I was walking there – all of us together – I was feeling as if I was walking in my own village back home . . . But I don't think I would go and take my children there. Maybe it's because I'm worried about the safety of my children and the safety of myself. Because no matter what I do, I can't change the colour of my skin. And I might be picked up . . . there might be a few gangs who all come to the woods, and I might be picked up for "fun's sake" or something like that. But I would like to go there and take my children out for a walk' (cited by Burgess, in Ainley, 1998).

3 MALCOLM MILES

Strange Days

INTRODUCTION

This chapter reconsiders dystopia in the literature of urbanism since the 1980s, taking as examples texts by Mike Davis (1990, 1998), Lebbeus Woods (1995) and Neil Smith (1996a). It asks how dystopian critiques might impact urban futures, and contextualises its enquiry through narratives of the Apocalypse, and the urban disaster scenario in film and popular fiction – a scenario brought into sudden and dramatic focus by the events of 11 September 2001.

The chapter asks to what extent tropes such as a cold war on the streets, architecture as war, or a new urban frontier offer alternative strategies for urban change, and compares the dystopian scenario with other approaches in the work of Doreen Massey (1994), Leonie Sandercock (1998a, 1998b), and Iris Marion Young (1990). The chapter argues that dystopian writing succeeds in exposing contradictions in the rhetoric of liberal, or neo-liberal, society but that its limitation is a reproduction of reductive aspects in the dominant narrative. In contrast, though their embrace of difference may shatter the liberal myth of social unity, insights from radical planning and identity politics offer more sustainable urban trajectories.

CONTEXT: STRANGE TIMES

Strange days haunt the city. The towers fall down again and again on television screens – the doom scenario. But is this anything new? Georg Simmel wrote of the 'overwhelming social forces' which threatened the individual's autonomy at the beginning of the last century (Simmel, 1903: 175); and Vita Fortunati writes towards its end that 'no end-of-the-century fails to assert itself as a time of crisis, as a moment of passage' (Fortunati, 1993: 81). It is, of course, not the end-of-the-century which asserts itself but commentators who interpret its significance. But if designation of the year 2000 as *the* rather than *a* millennium shows the ease with which ideologically determined perspectives are naturalised in figures of speech, so also a sense of crisis can be naturalised through reiteration of its link to an end of century or millennium.

Any point in the continuum of time is transitory between a perceived yesterday and tomorrow. The difficulty is that yesterdays and tomorrows become formalised as pasts and futures. These are generalised, like panoramas from a distant vantage point – which is detached but not disinterested (Massey, 1994: 232) – as specific histories. This gives rise to a notion of history as a series of turning points, awareness of which may intimate a potential for transformation, as in 1968, or deepen insecurity when models of society (or the world) which have hitherto seemed stable are threatened. Something like this is encountered in Elizabethan literature, for instances in Shakespeare's *Troilus and Cressida*, Elyot's *The Governor*, or Hooker's *Laws of Ecclesiastical Polity* (Tillyard, 1943: Chapter 2). Insecurity may give rise to nostalgia, as in the popularity of Elizabethan-style furniture suggesting Merry

England during the economic depression of the 1930s, or to incorporation of feelings of individual insecurity within established eschatologies.

A millennium lends itself to such apocalyptic fantasy, and events may support the incorporation. Fortunati lists the signs as she sees them today: war always somewhere; deterioration of the physical environment; failure of economic systems to alleviate abjection; overwhelming growth of scientific discoveries including genetic modification of organisms; and the unchecked concentration of power in a single state – a litany reminiscent of the account of the Last Days in *Revelation*, during which a terrible power is loosed on the world. Fortunati notes that popular novels in the USA at the time of the Gulf War cast Saddam Hussein as the Antichrist rebuilding Babylon (Fortunati, 1993: 81–2), an image coincidentally reinforced by a billboard in Baghdad depicting Saddam towering over the Ishtar Gate (al-Khalil, 1991: 53). But if the message of the billboard is a power it would be fruitless to oppose, and Saddam is not the only candidate for Antichrist, the appeal of popular culture trading on apocalyptic imagery has another dimension.

Disasters produce adrenalin. Like nicotine or alcohol their excitement is addictive, giving a short-term buzz and a longer depressive withdrawal. In late capitalism, short-term contracting and crisis management – management which generates constant crises – are means to disempower employees (Sennett, 1998). Similarly, a global crisis conveyed by news management acts as justification for curtailment of civil liberties and centralisation of power (Vulliamy, 2001). Addiction to crisis is fed by media coverage of disasters and by the disaster genre in popular fiction and the movies, and the ultimate fix, as it were, is given by the scenario of ultimate destruction in prophecies of nuclear holocaust, the toxic soup of pollution, or the threat of globalised terrorism. Questions such as militarisation and pollution do, emphatically, require urgent action, but the point is that from reiteration of doom – for instance in the news media – a sense of crisis is introjected in a process comparable to the self-coercion identified by Michel Foucault (1991a) in the operation of the panopticon prison. Fortunati sees evidence in the fiction of J.G. Ballard, for instance, of a desire for immolation: 'The Apocalypse is . . . no longer fought against; it is embraced' (Fortunati, 1993: 88). Ballard himself writes in *Empire of the Sun*, in the final scene set in Shanghai after Hiroshima: 'The crowds watching the newsreels . . . failed to grasp that these were the trailers for a war that had already started. One day there would be no more newsreels' (Ballard, 1994: 349). The writing makes it seem unavoidable.

Fortunati sees in the apocalypse a 'bipolar internal structure of positive and negative elements' such as the fire and water of cosmic annihilation, a Manichaean vision extended into a triad of destruction–judgement–regeneration which, she argues, betrays fears of separation, loneliness, and death:

> the Apocalypse functions internally rather than externally, as something close at hand . . . exciting a sentiment not unlike the morbid fascination with which, in much contemporary science fiction, one contemplates the end of one's own life and world.
>
> (Fortunati, 1993: 83)

If, then, visions of the world's end convey a depressive state of psyche, and are addictive, this breeds fatalism and limits the capacity for intervention. This, of course, is not the only possible reaction.

Utopia, however, may be more the other side of dystopia rather than an alternative to it. While dystopia is cataclysmic change, utopia rejects all change, its perfection as immutable as it is impossible to attain, hence set away in a non-place (Carey, 1999: xi). Both dystopia and utopia, then, are aestheticised; one negatively as sublime convulsion, the other positively as an equally sublime harmony. And both are imminent – soon-to-come, though in an elastic sense of soon. Imminence, like emancipation in Laclau's (1996: 1–19) analysis, is either radically separated from the present by a chasm, or is conceived within the present as a reaction which reproduces present contradictions in a fantasised future. But the question arises as to the source of the vocabulary in which the fantasy is articulated, which, however much a chasm is projected as wiping out the present world, must nevertheless be within its structures of oppression and repression. Similarly, notions of a golden age project a compensatory inversion of present lacks onto a past sufficiently remote as to accept any inscription. The concept of a world-to-come, however, differs fundamentally from the immanence, or pervasive inherence, of a liberation *now*. Imminence, if not purely aesthetic, requires struggle or sacrifice – seen by Irigaray as patriarchal and extending to 'the sacrifice of nature and the sexual body, particularly woman's' (Irigaray, 1994: 12), while realisation of the immanent comes through recognition, as, for Lefebvre, in moments of liberation within routine (Shields, 1999: 58–64).

The immanent is already here, its dimension space; radically different from normalised reality, it permeates it. Located in this world not the next, immanence can be depicted as a state of repose and is perhaps what Ernst Bloch means in his use of the term 'residual Sunday' in relation to the paintings of Cézanne (Bloch, 1986: 813). For Bloch, Cézanne 'transforms even his still lifes into places . . . in which happy ripeness has settled' and condenses 'whole worlds of repose into these small paintings'; in a landscape of L'Estaque, the 'repose of a settled nature appears' which is un-contemporaneous with capitalist modes of production (Bloch, 1986: 815–16). Hence a universal Sunday for all social classes, like Fourier's libidinisation of work in the phalansteries (Geoghegan, 1987: 20–1), or the representation of an ending of scarcity through factory production in Seurat's *Bathers* (1884, National Gallery, London), though Bloch sees Seurat's work as boring (Bloch, 1986: 814). In European history, however, imminence is privileged over immanence. Bloch sees the emotive appeal of fascism in the imminent aesthetic of a perverse millenarianism; fires in the night and tales of lost crowns in the Rhine project a mythicised past onto a sublime thousand-year future – designed even to ruin well (al-Khalil, 1991: 38–9) – to appeal to the insecurities of the petit-bourgeois class (Bloch, 1991: 56–63, 117–32, 191–4). This looming apocalypse, then, serves power while immanence disrupts the operations of power, as with the abolition of office and property in Joachim of Fiore's Third Kingdom (Bloch, 1986: 509–15).

The refusal of rent not in a celestial future but in the landlord's lifetime is suppressed ruthlessly for obvious reasons. For Bloch such moments of true millenarianism nonetheless stand as evidence of a perpetual desire for freedom: 'The tomorrow in today is alive, people are always asking about it'; and 'there arises in the world something which shines into the childhood of all and in which no one has yet been: homeland' (Bloch, 1986: 1374, 1376). But, if hope is grasped in the visibility of its arisings, fear is reproduced in bleak images in the prosaic violence of the news media, or in movies and popular fiction. When constantly reiterated, these images breed fatalism or cynicism. Meanwhile, ordinary yet immanent disruption, in strikes or the breakdown of computer systems (or hacking), offers a strange enjoyment.

Herbert Marcuse sees abstraction in art as disrupting the norms of perception: new forms including jazz as well as non-objective painting 'dissolve the very structure of perception' to render the familiar object impossible (Marcuse, 1969: 45). But concepts as well as modes of perception may be subject to rupture, including the liberal concept of *civitas* expressed in civic architecture, and the claims of governments to promote unity in social harmony. The popular currency of an apocalyptic image – such as the falling towers of woodcuts illustrating the *Book of Revelation* (Washton Long, 1975: 222–3) – denotes a desire to shatter the coherence imposed by the dominant culture. That coherence is a means to order the unruly, and in popular imagery of the apocalypse it is usually popes and princes, with wealthy merchants, who are trampled under the hooves of the four horsemen's mounts. Or, in high culture, the shattering may take the form of absurdity or degradation, or near-silence. For Adorno, in face of a grim reality, 'New art is as abstract as social relations have in truth become'; and Samuel Beckett charts in his writing a 'shabby, damaged world' which is a 'retreat from a world of which nothing remains except its *caput mortuum*' and in which, at 'ground zero . . . where Beckett's plays unfold . . . a second world of images springs forth, both sad and rich . . . the negative imprint of the administered world' (Adorno, 1997: 31).

Perhaps that imprint is found, too, in the everyday commotion of a metropolis or the rubble of a demolition site. Stephen Barber, writing from travels in post-liberalisation eastern Europe, sees 'exhilaration' in the ferment of cities in transition, in 'the multiplicity of voices passing between the transforming city and the transforming individual' (Barber, 1995: 9); and in demolitions:

> The periodic demolition of entire areas of the city makes its perspectives swing crazily, imparts a sense of exhilaration which is compounded from anticipation of a new 'coming into being', and from a lust for raw destruction.
>
> (Barber, 1995: 29)

It may be that, in a world of routine oppression, images of demolition subvert industrial society's principle of productivity; or that the rawness of breakage counters the inauthenticity of a highly mediated world. But this depends, also, on the destruction being either contained in a normalised reality, or safely distanced in a cultural medium. War ruins speak otherwise, too near. In Dresden, which has a new Hilton Hotel, ruin is being concealed. The Frauenkirche is being rebuilt after standing for fifty years as a blackened testament to the city's destruction in a firestorm caused by allied bombing on 13 February 1945 (Hertmans, 2001: 53–4). The event is still in living memory, but the city's symbolic economy has other demands than mourning.

What will happen at the site of Ground Zero?

THE DISASTER SCENARIO

The destruction of the World Trade Center, New York, on 11 September provided the most reproduced media footage of 2001, and enforces a new twist in the reading of disaster imagery. Reporter Stephen Evans, present in one of the twin towers at the time of the attack, retrospectively described the hourly repetition of the image and its continued use in news programme promotion as pornographic (Wells, 2001). The inference is less a likening of the repeated screening of a single scene to a fixation on body parts under the reductive selectivity of the masculine gaze, than to an obscenity in capitalising on a site of mass death. Perhaps what can be said is that reiteration of the image of Ground Zero reduced the attack to a view of its site. Like the sound bite, also given to repetition, visual encapsulation eliminates complexity. The event is translated to fit a scenario which is then read back into it, at which point the image comes to epitomise the event.

News footage claims objectivity. But while, unlike the movies, it is often produced in a single take, it remains selective of subject-matter and angle, and is edited. The newsreels in Ballard's Shanghai seem to have a function similar to film in splicing together a particular, historically determined and determining narrative. Walter Benjamin, as Esther Leslie (2000) notes, saw a distinction between a false claim for objectivity in photographic images which derive authority from their copy relationship to a world they represent as totality, and a recognition of the intervention of the means of photography into the world represented (Benjamin's notes for the second version of *Das Kunstwerk im Zeitalter seiner technischen Reproduzierbarkeit* cited in Leslie, 2000: 140). Overt intrusion allows, for Benjamin, possibilities of analysis and insights of the unsuspected, and is possible when the audience knows how films are made – that they are not real life. Because film is not only the movies, Benjamin's argument is not incompatible with Adorno's and Horkheimer's that 'Real life is becoming indistinguishable from the movies' (Adorno and Horkheimer, 1997: 126) when put in context. Adorno and Horkheimer write in the 1940s and refer to Hollywood, Benjamin more of the experimental film of Germany and Russia in the 1920s – though in *Moscow Diary* he sees Russian film as more limited by censorship than theatre, and asks whether film, 'one of the most advanced machines for the imperialist domination of the masses', can be used for purposes such as anti-bourgeois satirical comedy (Benjamin, 1986: 55). For Benjamin, however, film retains a democratic life at least in a visibility of alienation. Leslie paraphrases Benjamin:

> Film imprints on celluloid the alienated existences of humans. Simultaneous to the forfeiture of aura and the loss of presence of the 'here and now', a sense of shattered totality of personality is promoted by the stage actor . . . the film actor does not play a coherent role, but rather a disjointed series of fragments, a number of efforts and essays.
>
> (Leslie, 2000: 152)

Benjamin sees film production in repeated takes as equivalent to the supervision of workers in offices and factories. Despite a seamless visuality audiences can, he thinks, identify with this alienation which speaks their own.

The televised image of Ground Zero, sequentially reproduced countless times, offers less scope for intervention. The reproduction of the image and its translation into narrative instead provides legitimacy for security responses, and normalises the event. It implies required responses to a plot already scripted. Normalised through reiteration, news becomes more like the movies, or like Adorno's (1991) analysis of them: 'film provides models for collective behaviour. . . . The movements which the film presents are mimetic impulses which, prior to all content and meaning, incite the viewers . . . to fall into step' (Adorno, 1991: 158). But if news footage reduced the events of September 11th to the image of their site, the responsibility of critique is to draw out and reclaim historicity. Susan Buck-Morss writes that the events bring into focus a 'wild zone of power, barbaric and violent' occupied by the machine of national security (Buck-Morss, 2001: 6). She notes that on another September 11th in 1973 the US government engaged in murder in support of the military coup in Chile, and the similarity between the US School of the Americas – a training camp for armed insurgents – and al-Qaeda camps in Afghanistan. She reminds the reader that the national security state is served by the attack while applying selectively and unequally a concept of human rights which, for Buck-Morss, is indivisible. This leads to realisation that the US, in its global security mission, is 'the violent arm of global capital' (Buck-Morss, 2001: 7–9). Global capital, it could be added, was hardly disrupted in its operations, moving to back-up facilities as the mass public stood mesmerised in front of television screens. What emerges is the continuity of the wild zone of power.

If the media representation of September 11th conceals that zone, are other media more open? For Benjamin, film is a democratic cultural form because spectators can insert themselves into the plot to playfully imagine other ways in which it might end. Readers of academic texts might expect to do this through critical appraisal, as Buck-Morss assumes. And perhaps the disaster movie subverts both coherence and the dominant rhetoric of productivity, though on 12 September 2001 several disaster movies were put on hold as too hot to handle. This raises a deeper question, which is whether dystopian writing on cities, borrowing its tropes from the disaster genre, is not so much overshadowed by September 11th as brought into a new focus by the continuity exposed by Buck-Morss. Does dystopianism trade on the doom scenario, drawn into its adrenalin-producing attraction? If so, does it reproduce the fatalism of the plot it seeks to criticise?

THE DISASTER SCENE: LOS ANGELES

In *City of Quartz* (1990), Davis portrays Los Angeles as a city divided along lines of class and ethnicity, and turned by the siege mentality of its affluent class into a fortress. Recognising the internal need of the security state for enemies, he identifies a 'cold war on the streets' (Davis, 1990: 234) against homeless people and members of minority publics. Although written before the 1992 riots following the acquittal of four white policemen filmed beating up black driver Rodney King, the book charts a continuity of oppression and resistance in which the Watts rebellion of 1965 is a preliminary skirmish.

The assault on the poor is carried out, as Davis makes clear, by design rather than oversight, as the city authorities employ the skills of designers as a weapon:

> the city is engaged in a merciless struggle to make public facilities and spaces as 'unlivable' as possible for the homeless and the poor.
>
> (Davis, 1990: 232)

In the absence of mass deportations the street environment is hardened by 'bum-proof' benches and outdoor sprinklers that switch on randomly throughout the night. Cyanide is not added to the garbage but a restaurant has bought a $12,000 steel, bag-lady-proof trash cage (Davis, 1990: 233). Illustrating how the poor are forced through everyday situations to act the negative persona imposed on them, Davis notes that the number of public lavatories in downtown Los Angeles is lower than in any other US city; while middle-class citizens use the washrooms in art galleries and restaurants, the homeless, denied access to such plush surroundings, have nowhere but the street to piss.

The demonisation of the poor and evicted is complemented by the extent to which affluent neighbourhoods are protected against their intrusion by fortress architecture:

> The carefully manicured lawns of Los Angeles Westside sprout forests of ominous little signs warning: 'Armed Response!' Even richer neighbourhoods in the canyons and hillsides isolate themselves behind walls guarded by gun-toting private police and state-of-the-art electronic surveillance . . . . In Hollywood, Frank Gehry . . . apotheosizes the siege look in a library designed to resemble a foreign legion fort.
>
> (Davis, 1990: 223)

Looking to Foucault, Davis sees coercion in the panopticon-like malls built in inner-city districts to profit from the consumption needs of the poor, and echoed in the militarisation of housing: 'the counterpart of the mall-as-panopticon-prison is the housing-project-as-strategic-hamlet . . . fortified with fencing, obligatory identity passes and a substation of the LAPD' (Davis, 1990: 244). Fortress Los Angeles, for Davis, is a city of massive insecurity in the conditions of minorities, mirrored by a psychological insecurity on the part of the rich who fear the incursion of the poor and resort to increasing levels of security. The visibility of such measures reinforces the myth of insecurity: 'the market provision of "security" generates its own paranoid demand . . . becomes a positional good defined by income' (Davis, 1990: 224). Davis notes that while rising crime in black neighbourhoods is largely contained within them, white perceptions are heightened 'through a demonological lens' (ibid.).

The production of insecurity through technologies of security is described, also, by Zygmunt Bauman (1998). Citing G.H. von Wright (1997), Bauman sees a new world disorder:

> Thrown into the vast open sea with no navigation charts and all the marker buoys sunk . . . we have only two choices left: we may rejoice in the breath-taking vistas . . . or we may tremble out of fear of drowning. One option not really realistic is to claim sanctuary.
>
> (Bauman, 1998: 85)

The new world disorder entails a polarisation of wealth and deprivation (see also Polet, 2001: 4–6; de Rivera, 2001: 64–6); a diminishing of the power of states compared with that of global capital (see also de Rivera, 2001: 45–54); an introjection of consumerism which spawns a new divide between those for whom consuming is naturalised and those for whom it is beyond reach; and the production of new classes of migrant, both elite and abjected, as abolition of entry controls for some is matched by greater controls for others (Bauman, 1998: 85–9). Similarly, appeals to globalised fora such as the General Agreement on Tariffs and Trade (GATT) allow producers in the affluent world to enjoy a deregulation of markets in which poorer producers were hitherto protected by national laws; and law enforcement agencies ignore corporate crimes undetectable within a mass of ordinary dealings to target instead groups who reject or are denied the benefits of neo-liberalism. Foreseeing the present emphasis on security, Bauman concludes:

> Rejection prompts the effort to circumscribe localities after the pattern of concentration camps. Rejection of the rejectors prompts efforts to transform the locality into a fortress. The two efforts reinforce each other's effects . . . fragmentation and estrangement 'at the bottom' remain the twin siblings of globalization 'at the top'.
>
> (Bauman, 1998: 127)

Bauman's litany has much in common with Davis', and resembles Fortunati's in its bleakness. Unlike Fortunati's, however, it is rooted in political economy rather than apocalyptic myth, and so opens a possibility for intervention. For Bauman, understanding the operations of globalised capital is a prerequisite to critical positions which challenge it. In his introduction, Bauman writes that 'the trouble with the contemporary condition . . . is that it stopped questioning itself . . . . The price of silence is paid in the hard currency of human suffering' (Bauman, 1998: 5). Davis exposes the contradiction between affluence and deprivation, but while both accounts draw on research, and Bauman's on textual rather than experiential knowledge, it may be Davis who is more distancing. While Bauman analyses conditions, Davis attaches them to an over-arching scenario of dystopia – taken further in *Ecology of Fear* (1998), in which he conflates a narrative of Los Angeles with the scenario of the disaster movie – the impact of which is to close argument and comply with doom.

There are two difficulties: the concept of fortress Los Angeles, while in many ways insightful, limits observation to a vision of doom; and, from that, is marked by the absence of other voices. Minorities appear in the narrative only as objects of oppression, their voices mute and their organisation invisible. The latter point is made by Edward Soja in *Postmetropolis* (2000), in which he reproduces several extracts from the book. He sees Davis (1990) as appealing both to the hard left and to despairing liberals, but:

> as a framework for interpreting and acting politically . . . this piling of blame on neo-liberalism as the all-powerful right arm of the new capitalism has some serious shortcomings.
>
> (Soja, 2000: 302)

Soja identifies these as despondency in the face of overwhelming odds, failure to depict the range of oppositional movements within the situation, and closure of argument in a rhetoric of social warfare and no-longer adequate notion of class struggle. Similarly, James Duncan sees Davis' portrayal of the militarisation of urban space as accurate but lacking explanation: 'The city . . . becomes the stage upon which that well known morality play "class struggle" is enacted' (Duncan, 1996: 259). Davis himself writes of an 'explicit repressive intention, which has roots in . . . [a] history of class and race warfare' (Davis, 1990: 229). Suggesting that Davis falls into the trap of reproducing the scenario he criticises, Soja writes: 'Davis walls off his City of Quartz from the new cultural politics and the most insightful feminist, postcolonial, and postmodern critiques' (Soja, 2000: 303) – something of which Soja himself was once accused (Massey, 1994: 218).

In *Ecology of Fear* (Davis, 1998) the city approaches catastrophic failure. The real and the foreseen are conflated in a doom scenario framed by a worsening of conditions. Looking back on *City of Quartz*, he reflects:

> Events since the 1992 riots – including a four-year recession, a sharp decline in factory jobs, deep cuts in welfare and public employment, a backlash against immigrant workers, the failure of police reform, and an unprecedented exodus

> of middle-class families – have only reinforced spatial apartheid.
>
> (Davis, 1998: 361–2)

But rather than criticise these developments in terms of the economic and political mechanisms by which they are determined, or, as Soja suggests indicate resistance, Davis subsumes history to fantasy in the form of the disaster script. After a major storm and flood in January 1995, for instance: 'What was exceptional was not the storm . . . but the way in which it was instantly assimilated to other recent disasters as a malevolent omen' (Davis, 1998: 6). Which may be the case, but suggests a mystification in which Davis' account, too, is drawn into an adrenalin-producing excitement from which there is no exit.

From mass culture, particularly the movies, as well as the official blueprint *LA 2000: A City for the Future* – which has a final section by Kevin Starr on what might occur if the city approximates to its image in popular culture as a 'demotic polyglotism ominous with unresolved hostilities' (cited in Davis, 1998: 359) – Davis argues that the city incorporates the disaster movie in its imagined future. Fleshing out the disaster scenario, Davis cites Ridley Scott's *Blade Runner*, John Carpenter's *Escape from LA*, and Kathryn Bigelow's *Strange Days*, and the moral right's final solution in the form of Christian evangelist Pat Robertson's novel *The End of the Age* in which, as Davis puts it, 'God himself decides to flush Los Angeles down the toilet' with a meteor (Davis, 1998: 339).

If Los Angeles stands in this genre for the anti-matter of the American dream – the end of community along with cherry pie and the white picket-fence, recovery of which is a fantasy currently revived by the Disney Corporation in their venture into real estate at Celebration, Florida (MacCannell, 1999; Frantz and Collins, 2000) – the scenario is not exclusive to postmodernity. Of *Blade Runner*, Davis says 'peel away the high-tech plumbing . . . what remains is the same vista of urban gigantism and human mutation that Fritz Lang depicted in *Metropolis*' (Davis, 1998: 360).

Perhaps the underpinning theme of Davis' books is the failure of a Modernism which tried to engineer a new society but got widening deprivation in a city determined more by the demands of capital than planning. Like Bauman's analysis this derives from political economy and attacks neo-liberalism's pursuit of exploitative monopolies through deregulation. But the doubt which nags (this reader) while reading Davis' two narratives is whether they are so allied to dystopian imagery, presenting this as *the* narrative of Los Angeles, that they fail to offer any other trajectory. The doubt is magnified by a passage towards the end of *Ecology of Fear* in which Davis cites the thermal spot of the 1992 riot seen from space as a unitary geophysical phenomenon. He sees the riot as composed of individual human acts but there is a naturalisation and a distancing in his concluding sentence: 'the city that once hallucinated itself as an endless future without natural limits . . . now dazzles observers with the eerie beauty of an erupting volcano' (Davis, 1998: 422). The rioting city is thus aestheticised, its beauty dependent on a distanced viewpoint which allows reduction to a single visual image. This is more than a literary device or nice way to round off the discussion and says much about the narrative voice in which the account is given. For Massey, visuality is the sense which gives most mastery, epitomised by the eye-in-the-sky viewpoint of the city plan. Massey writes:

> The privileging of vision impoverishes us through deprivation of other forms of sensory perception . . . the reason for the privileging of vision is precisely its supposed detachment.
>
> (Massey, 1994: 232)

The distanced view is impersonal and allows an illusion of omnipotence, or what Fortunati describes as an 'awareness of his [*sic*] superiority' on the part of the speaker for the elect (Fortunati, 1993: 83). And the view from space merely extends the distancing aspect of a view of a city from a vantage point, be it the city walls or top of a high-rise tower. Arturo Escobar, similarly but from a different context, argues that the view from space 'was not so great a revolution . . . [but] only re-enacted the scientific gaze established in clinical medicine' (Escobar, 1996: 49). Escobar's concern is to critique the notion that sustainability involves global management of natural resources, which he sees as ensuring 'that the degradation of the Earth be redistributed and dispersed through the professional discourses of environmentalists, economists, geographers, and politicians' (ibid.). Perhaps the reduction of the 1992 riot to a visual image, like that of Ground Zero on the television screen, has a certain hypnotic fascination, but dissipates the critical case which Davis seeks to make in his writing, placing Los Angeles beyond intervention, and, crucially, in a narrative contained within either literature or the urban counterparts of those academic and professional disciplines to which Escobar refers.

If Los Angeles is for Davis reduced to a beautiful Armageddon, a similarly bleak scenario is donated by Lebbeus Woods (1995) employing the trope of architecture as everyday warfare. Everyday reality, he asserts, conceals violence. Likening this to military conflict – using images of war-torn Sarajevo as illustration – Woods re-presents the city as a site of the duality construction–destruction: 'in order to build, something must be destroyed' (Woods, 1995: 49). This has a certain naive reality but Woods delights in provocation, beginning his text with a confession that his trope 'has upset many people' (Woods, 1995: 47). He continues that 'Building is by its very nature an aggressive, even warlike act' (ibid.: 50), and that architects are engaged not in the creation of beauty but conquest of space (ibid.: 52):

> A building will only be beautiful and useful for those who benefit directly and tangibly from its existence. As for others . . . the building is little more than an instrument of denial.
>
> (Woods, 1995: 50)

Woods calls for a renewal of ethics in architecture, but it is difficult to resist the interpretation that he derives from his equation of architecture and war something of the excitation of the war comic. When he writes, citing Marshall Berman (1983), that 'The wrecking machines that levelled houses and urban blocks were no less destructive to culture than if they had been the tanks and artillery of an attacking army' (Woods, 1995: 50), he conceals the nuances of Berman's text, in which Moses ambivalently 'seemed to glory in the devastation', and was able to convince New Yorkers that his urban vision represented progress: 'Moses was destroying our world yet he seemed to be working in the name of values that we ourselves embraced . . . he genuinely loved New York – in his blind way – and never meant it any harm' (Berman, 1983: 293, 295, 307).

Something of the same capture by a resonant phrase may pertain to Neil Smith's treatment of gentrification in Manhattan. Smith writes of the new profitability of neighbourhoods such as the Lower East Side, once abandoned to the poor and migrant, associating new urbanism with a repolarisation of class linked to globalisation and expressed in a renaming of *Loisaida* as the East Village. He identifies a naturalisation of the process in advertising rhetoric, and describes its effect in producing an affluent bohemianism before a downturn in real estate values results in some neighbourhoods becoming part of what he terms a revanchist city of increasing attacks on minorities and violent removal of squatters.

Conflict in urban growth was previously naturalised by E.W. Burgess (1925) in his use of a biological model to locate the city outside history, that is, beyond human intervention. In this way the deregulation of a free market economy would be likened to a state of nature, and regulation – or planning – to some manner of curtailment. As Elizabeth Wilson argues, however, planning also served the needs of capital in producing orderly publics, while its radical potential is to replace its authoritarianism with acceptance of the specifically urban gratification of carnival (Wilson, 1991: 158). In a reconsideration of *Sphinx in the City*, Wilson (2001) alludes to Smith (in an earlier publication) as offering a 'strong critique of development' but sees his effort overtaken by a wider acceptance of consumption in postmodernity, and in 'the fatalism of nostalgia and the fascination of cities in decline' (Wilson, 2001: 66). Smith questions naturalisation, and quotes one developer's evident irresponsibility: 'To hold us accountable for it is like blaming the developer of a high-rise building in Houston for the displacement of the Indians a hundred years before' (cited in Smith, 1996a: 27). But does Smith yet subscribe to that fascination of decline in adopting the trope of a mythicised frontier?

The frontier metaphor is taken from a real estate advert proclaiming 'The Armory Celebrates the Taming of the Wild West' (Smith, 1996a: 14). Artist Martha Rosler, problematising the links between art-space and eviction in SoHo in her project *If You Lived Here*, cites another image from the real estate supplement of *The New York Times* showing three executives, one female, wearing business suits and western hats under the caption 'The top guns of New York are heading West' (Rosler, 1991: 18). Both adverts refer to buildings in midtown, but for Smith it is the Lower East Side where the conflict comes to a head in clearance of Tompkins Square on 6 August 1988 (Smith, 1996a: 3–29), and alongside reference to real estate advertisements the notion of a new frontier is borrowed from the words of Lower East Side authors: '[a] frontier where the urban fabric is wearing thin' and 'Indian country: the land of murder and cocaine' (Smith cites J. Charyn (1985) *War Cries over Avenue C*, New York: Donald I. Fine; and J. Rose and C. Texier (eds) (1988) *Between C and D: New Writing from the Lower East Side Fiction Magazine*, New York: Penguin – Smith, 1996a: 7–9). Smith's critical framework is drawn from political economy, and his analysis is compatible with Sharon Zukin's (1982) account of loft development in SoHo, and (1995) critique of New York's symbolic economy. What he adds is the overlay of the frontier as a description of an affluent bohemianism. Smith writes:

> Hostile landscapes are regenerated, cleansed, reinfused with middle-class sensibility; real estate values soar . . . . As with the Old West, the frontier is idyllic yet also dangerous, romantic but also ruthless.
>
> (Smith, 1996a: 13)

Smith cites *New York* magazine to the effect that 'A sort of wartime mentality seems to be settling onto New Yorkers affected by the housing squeeze' (cited in Smith, 1996a: 26), and justifies his use of the frontier image by reference to frontier motifs in everyday consumption, such as the growth of tex-mex restaurants, cowboy fashion, and appropriation of native American goods or styles, such as Navajo rugs and Lombak baskets, in interior design boutiques such as Zona (ibid.: 15). This follows a nineteenth-century use of frontier zones in the naming of buildings – The Dakota Apartments, for instance. Smith concludes that 'the frontier ideology continues to displace social conflict into the realm of myth' while reaffirming class divisions. It is not neutral 'but carries considerable ideological weight' (ibid.: 17). But if that is the case, does Smith himself subscribe to the myth in reproducing it as the presiding motif of his book? Rosalyn Deutsche (1991), too, addresses the naturalisation of eviction, and cites Mayor Koch:

> These homeless people, you can tell who they are . . . sitting on the floor, occasionally defecating, urinating, talking to themselves. . . . We thought it would be reasonable for the authorities to say 'You can't stay here unless you're here for transportation.' Reasonable, rational people would come to that conclusion.
>
> (cited in Deutsche, 1991: 159)

But Deutsche moves to a Lefebvrian analysis of spatial production as counter to naturalisation. Smith, producing his own mythicisation of the frontier superimposed on its history and 1980s re-presentation, on the other hand, tends to affirm conflict as a natural condition. His strategy for change is based on support for squatters in their fight for housing, and a retaking of the urban frontier as resistance to the dominant structures of wealth and power. Having begun from advertising images in which the new cowboys are affluent professionals, and residual or migrant publics presumably the Indians, Smith subverts the narrative so that the squatters are the pioneers – taking the place of the rich but perhaps assuming also some of their contradictions. This, however, has some historical accuracy. Though likening attacks on homeless people to the extermination policy of Custer's Indian wars, Smith points out that, prior to the Homesteading Act of 1862, the pioneers who appropriated the west were illegal squatters 'democratizing land for themselves', to whom the Act lent legitimation (Smith, 1996a: 232). The squatters organised themselves in clubs and provided a model of self-empowerment, as Smith sees it, for squatters' organisations today. Eliding the fact that the squatters took their land from a native population subjected to genocide, Smith urges a reinscription of the class base of the myth to embrace the city as frontier. As melodrama it works well; and as reassertion of a call for social justice it is refreshing. Although, unlike Davis, Smith gives considerable attention to mobilised resistance, like him he gives little room to other voices, and like him also takes class as a privileged category; in reproducing the violence on which the USA was built he depicts a trajectory in which means are determined by ends but ends repeat history. But if not that, what else?

OTHER VOICES, OTHER STORIES

Three issues are raised: the difficulty that dystopian texts may themselves be drawn into the impending myth with which they are aligned; the tendency of narratives to generalise, distancing the object of the story rather than engaging with the conditions in which it is produced; and the absence in much dystopian writing of voices other than the narrator's (which assumes the position of a speaker for the elect). These issues suggest a need for recognition of plurality, even, or perhaps especially, if this ruptures the coherence of the narrative – a rupture which may have the same attraction as the demolition site (Barber, 1995). If the dominant narrative is characterised by seamlessness, then a counter-narrative which offers only another foreclosure of argument is no alternative. So, Davis' (1990) account of Los Angeles as 'a deracinated urban hell', as Sandercock puts it (Sandercock, 1998c: 7), based on an understanding of its past through the genre of *noir* fiction, ruptures the notion of a land of opportunity, but *noir* mythologises, and Davis' narrative re-mythologises, dysfunctionality. For an alternative, it may be helpful to shift emphasis away from explanations and solutions towards interrogation of assumptions and the generation of processes which themselves embody liberation; that is, processes in which the new society, as it were, *is* the working through of the process as a liberating series of experiences, rather than processes which are seen as steps towards a greater, usually sublime, end.

The idea of such a process, as the end as well as means, sits well with Lefebvre's emphasis on the everyday, in his efforts both to reclaim value for experiential spaces and to see moments of liberation within routine. This implies that it happens regardless of interpretations from privileged positions, and exposes the structural failure of the modern concept of an avant-garde leading society in its new direction (Nochlin, 1968), thereby reproducing the power relations of the old regime. John Roberts writes:

> Political art . . . assumes that those whom the artwork is destined for (the fantasized working class) need art in as much as they need Ideas in order to understand capitalism . . . . There is never a moment's recognition that people are already engaged in practices . . . which are critical and transformative.
>
> (Roberts, 2001: 6)

In other words, those who interpret the world *for* others exclude the possibility that those whom they classify as other might interpret the world *for themselves*. Thinking of Benjamin's critique of film, this is as if to say a counter-narrative which retains a privileged voice allows its audiences as little play of reconstruction of the plot as the narrative it counters.

A field in which much work has been done on alternative approaches is development studies. In its expanding literature, the knowledges of local people have frequently been seen to be more appropriate in addressing environmental problems than external expertise (Pradervand, 1989; Peet and Watts, 1996; Guha and Martinez-Alier, 1997, for examples), and the governing principle of development itself has been questioned. Richard Peet and Michael Watts, for instance, apply discourse theory:

> Such reconceptualisations [as those of the World Bank] of power-knowledge . . . see development as perhaps *the* main theme in the Western discursive formation . . . the passage of time is understood developmentally, that is, 'Things are getting better all the time'. . . . This seizes control of the discursive terrain, subjugating alternative discourses which Third World peoples have articulated to express their desires for different societal objectives . . . . Through critique, post-structural theory wants to liberate aspirations.
>
> (Peet and Watts, 1996: 16–17)

Poststructuralism, then, recasts the concept of development as a colonial strategy as disempowering as any other, and the argument has much to offer critiques of urban planning – in which development is seen, also, as the overriding aim, and is generally conceived as a top-down process. Again, it is the exclusion of other voices, or the voices of those regarded as other, which characterises the dominant approach. In contrast, Paolo Freire, constructing a pedagogy of liberation from his experiences of adult literacy programmes in Brazil in the 1960s, writes:

> Those who have been denied their primordial right to speak their word must first reclaim this right and prevent the continuation of this dehumanizing aggression.
>
> (Freire, 1972: 61)

Accordingly, Freire sees revolutionary leaders not taking messages of salvation to the people but establishing a dialogic society, a process which is itself the new which begins to permeate the old.

But what is a dialogic society? Michael Mayerfeld Bell, drawing on Mikhail Bakhtin, sees culture as a dialogic process. He does this partly to avoid culture's reification and to recognise its creativity, and in part to see it 'as collective agency in face of frequently bad odds' (Bell, 1998: 52). Bell emphasises that the conversations envisaged are both predicted and unpredicted, and continues:

> Difference is central to the conversations we expect and hope to have . . . . Cultural understanding . . . depends on drawing boundaries, constructing categories and differences . . . . But it also depends on transcending those boundaries . . . . In conversation we discover our boundaries and transcend them as we interact with difference – that is with each other – in a collective act of dialogic improvisation.
>
> (Bell, 1998: 53)

This is close to Arendt's (1958) idea of natality, in which a mature self is produced in the interactive visibility of the perception(s) of others. For Arendt natality is aligned with interaction through speech and act in a public sphere, the essence of which 'is to arouse the impulse to freedom and let it shine' (Curtis, 1999: 73–4). Dorothy Smith, in the same volume as Bell, notes that for Bakhtin the novel 'as a literary form is of, and embedded in, a society of diverse forms of speaking and writing', and argues that social science, too, 'is embedded in the heteroglossia of a diverse society' (Smith, 1998b: 65). Buck-Morss (2001), in the conclusion of her commentary on the events of September 11th, reaffirms the need for a public sphere, in the conception of which she is indebted to Jürgen Habermas, as location of informed action. She sees the alliance of global power and global capital of the US national security state as a 'reactionary cosmopolitanism' lacking a radical sense of social justice, and Al-Qaeda as standing for a 'reactionary radicalism' lacking a cosmopolitan idea of a public sphere (Buck-Morss, 2001: 9). The means to liberation, this implies, is a radical cosmopolitanism – which Buck-Morss defines as a global public sphere.

PLANNING FOR DIFFERENCE

Sandercock, in her essay on the death of modernist planning (1998b), begins with an image of Los Angeles:

> Twilight. . . . The city is burning. As the smoke and glow from the fires begins to rise over the city, millions of horrified citizens huddle in front of TV sets which transmit images of rioting and looting.
> (Sandercock, 1998b: 163)

Her approach differs, though, from Davis' (1990). Sandercock, probing rather than being the *flâneuse*, problematises the image of the riots of 29 April 1992 as (black) race-riots, noting the ethnic diversity of the neighbourhoods involved and predominance in them of Korean and latino-owned shops, and that 1,200 of the 16,291 people arrested were illegal immigrants. Citing Davis (1990) and Soja (1989), she sees the image of 'a worst-case scenario of racial and ethnic conflict, social polarisation and residential segregation' as overused (Sandercock, 1998b: 163–4), arguing that the problem, as in the gulf between social inclusion in a networked society and exclusion for those who service it, is not unique to Los Angeles, and, from Seyla Benhabib, that the elision of voices in generalising accounts glosses over the multi-identity of communities in which 'every "we" discovers that it is in part a "they"' (Benhabib, 1995, cited in Sandercock, 1998b: 164). This is an antithesis of Davis' image of the hot spot seen from space. Sandercock adds that, while in the 1970s feminist planners 'swapped "war stories" and gave each other support at conferences', in the early 1980s 'The "we" of feminist urban analysis was challenged by "Other women" who argued that the "we" had never included "them"' (Sandercock, 1998b: 168). The challenge for radical planning, then, is to include such voices:

> The voices from the borderlands also inhabit and embody the new cultural politics of difference, complicating that politics with their intersectionalities of race, class, ethnicity, gender and sexual preference formations of 'difference'. Together they suggest that social justice . . . is inseparable from a respect for and an engagement with the politics of identity and difference.
> (Sandercock, 1998b: 169)

In a politics of difference which assumes the contestation and negotiation of identity, the understanding of identity produced cannot, according to Young (1990), be contained within the liberal myth of social unification; nor is it served by the liberal notion of equality as equal treatment under a law made in the image of one, dominant social group. A question which arises is how the marginalised gain the capacity to be heard; but another, not trivial, and which can be taken literally or metaphorically, is where they do this.

Arendt (1958), observer at Eichman's trial in Jerusalem, sees the oblivion of oppressed groups in enforced silence or invisibility as both acutely painful and a precondition of persecution. Kimberly Curtis writes in a study of Arendt's political philosophy:

> This form of injustice is . . . prior to social injustice in the sense that degradation of obscurity is a primary precondition of our capacity to inflict and sustain the suffering involved in the many forms social and economic inequality take.
> (Curtis, 1999: 68)

But if Arendt, drawing on classical philosophy, sees democracy as beginning in a public realm she contrasts in its possibility for freedom with a private realm of secrecy, the concept seems out-dated. Not only does privatised space, as in the mall, encroach on and erode public space (as space of open access and unplanned mixing), but residual public spaces such as the city square or boulevard are in any case both gendered and marginal to the operations of power. If, as Bauman (1998) observes, globalised capital is more influential than many states, or when parliaments become tourist attractions, what is the point of demonstrating outside their buildings?

The public realm, then, as where identity formations (or natality for Arendt) are negotiated, can no longer be considered only or primarily as a physical space. When lattices of communication, not just capital, are globalised (Eade, 1997), communicative action, too, may define new spaces which are, in a conventional sense, without boundary yet bounded by access, whether to technologies of communication or to influence. If the challenge for radical planning is inclusivity, this entails a new complexity and a new kind of engagement. It means, too, Young argues, a deconstruction of the ideals of liberty and equality:

> In recent years the idea of liberation as the elimination of group difference has been challenged by movements of the oppressed. The very success of political movements against differential privilege and for political equality has generated movements of group specificity and cultural pride.
> (Young, 1990: 157)

Young urges differential treatment for oppressed groups, and a reconsideration of the concept of representation in a democratic society. Because the equality granted in law is not reflected in social life, oppressed publics are more empowered by asserting difference than by accepting assimilation to a norm produced in the image of privileged publics: 'Social movements asserting the positivity of group difference have established this terrain, offering an emancipatory meaning of difference to replace the old exclusionary meaning' (Young, 1990: 169). Instead, then, of a society in which from all is produced One(ness), Young anticipates a society in which all remain all. This avoids the reductive generalisation which, in bringing everyone 'to the unity of a common measure', marks as deviant 'those whose attributes differ from . . . the norm' (Young, 1990: 169).

From Young's analysis emerges a political model of group representation: 'a democratic public should provide mechanisms for the effective recognition and representation of the distinct voices and perspectives of those of its constituent groups that are oppressed or disadvantaged' (Young, 1990: 184). This requires support for self-organisation, and the generation of policy within groups granted right of veto over the supposed consensus. What it does not mean, however, is the utopian notion of a localised, micro-scale or face-to-face communitarianism. Young argues strongly against such models, seeing their assumed immediacy as false – 'a metaphysical illusion' (Young, 1990: 233). She concludes:

> These arguments against community are not arguments against the political project of constructing . . . a positive group identity and relations of group solidarity, as a means of confronting cultural imperialism . . . [but] affinity cannot mean the transparency of selves to one another.
> (Young, 1990: 236)

Perhaps this suggests the diversity of experience which Wilson sees as exciting and specifically urban (Wilson, 1991: 158). But where does the interaction necessary for liberated identity formation and empowerment take place? If not the public square or common, and not the mall or the designer bar, then where? It would seem in a space the qualitative dimensions of which have yet to be constructed. But if it is not positively, pro-actively constructed, the possibility for democratic interchange may disappear and all determination be subsumed to the illusory and de-politicised choices of consumption.

If the location of an appropriately re-energised planning, in the widest sense, is problematic, then, so also is the methodology. For Sandercock (1998a), as for Young (1990), analysis of inequality has shifted from a class-based framework not only to one of gender and ethnicity but also of contestation within such categories – as indicated by her reference (Sandercock, 1998b: 164) to Benhabib above. Work with mobilised communities entails recognition of their complexities, then, but also of their specific, if un-codified, knowledges. These knowledges may be lent visibility by professional collaboration, but cannot be reproduced by professionals any more than the groups concerned can be represented by others more attuned to, or assimilated into, the dominant society. Sandercock cites Jacqueline Leavitt:

> Community-based groups who develop bottom-up programs are engaged in planning that occurs

> outside the local planning establishment. At some point the people with whom I work will interact with either the planning establishment or other political bodies. They will frequently need to use research I have helped produce with them for that purpose. I may or not be at the meetings: when I am, my role is to validate but not to be their voice. The overall intent of this type of practice is not to create a plan as much as it is to generate a political process that involves plans or programs.
>
> (Sandercock, 1998a: 98)

Such practices require time, and willingness to transgress: 'The identity of the radical planner . . . is that of a person who has, essentially, gone AWOL from the profession, has crossed over "to the other side", to work in opposition to the state and corporate economy' (Sandercock, 1998a: 99–100). For Young (1990), it is likely to be in opposition also to the myth of social coherence which is foundational for liberalism, and reconstituted as an exclusionary device in neo-liberalism and its application to real estate in new urbanism (MacCannell, 1999). Strange days . . . or days which strangers, too, enjoy.

CONCLUSION

Images of a twisted steel frame and accounts of a sickly smoke still rising two months on from what was in part a pile of human ash at Ground Zero add a further chapter to the dystopian story. Repeated broadcasting of the collapse of the twin towers reduces the event to the representation of its site while reiteration of the disaster scenario, not least in academic texts, breeds fatalism. Unifying concepts of society developed in liberalism and exploited in neo-liberalism are no less reductive, and limit the possibilities of liberation to assimilation in an already delineated public realm. The social reform programmes of liberal societies, progressive in their day, are inadequate to the needs of complex societies in a world of globalised communications and increasing migrancy, just as the old revolutionary category of class and model of an avant-garde are inadequate for the complexities of communities of interest articulated by gender, ethnicity, age, and many other variables down to the micro level. Neither will exposure of contradictions itself produce alternative trajectories for urban change. If, as Marx says in his 1845 eleventh *Thesis on Feuerbach*, the point is to change the world through human intervention, then the process must itself be liberating. Perhaps this suggests, following assertion of a common right to speak, a culturation of the planning process – not as aestheticisation but as recognition of the ways of life of dwellers whose lives, as Roberts (2001) argues, already entail transformative acts, and whose knowledges, as work in development studies shows, may offer more appropriate insights than external expertise. This will not produce a new society in place of the old, either descending from heaven or founded on the barricades. It might, however, recognise diverse, non-cohering but subversive and liberating energies within the crevices of the dominant society. In time, no doubt, these will be formalised as the new society, but that is not the point. There is, then, no plan. Only a high-risk handing over of the means in a dialogic space yet to be fully understood, and which is likely to be understood only in its active construction. Otherwise, as Buck-Morss dreads, an all-out war on terror which protects capital and further marginalises those already excluded from affluence (Buck-Morss, 2001: 9). This, of course, is not a choice. It may be both for a long time yet.

CODA: STRANGE DAYS HAVE FOUND US

The Doors sing on their album *Strange Days*, recorded in the 1960s in a climate of dropping out, turning on and being free but also quite often depressed and not unafraid of nuclear holocaust, as well as of the usual separation, loneliness and death:

> *Strange days have found us*
> *strange days have tracked us down*
> *threaten to destroy*
> *our casual joys . . .*

The cover of the album is illustrated in Debord's *Society of the Spectacle* (1977, not paginated – but only £4.95). It does not illustrate anything in particular, and is one of four funny pictures, but quite nearby the text reads, 'Where the real world changes into simple images, the simple images become real beings and effective motivations of hypnotic behaviour' (paragraph 18). Later, Debord writes, 'Urbanism is the modern fulfilment of the uninterrupted task which safeguards class power' (paragraph 172). Debord's statement needs updating in light of insights from Massey, Young and Sandercock relayed above, but if the question is how cities might not correspond to the terminal scenes of Kathryn Bigelow's *Strange Days*, The Doors have an answer at the end of the album:

we want the world and we want it now

Immediate gratification of all desires: it's like Marcuse's pronouncement of society as a work of art at the Dialectics of Liberation Congress at the Roundhouse, London in July 1967:

> What does this mean, in concrete terms? . . . creative imagination and play, becoming forces of transformation. As such they would guide, for example, the total reconstruction of our cities. . . . The sensitivity and the awareness of the new transcending, antagonistic values – they are there. And they are there, they are here, precisely among the still non-integrated social groups and among those who . . . can pierce the ideological and material veil of mass communication and indoctrination.
>
> (Marcuse, 1968: 186–7)

Well . . .

PART II
Moving

A loft shop sign, Manchester, UK (Photo: Malcolm Miles)

INTRODUCTION

The image presiding over Part I is Klee's angel at the momentarily still threshold. Part II crosses the threshold into a city of movement, a city such as Simmel's Berlin (1903) always in flux, disconnected from origins and with no predetermined end. Simmel describes the conditions of metropolitan cities which to him seemed radically new, but as the trams are housed in museums and the department stores give way now to malls and franchises, the vocabularies as well as the categories of urban discourse have changed. Instead of movement, or the Futurists' celebration of the speed of a racing motor car, the key word might be mobility: geographical mobility in business travel and tourism; social mobility through education and fluid patterns of employment; cultural mobility in contrasting networks; political mobility as electorates switch allegiances more frequently, adhere to single-issue campaigns, or switch off altogether; economic mobility for some and dreams of winning the lottery for others . . . but always moving on. Identities, too, shift, produced in formations which are themselves in motion. Fashion demonstrates something of this flux when styles, represented by images in magazines, condition what people wear but are also themselves conditioned by street-level appropriations. The relation, however, is not entirely reciprocal because power remains with capital while consumers negotiate their space as best they can within what is on offer as well as in relation to other groups. Perhaps a key question of urban mobility is how they do that, how much they get away with.

Steven Miles asks whether urban youth are rebellious or complicit in consumerism, or whether they have an ambiguous relation to it, taking what they want and rejecting the rest, appropriating space as they can. Seen as driving much inner city renewal through their spending power in boutiques, bars and clubs, as well as fuelling industries such as popular music, young people may seem to enjoy an unprecedented sense of their own empowerment, colonising urban spaces for their (sub-)cultures. Miles argues, however, that a more critical appreciation of the data reveals that while young people do consume extensively and their identities may as manufacturers and advertisers know be defined in terms of highly specific, branded forms of consumption, they are also, in some cases, able to play games with the conditions they encounter. Against the presiding image of free-wheeling, free-spending youth, Miles cites a study of unemployed youth in Australia who use the mall as social space, performing themselves there rather than in urban wastelands, without spending money on things they have no means to afford. He questions the assumption that youth and rebellion go together – they might be chalk and cheese rather than bread and cheese, so to say – and the conventional dualism of youth and mainstream society. Young people should not, then, be expected to lead revolt against consumption, even though the way youth consumption shapes some city centres has little reflection of needs they might articulate themselves. As Adorno writes, looking at the culture industry: 'The customer is not king, as the culture industry would have us believe, not its subject but its object' (Adorno, 1991: 85). The same may apply to youth consumer culture, but to state it is an overview from academia; many young people revel in rather than rebel against the shiny world on offer to them, just as their parents do in their sphere of consumption. Similarly, Wilson remarks (2001: 71, cited in this volume, page 6) that most people's relation to consumerism is today ambivalent, neither impervious to its delights nor uncritical of excess. And, as Miles points out, not all young people belong to a subculture; many are able to pick and mix their consumption of both dominant and subcultures, perhaps in a post-modern eclecticism.

In cities such as Manchester or Newcastle, youth consumption drives much inner city development, and the emergence of a new, affluent young professional class is reflected in the conversion of redundant industrial buildings to loft-living spaces. But one person's cultural arena is another's no-go area, and the colonisation of spaces for specific marketing opportunities can be exclusive. The bar with live music and fifty-seven kinds of beer has little appeal for the elderly resident, member of a residual public, who remembers it as the corner pub. Similarly, schemes for city-centre redevelopment, often led by commercial interests, can be exclusive. The case of Birmingham shows that even when disguised as a cultural zone, with ubiquitous public monuments by well-known contemporary artists, its remodelled central business district offers uneven access to the city's many population groups.

Patrick Loftman and Brendan Nevin, both working for several years in Birmingham and critical of its redevelopment policies, write of the shifts from manufacturing industries to service economies, and efforts to improve their competitive position and attraction for inward investment, of cities such as Birmingham where public policy (and the public purse) has favoured flagship schemes in central zones over wider infrastructural improvements. Cities may raise their national, even international, profiles, while excluding some or many of their citizens (or council tax payers, and those exempt because on benefit) from the city's parade of new delights. As they argue, flagship schemes are aimed at facilitating the physical, economic and cultural rebuilding of a downtown area but offer unevenly distributed benefits. As social inclusion rises on political agendas across the spectrum, downtown redevelopment may be at odds with it.

Loftman and Nevin focus on Birmingham's city centre strategy, which includes the £180 million International Convention Centre at Centenary Square (in which a 'Percent for Art' policy was applied to commission an array of sculptures and a light-work) and £500 million proposed redevelopment of the 1960s Bullring shopping centre near the central New Street Station. Postcards sold at New Street Station carry the text 'Birmingham – City of Culture'. It never was officially, like Glasgow or Dublin, but the message indicates an ambition illustrated with a mix of Victorian civic architecture and contemporary public art in Centenary and Victoria Squares. The City Council claims huge benefits in terms of commercial and leisure investment from such highly visible schemes but Loftman and Nevin remain sceptical, seeing a boom in luxury housing close to the newly redeveloped areas but little wider spread of either rising quality of life or stable employment. As their 1998 study showed, many jobs created in such schemes are temporary, part-time and low paid. Drawing on experiences in other European and North American cities they reconsider the broader trends of city centre regeneration.

Tim Hall cycles between home and work, and around the several sites of his University. In his chapter he dwells on the uses of the car. He begins by observing the lack of studies of mobility in urban geography, despite its ideological as well as spatial implications and wide impact on other areas of urban policy and planning. In contrast to the well-funded position of the road (for which read also petrol and automobile manufacturer) lobby, Hall sees car use as far from benign, producing toxic emissions but also emphasising social deprivation in as much as car use corresponds largely to income category. Just as flagship development is linked to social exclusion, so a privileging of roads may disadvantage some groups; if private cars replace public transport as a policy priority, non-drivers are excluded and their employment prospects worsened. From this, Hall argues that understandings of sustainable urban mobility need to take into account multiple social and cultural geographies.

Drawing on a broad range of empirical data, Hall charts the footprint of car use in global terms as well as those of localised economic impact and health. Noting that since 1980, for instance, motor vehicles have dispatched 31,000 North American citizens to the threshold of the angels, and that this is far more than the fatalities of violent crime (or September 11th), Hall argues that new urban strategies are needed to limit car use.

This argument is perhaps appreciated by many car users for whom public transport does not yet offer a good alternative, and by an increasing number of city authorities. As to action on sustainable, low energy, public transportation, though this is outside Hall's commentary, a notable case is Curitiba, Brazil. Cities in Germany and Scandinavia, too, have invested in coordinated public transport systems with interchanges between rail and tram, harmonised timetables, and low fares. And the New York Subway has undergone a major refit and removal of graffiti. In contrast, Birmingham's public transport system is confusing, while the city centre is almost fortified by major roads. Little opportunity there for the appropriation of space Steven Miles hints at, nor in the buildings round Centenary Square: unemployed youth wandering in the Convention Centre, or the adjacent Hyatt Hotel?

4 STEVEN MILES

Resistance or Security? Young People and the 'Appropriation' of Urban, Cultural and Consumer Space

Are young people servants of urban consumer society? Do they simply accept the dominant ideologies of the status quo or is their relationship with the status quo a critical and reflective one? As observers of the consuming city it might be tempting to come to the conclusion that young people positively embrace the city as a site of consumption and that they are, more so than any other age group, happy to indulge in the superficial attractions of consumer lifestyles. Yet despite young people's apparent willingness to positively embrace the principles of consumerism, their relationship with urban culture remains ambiguous. They might be said to be at one and the same time upholders of the consumerist ideal and usurpers of urban consumer space.

Far from being at the forefront of changes to the nature and character of the urban fabric, in this chapter I want to suggest that young people's 'consumption' of urban life is symptomatic of a broader melodramatic and pathological model of what it means to be a young person. In other words, young people may well appropriate components of urban life, and that appropriation may well provide them with a powerful sense of meaning but, potentially at least, that meaning is inevitably taciturn.

In many respects, young people's use of urban space, however diverse in nature, may actually serve to reinforce the status quo. This contention is not problematic in itself. What is problematic is that there remains an overriding assumption that young people *should* be resistant. In contrast, and however disappointing it may sound, there may be an argument for suggesting that young people simply do not *want* to be at the vanguard of urban change.

Until recent years the impact of space and spatial location has had a relatively limited impact on sociology. This reflects a concern that the process of modernisation brought with it a proclivity to expose human beings to the same material conditions and ideas which would therefore undermine cultural diversity and promote similarity in lifestyles and values. It was therefore assumed that spatial location would have a limited impact in determining social lives (Sekhon, 1997). It might well be argued that at the beginning of the twenty-first century the continued dominance of consumerism as a way of life (Miles, 1998) has reinforced this state of affairs to the extent that any appropriation of social space in urban culture does not constitute 'appropriation' at all, but simply serves to reinforce the status quo. However, there is a compelling argument that space plays a crucial role in the construction of youth identities (Ruddick, 1997; Toon, 2000). This is a key debate and one that I seek to address throughout the remainder of this chapter.

However, before I can begin to effectively outline young people's relationship with urban space I need to consider the overwhelming (though often unintentional) tendency on the part of the literature to portray young people in a negative light. I will go on to argue that academic and popular portrayals of young people are, in some respects, themselves to blame for the possibility that the creativity of young people continues to be largely suffocated within the urban environment, precisely because such portrayals put unrealistic expectations on those young people.

REBELLIOUS YOUTH?

The abiding image of young people in urban environments is perhaps that of the rebellious and potentially dangerous troublemaker 'hanging out' on street corners. This is an image of young people as unsavoury deviants. It is an image of young people as lost, hopeless and potentially threatening. But perhaps most importantly, this is an image of young people as victims. This popular image is reflected by the sociology of youth which itself continues to be dominated by an orthodoxy in which the agency of young people is largely neglected (see Ettorre and Miles, 2002). Discussions of youth transitions into adulthood adopt an almost entirely (although perhaps at times unintentionally) negative conception of what it actually means to be young. From this point of view, being 'youthful' is about being unemployed, homeless and unwanted. This suggestion might appear somewhat excessive in nature, but it is an approach that is implicit in the way in which youth research is conducted. Discussions of young people's transitions are largely concerned with aspects of housing, unemployment, education and training (for example, Roberts, 1995). In doing so, these discussions tend to focus on broad economic trends which determine how easy it is for young people to reach adulthood. Here, the argument is that traditional routes into adulthood have, in recent years, been problematised. Discussions of the 'transition' focus on the ways in which social structures affect how young people grow up. Within this discourse there is very little room for discussion of the ways in which young people actively negotiate with dominant power structures.

It is true to say that there are examples of research that engage with young people's experience of the above trends (for example, Griffin, 1993), but the fact that these trends are prioritised perpetuates a vision of young people as reactive, as opposed to proactive; as controlled rather than in control. The suggestion that young people are amongst the most vulnerable of social groups is not without foundation: young people are indeed subject to the ups and downs of socio-economic change. But this approach to what it means to be a young person is very much a by-product of the 1980s. Cashmore (1984), for instance, describes a world in which young people have 'no future' and in which they are resigned to having few ambitions, limited horizons and next to no prospects. Young people are in a crisis inasmuch as rising unemployment, welfare and education cuts and a decade of right-wing government are apparently taking their toll (Griffin, 1997). From this point of view young people are wounded by, if not victims of, social change.

The question remains, however: does this really tell us the whole story? Perhaps we *could* describe young people as an index of social *ills* (Jones and Wallace, 1992) but to accept this position uncritically serves to underestimate the complexity of young people's experience. Young people are more than passive casualties of unemployment trends, drug misuse and problems associated with teenage pregnancy. Young people are not simply all about the melodrama of subcultural life or the terrors of youth crime, drug addiction and alcohol consumption. In many ways young people are an index of social *norms*, and their patterns of consumption constitute the playing out of such norms. A discussion of young people's relationship with the city should help to illustrate the fact that their consumption of city space represents a prime means by which they negotiate the ups and downs of life in a so-called risk society.

In light of the above it is absolutely crucial that in understanding young people's use of urban space we adopt a differentiated notion of 'youth'. Many authors have questioned the validity of a sociology of youth and the tendency, in particular, to over-generalise about young people's experience. Young people clearly do not relate to such space in a uniform manner. Gender, ethnicity, class, and indeed age, all play a crucial role in young people's fragmented urban experience. For instance, Watt and Stenson (1998) studied the use of public space by South Asian, Afro-Caribbean and white youth living in a medium-sized town in the south-east of England. They found that parts of 'Thamestown' were associated with particular racialised perceptions of 'danger' and 'trouble'. Such associations often contrasted with the more personal sense of safety and security expressed by many of the young people Watt and Stenson interviewed.

In many respects urban spaces are contested spaces, not least through the exclusion of ethnic minority groups from particular areas of the city. But conceptions of urban space are not universal. For instance, as Watt and Stenson point out, a young middle-class woman from a suburban area may feel less safe in certain areas of the city than a young working-class man who may have personal knowledge and relationships that serve to counteract place-based stereotypes. As such, socio-spatial inequalities can be said to exist uneasily alongside 'postmodern fluidity'. The impact of consumption in the experience of urban space may well be to help provide some semblance of security at a time when young people are having to balance the old trajectories of inequality alongside new risks, uncertainties and opportunities (Furlong and Cartmel, 1997).

Bearing in mind the diversity of young people's backgrounds I want to suggest here that young people's relationship with urban space is essentially paradoxical. Consider, for example, Presdee's (1986) work on the activities of young unemployed people in an Australian shopping mall. Presdee points out that significant ideological interests underpin the experience of the shopping mall, to the extent that the politics of the production and consumption of products is made invisible. But as Presdee indicates, despite and perhaps because of these ideological dimensions, 80 per cent of the unemployed young people he researched visited that mall at least once a week and nearly 100 per cent of unemployed young women also considered themselves to be 'regulars'. Young people apparently use the fact they are economically excluded to usurp the dominant ideology of consumerism:

> For young people, especially the unemployed, there has been a congregating within these cathedrals of consumption, where desires are created and fulfilled and the production of commodities, the very activity they are barred from, is itself celebrated on the altar of consumerism. Young people, cut off from normal consumer power are invading the space of those with consumer power.
>
> (Presdee, 1986: 13)

As far as Presdee is concerned the key thing here is that young people possess the space available to them in a shopping mall. Young people parade, 'not buying, but presenting, visually all the contradictions of employment and unemployment, taking up their natural public space that brings both life and yet confronts the market place' (Presdee, 1986: 16). This apparently therefore represents what Fiske (1989: 17) describes as an 'oppositional cultural practice'. Regardless of the diverse nature of young people's backgrounds, they are often perceived as oppositional for no better reason than that they are young people.

As Fiske (1989) points out, these young people aren't actually behaving any differently from any other shoppers who happen to be browsing. Malls have always been places of refuge; places on the cusp of the public and the private that provide an escape from the hostile world outside. But how do young people actually use this space? Do they use it any differently from any other social group? Or are we as guilty of making assumptions about how young people use such space as we are of assuming that young people constitute something of a threat? Need young people's use of the shopping mall, for example, be oppositional at all? Are young people really the oppositional breed that sociologists of youth like to think they are?

YOUTH SUBCULTURES

In order to begin to address this question I want briefly to outline the value of traditional interpretations of youth culture, the legacy of which, I argue, has been to perpetuate a misleading image of youthful urban rebellion. In this context, Wyn and White (1997) point out that there is a tendency to essentialise youth cultural formation: to focus on superficial aspects of youth cultural style, whilst ignoring the differences that exist both between different young people and between different age categories. This approach has simply served to pigeon-hole young people as oppositional, when in fact young people's experience of urban cultures may well be far more to do with conformity than it is rebellion.

In other words, the legacy provided by members of the CCCS in Birmingham in particular has created a mindset in which the assumption is that young people are or should be rebellious when perhaps they are not rebellious at all. In 1975, for example, John Clarke argued that skinheads have a preoccupation with territory. The skinhead style represents an effort to recreate (or indeed substitute) through the 'mob' traditional working-class communities. This provided skinheads with a sense of collective solidarity that served to perpetuate an 'us and them' attitude. However, what is perhaps most important here is that skinheads cannot and do not, as Clarke points out, revive the community in any real sense. Any genuine sense of community had been deprived of its social bases long before. At this time skinheads were organised on a territorial basis insofar as they defended their own 'patch', which they identified with painted slogans. The street corner, the pub and the football ground became the 'mob's' local identity and a focal point around which they could organise themselves. However, ultimately, this sense of community was illusory,

> We may see these three interrelated elements of territoriality, collective solidarity and 'masculinity' as being the way in which the Skinheads attempted to recreate the inherited imagery of the community in a period in which the experiences of increasing oppression demanded forms of mutual organisation and defence. And we might finally see the intensive violence connected with the style as evidence of the 'recreation of the community' being indeed a 'magical' or 'imaginary' one, in that it was created without the material and organisational basis of that community and consequently was less subject to the informal mechanisms of social control characteristic of such communities.
>
> (Clarke, 1975: 102)

The key phrase here is that which refers to the experience of oppression. Young people's relationship with subcultures and thus the forms of urban expression in which they partake have in the past twenty to thirty years been radically altered in a world in which young people's feelings of oppression are at least partly offset by ready access to urban consumer culture. In other words, young people simply no longer appear to *feel* the need for mutual organisation and defence. Their experience of the urban fabric, although communal in some senses of the word, is now primarily individualised.

It is true that young people are less dependent upon the all-encompassing lifestyles we might associate with traditional subcultures. However, it is still important to recognise that young people also attribute particular significance to urban space and that their immediate neighbourhoods are less likely to play a key spatial role in their lives at least as far as them constructing their identities is concerned:

> There are a number of important reasons why these teenagers gravitate to the urban centre. It not only constitutes space in which they feel free from parental jurisdiction but it also provides them with a means of finding and creating a different spatiality, more exciting than that offered by the impoverished landscape of the neighbourhood. This distancing from the 'local' is part of an important struggle for self-definition for these teenagers.
>
> (Toon, 2000: 144)

For young people, then, the urban is a centre of sociability where chance encounters and unexpected events add a dimension to their lives that their locality cannot. In discussing the impact of CCTV surveillance on young people's use of urban space, Toon (2000) argues that CCTV's impact on the urban environment is not all-encompassing. Young people develop what Toon describes as 'narrative footsteps' insofar as they identify and move through concealed urban spaces and invisible routeways that allow them to re-appropriate such space for their own ends.

Of most interest here is Toon's recognition that such resistance is limited in its scope. Indeed, authors such as Edensor (1998) have commented that the commodification of the high street has effectively depoliticised the street. Whether or not we accept this argument, the contention that by concentrating their sociability in unmonitored gathering places young people are effectively policing themselves is worth considering. This is an interpretation that Toon (2000) dismisses in arguing that young people have a nonetheless highly visible presence in public space. In this sense Toon's analysis is a slightly misleading one. Toon does point out that there are negative implications to this process insofar as young people could be said to be creating a risky environment where groups of young people seek to assert, somewhat destructively, their territorial presence. From this point of view young people deploy tactics that ensure, at least to a degree, their urban imperceptibility:

> They have disrupted the strategies of control which deny them access to public space and manipulated and diverted the meaning of dominant space, re-inscribing it as a site of possibility and meaning, creating the ground for the re-location of their identities through a 'hidden' mode of resistance.
>
> (Toon, 2000: 159)

Toon does, however, go on to recognise that such activity on the part of young people does not necessarily equate with power and that such territorial behaviour can have serious and potentially violent implications. As such, I would echo Toon's contention that we need to avoid romanticising the condition of young people, in particular in ways that demand no further action. Young people need urban space that is their own. Toon recognises this, but what he doesn't recognise is the fact that young people do not necessarily feel the need to resist. What Toon describes is *not* resistance at all. Young people's use of urban space often represents an effort to secure space within dominant power structures as opposed to outwit them, as he suggests.

YOUNG PEOPLE, RISK AND CONSUMPTION

One means of coming to terms with the above contention is through consideration of current debates on young people's experience of the risk society. Authors such as Furlong and Cartmel (1997) have pointed out that the process of individualisation represents a key aspect of young people's experience of social change. In this context, social change is such that the individual has effectively been removed from traditional commitments and support relationships to the extent that:

> The place of traditional ties and social forms (i.e. social class, nuclear family) is taken by secondary agencies and institutions, which stamp the biography of the individual and make that person dependent upon fashions, social policy, economic cycles and markets, contrary to the image of individual control which establishes itself in consciousness.
>
> (Beck, 1992: 131)

Young people, of whatever background, have therefore been depicted as especially vulnerable to the heightened sense of risk and the individualisation of experience that has characterised the move towards 'high modernity'. In short, journeys into adulthood are becoming increasingly precarious.

Whereas in the past young people would have navigated a way through to adulthood with the help of traditional support mechanisms such as the family, community and religion, nowadays any tough choices or decisions about the future fall squarely on young people's own individualised shoulders (Furlong and Cartmel, 1997). In this context, it would make sense to suggest that this process is most readily expressed in the context of market dependency. This argument is developed most effectively by Ulrich Beck (1992) who identifies a new mode of socialisation, a 'metamorphosis' or 'categorical shift' in the make-up of the relationship between the individual and society. In particular, there is apparently an increasing focus on the individual as a reproduction unit for the social in his or her life-world. In other words, the traditional forms of social support upon which young people used to be able to depend are no longer available. Young people do not share common biographies as much as they perhaps would have done in the past.

Young people apparently live in a world of uncertain diversity, a world of contradictions: of opportunity and risk. As such, they are frantically looking for a means by which they can feel that they belong. The symbolism of traditional subcultures no longer represents a viable option in a world where you are under pressure to be an individual and yet where the material resources by which you do so are liable to promote conformity. The vast majority of young people do not belong to a particular urban subculture (if they ever did). They apparently pick and choose aspects of their identity from one minute to the next. But they do not do so freely. They are subjects to the whims of the market and they collude with that market and accept its limitations in order to establish a sense of stability in their lives. As such, the market is extended into every aspect of social life which serves further to undermine traditional sources of social support, to the extent that the individual becomes as standardised as he or she is individualised. Young people therefore express themselves through their lifestyles which are simultaneously individualised and communal (Miles, 2000). In this world the young people's safety net is consumption and, as such, they will inevitably, in some shape or form, reinforce the ideological underpinnings of that market.

In the next section I want to consider how this reinforcement is manifested in the urban environment and whether or not in particular this should mean that young people's urban 'voice' is necessarily mute. In particular, I want to suggest that at the beginning of the twenty-first century young people's relationship with the urban fabric is primarily established through their experience as consumers. As such, consumption represents the primary force in the conduct of social life in urban contexts at the beginning of the twenty-first century. This is particularly true of young people insofar as if they are valued as anything in contemporary society they are valued as consumers (Palladino, 1996). As such, Wyn and White (1997: 86–7) go as far as to argue that 'today youth itself is a consumable item, in that the superficial trappings of youth are now part of the consumer market'. This represents part of a long-term process that can be charted back to the aftermath of World War II, which authors such as Osgerby (1998) argue provided a major turning point in ensuring that youth emerged as an influential social group in its own right. Within the boundaries provided by the culture industries young people were actively able to carve out their own autonomy, an autonomy that was expressed through distinctive purchasing styles and patterns of consumption (Furlong and Cartmel, 1997). In effect, young people continued to take advantage of an economic situation in which manufacturers and service industries demanded their labour, as well as the money being earned by their parents.

It is worth pointing out that regardless of the increased spending power of young people it would be naive to suggest that young people's ability to consume was limitless, notably during the recession of the 1980s: far from it. Often, it seems to be the case that the opportunity to dream about consuming, to 'hang out' at shopping centres for example, was more important than the act of consumption itself, precisely because the ability to dream about consumption was grounded on communal ideals about what constituted a consumer culture:

> it's worth explaining what a teenager means by 'shopping'. They may eventually make a purchase, but most of the time they're just looking around,

> sizing up one thing against another, asking for their mates' opinions, deciding just how much they want to buy it. . . . They also get a lot of pleasure first by simply browsing around shops – looking at toiletries, magazines, books, stationery, records, fashion accessories, models, kits, toys, games, sweets etc. The kids felt free to poke around, see what was for sale, try out the testers, browse without being harassed and hustled: 'It's trendy to go to a place like Virgin Records on a Saturday morning – all the punks sit around outside so I really enjoy going there – it's just the getting over the fear of going through the door.'
>
> (Fisher and Holder, 1981: 170)

Today's teenagers apparently take it for granted that they were 'born to shop' (Palladino, 1996: xix; Starkey, 1989); the problem being that resource-wise, the ability to shop is easier said than done. Palladino implies, however, that young people's relationship with consumer culture is far from being the radical rebelliousness which may have been accurate in the past. Nowadays young people are compelled to find a job and thereby to make money in order to belong to the consumer culture which they so fervently crave. Young people are worried about their economic future and as a consequence their relationship with consumption is ever more tenuous and yet ever more fundamental to who it is they are.

Stewart (1992) points out that consumption has become increasingly fragmentary as far as young people are concerned, in that the status young people procure from what they consume is not a result of the ability to buy goods in a general sense, but more dependent upon buying goods that are different from what others buy. I tend to diverge from Stewart on this point in the sense that although young people evidently adopt fragmented forms of consumption, they consume in a broader framework dependent upon a cultural capital of consumption which is omnipresent amongst their friendship circle. They simply cannot afford to consume individualistically. I do, however, agree with Stewart's (1992) suggestion that what characterises young people in terms of their core values is that, above all, they are being incorporated into the mainstream. For the majority of young people, as Stewart (1992) suggests, there appears to be very little left for them to rebel against. Give a young person the opportunity to drive his or her first car and he or she will be happy (Veash and O'Sullivan, 1997). In this context, young people are independent of mind and yet paradoxically conformist in a world where the individual experiences and the influences individuals are exposed to are becoming 'increasingly homogenous' (Stewart, 1992: 224). The question that needs to be asked therefore is: are young people in control of their consumption patterns, or are they being controlled?

YOUNG PEOPLE AND SITES OF CONSUMPTION

The impact of consumption on city life is well documented (see Miles and Paddison, 1998) to the extent that it could be argued that our cities are no more and no less than sites of consumption. In this light Knox (1991) goes as far as to suggest that the whole of the contemporary landscape is being geared towards consumption, as a direct result of the emergence of an increasingly flexible form of production which is itself more and more dependent upon the demands of consumers. Of more interest, perhaps, is Chaney's (1990) suggestion that contemporary culture has, in fact, been *de*territorialised. That is, with the development of metropolitan-based mass communication and entertainment networks which disseminate the wares of consumer culture, what has emerged is a cultural homogeneity that overlays the diversity of places and spaces. Meanwhile, authors such as Zukin (1992) and Urry (1995) have commented upon an apparent postmodern reconstruction of the city that has in fact served to undermine people's sense of identity. More than anything else the city has become a spectacle:

> postmodern landscapes are all about place, such as Main St in EuroDisney, world fairs or Covent

> Garden in London. But these are simulated places which are there for consumption. They are barely places that people any longer come from, or live in, or which provide much of a sense of social identity. . . . Such spaces are specifically designed to wall off the differences between diverse social groups and to separate the inner life of people from their public activities.
>
> (Urry, 1995: 21)

It could be argued that young people are particularly prone to the sorts of processes Urry describes above. I do not wish to imply that young people are somehow powerless in these circumstances, nor that they engage with consumption in a uniform fashion, but rather that they are more familiar with and therefore comfortable about the opportunities that consumption provides for them. They are therefore more amenable to escaping from the ups and downs of a risk society *through* consumption. They are sophisticated enough consumers to adapt to the consumer society and the spaces that promote that consumer society. They use the opportunities urban cultures of consumption provide them. They are therefore prepared to make the sacrifices involved in losing a certain degree of control over who it is they are, for the joys and satisfaction that consumption can provide them with in the short term. Young people are not therefore controlled by the consumer ethic. They are not consumer dupes. But they will partake in the pleasures and the pleasurable spaces of consumerism in a reflexive way that subjects them to the ideologies of consumerism, whilst allowing them to maintain a critical distance (or at the very least the sense of a critical distance).

Many authors have also commented on the way in which young people use sites of consumption in a reflexive constructive fashion. For instance, Presdee (1986), who I mention above, notes that young people are attracted to shopping malls as 'cathedrals of consumption', often asserting their right to claim space as non-consumers in a consumer-oriented world. Similarly, Langman (1992) argues that mall culture has a particularly influential role as far as young people are concerned, in as much as it provides a hang-out in which they can be free of parental pressure, a community of peers, which ironically confirms to young people the legitimacy of a consumer lifestyle. In effect, malls play the role of harbingers of pseudo-communities, providing a means or focus by which young people are integrated into subsequent life trajectories. Depending on young people's age, ethnicity, gender, and class, malls give young people the opportunity to contest, reinterpret and mobilise particular meanings; they provide a source of empowerment through 'hyper-real gratification' (Langman, 1992: 60). But how far does this represent an authentic means of empowerment?

> Adolescent groups with common cultural capital become the socialization agents for, and models of, what Bellah *et al.* (1985) has termed lifestyle enclaves that reflect current life chances and subcultures of particular consumer tastes which offer identity packages that typically endure through subsequent life careers. But these are really more 'proto-communities' of shared patterns of cultural consumption and communication through shared tastes and fashion than the more traditional forms of community and face-to-face verbal communication.
>
> (Langman, 1992: 60–1)

It would be going too far, however, to claim that young people use consumption purely for their own ends. This would to underestimate the inherently paradoxical nature of consumer lifestyles (see Miles, 2000). As I have already said, far from being characterised by rebellion against the dominant values of society, young people often actually positively embrace those values wholeheartedly. This reflects the concomitant trend, for sociologists of youth to exaggerate the rebellious nature of young people. As McGuigan (1992: 91) notes, 'There is a tendency to overstate the cultural power of youth in the sphere of consumption.'

Focusing more specifically on the relationship between the construction and management of 'troubled teens'; apparently subject to specific disorders of consumption and transition, Griffin (1997: 5) identifies many young people, as 'disordered consumers in racially-structured patriarchal capitalist societies'. She goes on to point out that rising youth unemployment and increasingly complex global cultures and technological forms have actively undermined some of the certainties which formed a basis for an assessment of youth cultural styles.

In their study of consumers' use of two London shopping centres at Brent Cross and Wood Green, Miller *et al.* (1998) found, perhaps not surprisingly, that young people gained considerable pleasure from visiting shopping malls with their friends. In the context of this research, it is certainly true to say that the act of purchase was not pivotal to the experience. As one parent Miller *et al.* interviewed put it:

> My daughter's thirteen and her idea of an outing on Saturday is to go on the bus with one of her friends, and they don't spend much but they look at everything in the Body Shop and Miss Selfridge and Top Shop and Gap and all those places, and it's fun. Nightmare [to me] but to them it's fun.
> (Miller *et al.*, 1998: 98)

Rob White has also written extensively on young people's use of urban spaces in an Australian context. He points out that young people are increasingly marginalised from both the processes of production and consumption and community life more generally, to the extent that they are alienated and disaffected from a social system which does not provide for their specific needs (White, 1993: 87). Indeed, in many instances, public space is actively closed off to young people. Much of this space is constituted first and foremost as commercial space, but it also provides a place where young people can get together to spend their time among peers and friends, 'in an atmosphere of relative autonomy and excitement of the senses'. Young people use shopping environments as places to socialise first and consume second (White, 1996). But a key problem as far as White is concerned is that young people's marginalisation can be masked by youth fads, fashions and subcultural styles. More often than not young people's use of commercial space in Australia is highly regulated to the extent that their 'worth' as human beings is defined purely in commercial terms. White goes on to discuss crime prevention strategies, which are far too often primarily concerned with excluding people, which therefore heightens the perception of young people as a threat and thereby quite explicitly presents them as non-members of the community (White, 1993: 92). White (1996: 44) therefore contends that the city landscape has been transformed ideologically into a 'moral battleground'. Young people therefore constitute part of an unsightly and socially threatening urban underclass. To illustrate this point White (1996) discusses 'Operation Sweep' in Western Australia which allowed the police to pick up every teenager on the streets of Northbridge and Freemantle (the main entertainment districts) after 8 p.m. and take them to the police station for processing. This legislation was rationalised insofar as it was said to help young people 'at risk'. Apparently, good citizens stay off the streets in occupying their high-security consumption spaces, whereas bad citizens are identified on the basis that they occupy the streets (see White, 1996; Davies, 1995). For White (1996), what he describes as a 'militarisation' of the landscape, through extensive fear campaigns and a 'security at all costs' mentality, creates a climate in which commercialism outweighs sociological or political considerations and young people are the primary losers in this equation. In this context, White returns to the question of youth rebellion, 'For without income and space of their own, without social respect and personal self-esteem, young people have nothing to lose in rebelling in whatever fashion or manner they may choose against a system which for all intents and purposes has forsaken them' (White, 1992: 199). But I would argue, ironically, that young people actually choose *not* to rebel and that the vast majority of young people are happy to partake in the pleasures of consumption. Access to resources is obviously a key issue here, but even if such resources are scarce the day-dreaming quality of consumerism will come into play. Consumption, or at least the idea of consumption, offers young people a way out. It is able to do so precisely because it at least appears to offer a means by which young people can express their sense of difference, whilst simultaneously feeling that they belong.

Similar work on shopping malls has also been undertaken in the American context, where there is evidence to suggest that such centres have become central components of many young people's lives (see Vanderbeck and Johnson, 2000). What is interesting about this work is that it focuses not on young people as rebels or on the potential for rebellion within the context of the mall, but on attraction to young people of the mall as a primarily 'safe' place. The argument here then is that young people actually desire to enact relatively mainstream goals in urban shopping environments. Shopping spaces provide young people with a sense of choice and independence that are often missing in other aspects of their lives. For this reason, Vanderbeck and Johnson argue that shopping spaces must be considered in relation to what they provide for young people relative to other possible alternatives. For instance, young people may congregate in shopping spaces precisely because their options elsewhere in the neighbourhood are otherwise limited. In other words, shopping spaces play an important part in young people's broader geographical imaginations. They may not even visit the shopping mall very often, but 'the appeal of the mall can be understood in terms not just of what malls offer intrinsically, but what it provides relative to . . . other spaces' (Vanderbeck and Johnson, 2000: 145). In short, Vanderbeck and Johnson argue that the shopping mall fulfils a function for young people as both a real and imagined social space.

CONCLUSIONS

The ways in which young people 'hang out', whether in the shopping mall or on the street corner, may well be largely imaginary, in the sense that it is not on the surface, at least, resistant. The time has come to refrain from assuming young people constitute a progressive social force and to ask why it is young people often do not feel it necessary to be such. As Aitken (2001) notes, the 'street' represents an important part of a substantial number of young people's everyday lives despite the fact that young people's lives are currently so commercialised. It is also important to recognise that the street is not, as is often implied, a universally male (and therefore by misguided implication more resistant) domain (Toon, 2000). Young men and women alike elaborate 'a vibrant cultural capital based on the streets' (Aitken, 2001: 159). In other words, there should not be an assumption that young people relate to the streets in prescribed ways.

Young people's experiences of urban life are clearly diverse in nature. However, having said that, public urban space also provides young people with an arena within which they can maintain social capital through friendships and networks and hence manage such diversity on a day-to-day basis. Young people have very few places to go. The sheer use of space does not necessarily constitute subversiveness. As such Toon (2000) discusses young people's experience of CCTV surveillance and exclusionary youth policing. In developing De Certeau's (1984) notion of the 'tactic' Toon argues that young people re-appropriate public space in order to create meaning and context in their everyday lives. This is all well and good, but as a theory it tends to confuse the determination of young people to secure their own safe spaces where they are able to be themselves, with the intentions of such actions, which I would suggest are as much about delineating secure space within the consumer society as actively protesting against it. There is a need to allow young people to be themselves in an urban environment. But 'being themselves' is arguably more about accepting the status quo and asserting their rights as consumers than it is about usurping a system which it is in their interests, as sophisticated consumers of social change, to maintain. Young people will interact with the urban environment depending upon their age, class, ethnicity and gender, but ultimately that engagement is bounded by consumer society that allows young people the sense of security and the sense of freedom through which they can construct their own meanings. In this regard, young people use urban consumer space to their own ends; but those ends will also often fulfil the needs of the dominant social order.

As far as the future of urban life is concerned, a wide variety of authors have called for a more sensitive approach to the needs of young people and one that recognises young people as participants of public urban life (for example, White, 1992, 1993). Such sentiments are worthy ones, but they assume that young people want something radically different to what the urban environment currently offers them. My conclusion is that young people can be creative but that their creativity is more liable to be expressed within the parameters laid down for them by consumer society. Consumerism, however, can be used as a force for change. The vast majority of young people do not partake in anti-capitalist demonstrations but are nonetheless aware of capitalism's limitations and use its advantages as pro-actively as they can, largely as a motor for the maintenance of peer relationships. What is most urgent, then, is a change in the mind-set of those commentators who assume young people are necessarily unhappy with the urban present. Young people use urban environments in pro-active ways, but just because those ways are not as resistant as anticipated, or indeed wished, does not mean to say they are not valid. The urban future may well be a consumerist one. One thing is sure: if that is the case young people will be better equipped than most to cope with it.

SUGGESTIONS FOR FURTHER READING

Bell, D. and Haddour, A. (eds) (2000) *City Visions*, Harlow: Prentice Hall.

Miles, S. (2000) *Youth Lifestyles in a Changing World*, London: Sage.

Miller, D., Jackson, P., Thrift, N., Holbrook, B. and Rowlands, M. (1998) *Shopping, Place and Identity*, London: Routledge.

Skelton, T. and Valentine, G. (eds) (1998) *Cool Places: Geographies of Youth Cultures*, London: Routledge.

White, R. (1996) 'No-go in the fortress city: Young people, inequality and space', *Urban Policy and Research*, 12 (1), 37–50.

5 PATRICK LOFTMAN AND BRENDAN NEVIN

Prestige Projects, City Centre Restructuring and Social Exclusion: Taking the Long-Term View

INTRODUCTION: INTER-URBAN COMPETITION AND PRESTIGE PROJECTS

In response to the global restructuring of industry in the 1970s and 1980s, and the increased internationalisation of economic activity, there has been an intensification of the global inter-urban competition for investment and jobs. Cities now compete for mobile investment, public sector funds, hallmark events such as the Olympic Games and Commonwealth Games, and G8 summits. In addition they compete for, and seek to retain or attract, higher-income and skilled workers to live and/or spend within their jurisdictions (Porter, 1995, 1996; Begg, 1999; van den Berg and Braun, 1999). In the context of the increased inter-urban competition, many Western European and North American cities have adopted pro-growth local economic development policies as a means of securing their economic futures (Kotler *et al.*, 1993; Fainstein, 1994). A key component of such pro-growth economic development strategies has been the formulation, development and promotion of prestige projects (Loftman and Nevin, 1995, 1996). These prestige physical development projects are largely geared at facilitating the physical, economic and cultural restructuring of city centre or downtown areas.

A number of British cities (particularly those previously dependent on manufacturing industries) have increasingly sought to harness cultural, business and sports-related tourism as a vehicle for: enhancing or reshaping their city image; securing much needed visitor expenditure; retaining local business; attracting new business investment (particularly mobile service sector investment); and diversifying their economic base. During the 1980s and 1990s, flagship development projects have become a prominent, and possibly essential, component of UK urban regeneration initiatives. Many local authorities (in particular cities such as Birmingham, Manchester and Sheffield) have sought to utilise high-profile prestige developments as a means of re-imaging their respective areas; attracting private investment and visitor expenditure; and securing the physical transformation of previously neglected or declining areas (Bianchini *et al.*, 1992; Loftman and Nevin, 1995, 1996). In particular, cultural prestige development projects (such as concert halls, sports stadia, and themed entertainment complexes) have become increasingly common features of city regeneration strategies in the UK. Such complexes are also geared at reshaping the leisure cultures of their localities. Many prestige projects are found in or adjacent to city centre locations (Bianchini *et al.*, 1992; Loftman and Nevin, 1995). Furthermore, it is argued that the dramatic postmodern architecture often associated with grand prestige schemes plays an important role in place promotion and in 'mediating perceptions of urban change and persuading "us" of the virtues and cultural beneficence of speculative investments' (Crilley, 1993: 231).

The rationale behind the development of flagship projects as a tool for city promotion and redevelopment is not 'new' to UK regeneration policy. However, what was new in the

1980s and 1990s was the extent of public and private sector support for flagship schemes in the process of urban regeneration, and the desire of British city governments to import US city revitalisation models focusing on prestige development projects (Loftman and Nevin, 1995). This consensus was reflected in the commonly held view amongst policy and decision makers that 'a city without a flagship did not have a regeneration strategy' (O'Toole and Usher, 1992: 221). In this context a number of British cities have, during the 1990s, invested significant public sector resources in new flagship cultural visitor attractions. In particular, a number of high-profile cultural flagship projects have been developed (partially funded by the UK National Lottery) to mark the year 2000 and the arrival of the third millennium. It was originally anticipated by Peter Brooke (then Secretary of State for National Heritage) that these landmark projects would be 'at the centre of our Millennium celebrations' (Brooke, 1994: 8) and would provide a means of capturing 'the spirit of our age in enduring landmarks that symbolise our hopes for the future' (ibid.: 5). The most striking example of the current obsession of British cities with flagship cultural developments was the £800 million Millennium Dome project located in London. This development was secured by London following a national inter-city competition for the right to become the focus of the nation's Millennium celebrations and secure £400 million of National Lottery funds towards the construction of a flagship cultural visitor attraction of international standing – which would represent the nation's future and past.

This chapter focuses on the experience of Birmingham City Council's formulation and implementation of a pro-growth city centre focused regeneration strategy.[1] This strategy, since the early 1980s, has facilitated massive public and private investment in the development of prestige projects focused on two discrete zones of Birmingham's city centre – the Broad Street Redevelopment Area (in this chapter referred to as 'Westside') and the Eastside Regeneration Initiative. The 'Westside' redevelopment area (which was first started in the late 1980s) has included the development of the £180m International Convention Centre; the £57m National Indoor Arena; the development of the £300m Brindleyplace office and leisure complex; and the £100m Mailbox development. The Westside area also includes proposals for the development a flagship £300m Arena Central mixed-use complex. The more recent 'Eastside' development area (which was launched in the late 1990s) includes the £111m Millennium Point project; the £500m redevelopment of the Bull Ring shopping centre; and the development of a new City Centre Park on the eastern edge of the city centre area.

Birmingham City Council has estimated that as a consequence of the Westside and Eastside developments, and its aggressive programme of pro-growth local economic development, the city centre as a whole has benefited from £1.245 billion of public and private investment in leisure, office and retail schemes during the year 2000. Furthermore, it is argued that Birmingham's array of impressive prestige projects have been subsequently followed by a massive boom in luxury housing developments within Birmingham's city centre area.

This chapter reviews how Birmingham has sought to improve its relative position in the international 'league table' of major cities and raise its national and international profiles via locally driven strategies focused on the physical, economic and cultural restructuring of its city centre area. First, the chapter reviews the development of Birmingham's city centre focused pro-growth local economic development strategy between 1984 and 2001 – a period that encompassed three different phases of city council leadership. Second, the chapter examines the scale of physical and cultural restructuring that has occurred within Birmingham's city centre since the mid 1980s: focusing on the Westside and Eastside developments and the promotion of city centre residential development for higher-income groups. The chapter then concludes with a review of the potential longer-term policy implications and distributional consequences of such investment for disadvantaged groups living within Birmingham – particularly in terms of access to public and private sector services, and quality education and housing.

A REVIEW: BIRMINGHAM'S CITY CENTRE FOCUSED PRO-GROWTH POLICIES 1984–2000

Birmingham, located in the heart of England's industrial West Midlands region, is Britain's second largest city with a population of just under one million people. As such, it is generally considered to be the nation's leading provincial city and is often referred to as Britain's 'second city'. During the 1970s and early 1980s Birmingham's economy (and that of the West Midlands region) was traumatised by national and global economic change. Between 1971 and 1987 Birmingham lost 191,000 jobs, amounting to 29 per cent of total employment in the city. The city's manufacturing base was particularly devastated, losing 149,000 jobs over the same period. This loss represented 46 per cent of total manufacturing employment in Birmingham between 1971 and 1987 (Champion and Townsend, 1990). In addition to the immediate economic problems facing Birmingham, local residents and outsiders (particularly potential inward investors and visitors) generally perceived the city as having a poor image (Loftman and Nevin, 1998).

In order to address Birmingham's twin problems of economic decline and poor image, the city council sought to promote a new national and international image for Birmingham and provide an attractive climate for private investment within its boundaries. Birmingham's city centre was considered by the city council to be the major asset upon which public and private sector investment should be focused. Indeed, the city centre would provide the platform to build Birmingham's economic future – providing the primary vehicle and catalyst for promoting Birmingham as an international centre for business tourism, leisure and culture, and acting as a magnet for attracting footloose inward private sector investment and jobs (Loftman and Nevin, 1992, 1998). In this context, the city council adopted a pro-growth local economic development strategy focused upon its city centre area. The development of Birmingham's current pro-growth strategies and city centre prestige projects can be traced over two decades. This period encompassed three distinct phases of leadership of the Labour Party controlled city council: under the leadership of Richard Knowles (1984–93), Theresa Stewart (1993–9) and Albert Bore (1999–present). Each of these periods of Labour Party leadership of the city council are reviewed below.

The Dick Knowles era: genesis of the pro-growth agenda

During the 1980s and early 1990s, under the leadership of Richard Knowles, Birmingham City Council sought to engineer a new future for the city, based on grand development projects and aggressive city promotion initiatives (for a more in-depth review see Loftman and Nevin, 1992, 1998 and Loftman and Middleton, 2001). Under Knowles' leadership the city council devised the basis of Birmingham's longstanding aggressive pro-growth local economic development strategy and embarked on a period of massive city council investment in Westside prestige project developments. This phase of Westside prestige development (evidenced by the formulation and building of the International Convention Centre (ICC), National Indoor Arena (NIA), Hyatt Hotel and Brindleyplace developments) was underpinned and supported by Councillor Albert Bore, the influential Chair of the council's powerful Economic Development Committee (Loftman and Nevin, 1998). Over the period 1986/7–1991/2, Birmingham City Council invested £331.1 million in its city centre prestige projects (Loftman and Nevin, 1992).

Birmingham's investment in prestige projects, during the period of Knowles' leadership, has been further supported by a plethora of high profile city council civic 'boosterism' initiatives, which were seen as crucial in underpinning the local authority's projection of a new and positive image of an 'international city' (Loftman and Nevin, 1998). Examples of these boosterist activities include: the city's unsuccessful bids (made during the 1980s) to host the 1992 and 1996 Olympic Games; the transfer of the Sadlers Wells Royal Ballet company and the D'Oyly Carte Opera company from London; and the running of the Birmingham Super Prix, an annual formula two motor race (now defunct), held within Birmingham's city centre area. These place promotion initiatives were complemented by an aggressive city council marketing campaign portraying Birmingham as 'The Big Heart of England' and 'One of the World's Great Meeting Places' (Loftman and Nevin, 1998). The implementation of these civic boosterism initiatives, however, entailed considerable city council financial investment and costs (Loftman and Nevin, 1998).

Wider private and public sector support for the city council's pro-growth strategies and city centre prestige projects was gained via the hosting of two city centre symposia held in 1988 and 1989 (called the Highbury Initiative Symposium 1 and 2 respectively). The two symposia were largely made up of senior city council officers and politicians (including Councillors Richard Knowles and Albert Bore), central government officials, private sector interests and a group of national and international experts in the field of urban design and regeneration. However, the initiative was marked by the absence of any members of the general public or lay community representatives in Birmingham (Loftman and Middleton, 2001).

Table 5.1 provides a breakdown of the 'invited' participants in the 1988 and 1989 Highbury Symposia. From the table it can be seen that senior Birmingham City Council elected members and paid council officers accounted for 21 per cent of all the participants in the 1988 Highbury Symposium and 18 per cent of all participants at the 1989 event. Furthermore, senior Birmingham City elected members and officers accounted for 40.5 per cent of the Birmingham-based Highbury Symposia participants in 1988 and 27.5 per cent in 1989. In contrast, no representatives from any of the city centre residents' associations or tenants' organisations were present at either of the Highbury Symposia. However, one representative from a Birmingham-based voluntary organisation was present at the 1988 Symposium (amongst a total of 71 participants) and representatives from two Birmingham-based voluntary organisations were present in 1989 (amongst a total of 104 participants) (Loftman and Middleton, 2001). The absence of representatives from Birmingham's significant black and ethnic minority populations at the Highbury events was noted by a local journalist that attended the 1988 Symposium:

> The symposium came up with a powerful consensus which largely endorsed the direction taken by the city council while adding some fresh insights of its own, but more local voices need to be heard.
>
> For example, the contribution to be made by Birmingham's ethnic minorities was mentioned many times over the weekend, but none were represented at the symposium.
>
> (Grimley, 1988)

The Stewart years: 'Back to Basics'

In 1993, the political tensions and external criticisms around the prioritisation of scarce city council resources expenditure for prestige projects and city centre development (which had been evident for much of Knowles' period of leadership) surfaced within the city council's ruling Labour Party group. The fracturing of the all-party political 'consensus' around the utility of the city's prestige projects (which was carefully nurtured by Councillors Knowles and Bore) became evident in May 1993, when a veteran left-wing councillor Theresa Stewart was elected Deputy Leader of the council's Labour Party by her peers, and the then Leader (Councillor Richard Knowles) announced his intention to resign from his position as council leader. The election of Theresa Stewart as

Table 5.1 Profile of Highbury Initiative Symposia (1988 and 1989) participants

	No. and (%) of participants	
Highbury Initiative Symposia (1988 and 1989) participants	*1988*	*1989*
Birmingham-based		
City council elected councillors	4 (5.7)	5 (4.8)
City council officers	11 (15.7)	14 (13.5)
Birmingham-based community/voluntary organisations	1 (1.4)	2 (1.9)
Other Birmingham-based participants	21 (30)	48 (46.2)
Birmingham subtotal	**37 (52.8)**	**69 (66.4)**
Other UK (non-Birmingham) participants	23 (32.9)	26 (25)
International		
USA	5 (7.1)	5 (4.8)
Netherlands	2 (2.9)	1 (1)
Spain	0 (0)	1 (1)
Germany	1 (1.4)	1 (1)
Japan	2 (2.9)	1 (1)
International subtotal	**10 (14.3)**	**9 (8.8)**
Total number of participants	**70 (100)**	**104 (100.2)**

Source: Loftman and Middleton, 2001
Note: Figures are rounded

deputy leader facilitated a frank political debate within the local authority and in the local media about the financial costs and impacts of the city's prestige projects. In October 1993, Theresa Stewart was elected *Leader* of the city council's ruling Labour Party group.

The new political priorities of the Stewart administration were reflected in the city council's 1994/5 budget, which protected housing capital expenditure (despite a decline in central government financial support) and set revenue budgets for education and social services which were £2.9 million and £9.5 million above the level recommended by central government (Loftman and Nevin, 1998). The 'back to basics' philosophy of the new council leadership led to local services such as education and social services receiving priority in council resource allocations over prestige development and civic boosterist activities. In the 1994/5 council budget the National Exhibition Centre, ICC and NIA were collectively allocated a cut of £2.6 million in their revenue budget. Additionally, the city council vetoed plans formulated under the Knowles administration for a bid to host the 2002 Commonwealth Games (Loftman and Nevin, 1998). Furthermore, in 1995, the Stewart-led administration stated its reluctance to ratify a £2.5 million subsidy for hosting the 1998 Lions Club Convention (considered one of the largest in the world) – previously agreed by the Knowles-led Council (Loftman and Nevin, 1998).

Nevertheless, the Stewart-led council administration did not turn its back completely on the city's pro-growth and prestige project focused strategy – rather it adopted a more diluted form of pro-growth local economic regeneration in tandem with a more redistributive policy agenda. Under Stewart's leadership, the city council hosted and subsidised the hosting of the G8 Summit and the Eurovision Song Contest in 1998 and continued to pursue the development of prestige projects, but only where external resources could be secured to finance them (thus minimising the impact on council resources). Examples of Birmingham's attempts to secure external resources to finance its revised and diluted pro-growth strategy included: the unsuccessful attempts in 1995 to secure National Lottery and Sports Council funding for a new (then) £200 million English National Sports Stadium (losing out to London) and its bid to stage the prestigious £500 million Millennium Exhibition (again losing out to London, which proceeded to build the Millennium Dome). However, the city council was successful in its bid for one of the fourteen national Landmark Millennium Projects (funded through the National Lottery Millennium Fund), which facilitated the development of an education, science/technology, and leisure development – Millennium Point.

Albert Bore: return of the pro-growth agenda

The pragmatic strategy adopted by Stewart prevented the total collapse of Birmingham's public–private sector consensus around its pro-growth strategy in the short term, but this fragile peace finally collapsed in May 1999, when Councillor Stewart's leadership of the Labour group was challenged by a number of her Labour Party peers. At a Labour Group meeting held on 8 May 1999, Councillor Albert Bore (the main architect behind Birmingham's city centre focused regeneration strategy during Knowles' period of leadership) was elected as the new leader of the city council. This change in leadership marked a shift back towards the aggressive city centre regeneration and promotion strategy implemented by Knowles' city council administration, as noted in the local press:

> Although she [Councillor Stewart] poured millions of pounds into city schools in a successful attempt to raise educational standards, the vision that provided Birmingham with the International Convention Centre, National Indoor Arena, and National Exhibition Centre was missing.
>
> Mrs Stewart dismissed them sneeringly as 'prestige projects'.
>
> Ironically, Coun [*sic*] Bore was heavily involved in all of them, raising the European grants that enabled them to be built.
>
> (Bell, 1999: 7)

> To her [Councillor Stewart's] credit, she insisted on pouring money into the city's run-down education system and was probably as sincere a leader as the city has had for decades, but many felt her style was unsuited to the demands of a millennium city
>
> Coun [*sic*] Bore wants Birmingham to be the best-run and innovative local authority in Britain, and cites the way in which Joseph Chamberlain made the city a world-beater.
>
> (Bell, 1999: 10)

The period since Bore was elected leader of the council, has seen the formulation of a number of bold new strategies for the resumption of large-scale development within the city centre – most notably the redevelopment of the 'Eastside' of Birmingham; the £14 million downgrading of the inner ring road[2] within Birmingham city centre; the aggressive promotion of city centre luxury housing developments; and the formulation of new bids for the location of the new English National Stadium and for the European Capital of Culture 2008. Nevertheless, the pro-growth strategies developed under the Bore administration have been tempered by recognition that the needs of Birmingham's disadvantaged groups and most deprived areas were not being met by the rejuvenation of the city centre area.

This recognition was most evident in the Highbury 3 Symposium, which was held in February 2001. This symposium, in contrast to its two predecessors held in 1988 and 1989, considered issues beyond the city centre and examined those affecting the whole of the city, particularly its deprived neighbourhoods:

> The regeneration of local neighbourhoods emerged as a key message, inextricably linked with strands around local identity, devolution, and urban experience. More investment of resources and attention is urgently needed in local areas, as the success of the centre is not filtering out extensively or quickly enough, and local centres need to be revitalised to provide natural 'hearts' for local communities.
> (Birmingham City Council, 2001: 8)

Furthermore, the extent of local community participation in the Highbury 3 event was considerably expanded compared with the two previous symposia (see Table 5.1). At the Highbury 3 event, Birmingham-based community/voluntary organisation representatives accounted for 14 per cent of the 140 listed participants (Birmingham City Council, 2001), compared with less than 2 per cent in each of the 1988 and 1989 Symposia. Nevertheless, despite the broader scope and content of Highbury 3, the city's pro-growth agenda was strongly reaffirmed by the council leader Albert Bore via the symposium:

> There is no doubt that the city has been in a period of renaissance over the last decade. The tremendous physical transformation of the city centre and the exciting developments coming to a conclusion at Eastside and the Bull Ring are there for all to see. Birmingham's success in shaping a new role and self image has been widely acknowledged, not least when we staged the G8 Summit in 1998 which saw the city take a prominent place on the international stage. All the feedback shows that local people are proud of their city and what is being achieved.
>
> This process of transformation can be traced back to the first 'Highbury' initiative in 1987. This proved to be a pivotal point in the city's history – shaping the strategy for breaking down the restrictive inner-ring road, developing a series of distinct city centre quarters, reinventing the city in the eyes of the world and its own citizens, and strengthening further the already close partnership working between the public and private sectors in the city.
>
> This original vision statement served the city well but, in this era of intense global competition, no city can afford to stand still – it must continually adapt and evolve in response to changes in the global economy and patterns of social and community life. It seemed timely to reflect on the progress that has been made and begin to look ahead to the next decade and beyond, to the future for a new Birmingham; a modern European city.
> (Birmingham City Council, 2001: 6)

BIRMINGHAM'S CITY CENTRE FOCUSED PRO-GROWTH POLICIES: 2002

At the start of 2002, Birmingham's city centre pro-growth strategy was largely comprised of three key elements: the continued development of the Westside area of the city (which was begun in the 1980s); the new Eastside area which is currently the main focus of flagship development in the city; and the promotion of residential development for higher-income groups across the city centre – called 'City Living'. Each of these three elements of the city centre pro-growth strategy are considered in turn below.

'Westside': The Convention Centre Quarter

Four prestige projects located within the Broad Street area of the Convention Centre Quarter (situated on the 'west side' of the city centre) have provided the main focus and linchpin of Birmingham City Council's pro-growth strategy since the late 1980s. These four city centre prestige developments are:

- the £180 million International Convention Centre (ICC), opened in April 1991, with a maximum conference capacity of 3,700 delegates and inclusion of a 'world class' symphony hall;
- the £57 million National Indoor Arena (NIA), which was originally conceived as part of a city council bid to host the 1992 Olympics, constructed to enhance Birmingham's position as an international centre for sport, and seating up to 12,000 people;

- the £31 million Hyatt Hotel, built as an integral part of the ICC development;
- the £250 million privately financed Brindleyplace scheme (located adjacent to the ICC and NIA), which currently includes: two civic squares; 850,000 square feet of offices; 123,000 square feet of retail and restaurant space; a new theatre; a modern art gallery; a high-profile national aquarium; and 140 units of luxury residential accommodation. The 17 acre development is currently the UK's largest mixed-use scheme, and is now approximately 75 per cent complete.

These projects have been supported by local authority investment via the creation of a new civic square (Centenary Square) at a cost of £3.4 million and a £3.3 million pedestrian footbridge that links the Convention Centre Quarter with the city centre. In order to finance the development of the ICC and NIA prestige projects, Birmingham City Council applied its financial and organisational skills to securing private sector and European Community funds, in a bid to offset their massive construction costs (Loftman and Nevin, 1996, 1998).

The perceived success of Birmingham's four prestige projects has generated further private sector interest in other development sites located in close proximity to the Broad Street area. Two such developments are the Mailbox and the proposed Arena Central complex. The £100m Mailbox private sector mixed-use development (an 80,000 sq. m. former Royal Mail sorting office) includes fashion design retailing shops (such as Harvey Nicholls, Emporio Armani, and DKNY), restaurants, bars, offices and a new 180-bedroom hotel. In addition the development includes 200 high-quality residential apartments, 140 of which are rooftop apartments 'offering superb views across the city with private terraces, creating a residential setting unlike anything seen so far in the city' (Birmingham Marketing Partnership, 1999a: 12). The first retail units in the Mailbox complex opened in December 2000.

In 1998 an ambitious proposal to develop a 14 acre '24 hour' mixed-use (leisure, office and residential) scheme within the Broad Street area was submitted by private developers. The proposals for the £400 million Arena Central complex include 100,000 square metres of leisure facilities, office accommodation, a luxury hotel, 100 penthouse-style apartments and the development of a fifty-storey 'City Tower', the second tallest tower block in the UK. The Arena Central proposal was rationalised by the city in the following terms:

> The scale of development uses proposed, including high levels of leisure and entertainment floor space, a large hotel and housing, is consistent with aims to consolidate Birmingham as the regional centre for the West Midlands and is of such a scale as to promote a 'landmark' complex of national and international significance. This could provide both a strong central attraction for potential visitors from within the region and also a combination of commercial leisure, hotel and related facilities complementing Birmingham's role as a focus for business and visitor tourism in Europe.
>
> (Birmingham City Council, 1998: 2)

The scheme is considered to form an integral part of the city's plans for the promotion of Birmingham city centre and the Broad Street area as a nationally and internationally significant visitor attraction and also enhance the business and cultural tourism role now firmly established in the Broad Street area.

In particular, it is anticipated by the city council that the development of the fifty-storey City Tower will serve as landmark development for the city centre and will significantly enhance the city's skyline: 'it would serve as a landmark role establishing strong site identity and relating closely to the regional and national promotion of the site as a focus for visitors' (Birmingham City Council, 1998: 2). Similarly the current leader of Birmingham City Council – Albert Bore – stated:

> Arena Central is the last and most important piece in the Broad Street area regeneration jigsaw, and will provide a magnificent centrepiece for the city centre. It will be a magnet for local people and for visitors, and put Birmingham on the map as a major international city. It will also have a major

> impact on the city's prosperity, creating thousands of jobs, both during construction and afterwards.
>
> Together with the other major developments planned for the Eastside of the city, it will make Birmingham a vibrant place for living, visiting and shopping.
>
> (Birmingham City Council, 2000: 1)

Eastside Regeneration Initiative

Under the leadership of Albert Bore, Birmingham City Council announced the establishment of the Eastside Regeneration Initiative, which seeks to secure the revitalisation of the hitherto neglected eastern side of the city centre. The ten-year Eastside project covers 420 acres of a relatively underdeveloped area, which was the 'original birthplace' of Birmingham. The regeneration initiative is comprised of three key elements: the £113 million Millennium Point project; the £800 million Bull Ring/Martineau Galleries redevelopment; and the creation of a new City Park. Indeed, it is argued that the landmark regeneration initiative will act as a catalyst for the wider regeneration of the eastern side of the city centre, thus further extending the city council's regeneration work and investment into the western and central areas of the city centre. The Eastside initiative also includes proposals for the development of offices, retail, leisure and hotel space and around 3,500 homes and the city council is considering the possible relocation of Birmingham's Central Library to the Eastside area (Birmingham City Council, n.d. a). The importance of the Eastside initiative, the restructuring of the city centre, and the regeneration of Birmingham as a whole was clearly articulated by the City Council Leader Albert Bore at the launch of the initiative in 1999:

> At the heart of Birmingham's continuing renaissance is the phoenix of Eastside – where a highly innovative approach to development and investment opportunity will create a new City Quarter worthy of international recognition.
>
> Eastside is destined to be a role model for future European urban regeneration – a classic example of how to transform a tired industrial zone into a pace setting quality development which will establish new standards.
>
> But the vision cannot become a reality over night. Development will take place over a number of years. However, by 2010, Birmingham's vision-makers envisage that Eastside will enhance Birmingham's profile as a learning city, having already gained a reputation for its contribution to the city's continuing renaissance.
>
> Eastside will be a centre where international companies will operate alongside indigenous business.
>
> It will be home to a thriving community and it will be a bustling, vibrant, location where families will also choose to spend their leisure time.
>
> Above all, Eastside will demonstrate to the world that Birmingham – the 'Can Do' City – continues to set the standards to which others aspire.
>
> (Birmingham City Council, n.d. a)

The three key components of the Eastside Regeneration initiative are considered in turn below.

Millennium Point

This science and technology visitor attraction project has a key role in the regeneration of the Eastside area. The regeneration project represents the largest of Britain's fourteen Millennium Landmark Projects – of national and regional significance, which are part-financed by UK National Lottery funds allocated via central government's Millennium Commission. The Millennium Point project has four main interlinked components:

- The University of the First Age: a new educational institution aimed at providing out-of-school learning in the fields of science, technology and design. It is anticipated that up to 90,000 young people from across the West Midlands region will benefit from the new learning opportunities generated by the university.
- The Technology Innovation Centre: a national centre for excellence for technology and innovation, designed specifically to encourage access and participation by young people, the general

public and local industry – combining technology transfer, research and development and learning activities. The centre is now also the new home of the University of Central England's Faculty of Engineering and Computer Technology.

- The Discovery Centre: a major visitor attraction aimed at a regional, national and international audience (designed to attract more than half a million visitors per annum) – which will comprise a 'hands-on' twenty-first century science and technology museum, focused on new scientific developments and the region's technological history.
- The Hub: a major leisure and entertainment complex, including a 400-seat IMAX cinema, conference facilities and technology-themed restaurants, cafés and retail units.

The Millennium Point project is expected to attract one million visitors each year, generating an annual gross visitor spend of £4 million. Furthermore, it is anticipated the landmark project will attract up to £500 million of private investment and generate up to 4,000 jobs (Birmingham City Council, n.d. b). Additionally, the regeneration project is regarded by the city council as a vehicle for addressing two wider problems facing Birmingham – first, raising the educational achievement levels of the city's young people and its workforce, and second, tackling the under-representation of high technology, high value-added industries within the city, as 'the key to Birmingham's prosperity is creating a new focus for the city as the innovation capital of the region' (Birmingham City Pride, 1994: 22).

Bull Ring/Martineau Galleries Retail Complex

This £800 million development is comprised of two major retail development schemes that are being undertaken by Birmingham Alliance (a partnership between three of the UK's largest property investors). The Bull Ring development project, Europe's largest city centre retail development, entails the demolition of the existing outdated shopping mall and the development of 1.2 million square feet of retail space (100 new shop units and two major departmental stores). The key elements of the scheme proposals are:

- the demolition of the existing Bull Ring;
- a new 110,000 sq. m. shopping centre on three levels, including two department stores;
- a new indoor market and the refurbishment of the Rag Market and the Row Open Market;
- a refurbished fifteen-bay bus terminus;
- safe and secure parking for 3,000 cars;
- a major public open space, greatly enhancing the setting of St Martin's Church;
- the creation of more than 5,000 full- and part-time jobs, together with 1,500 construction jobs.

It is intended that the proposals will connect the Bull Ring with the rest of the city centre by breaking the 'concrete collar' of the inner ring road and subways, with the extension of the city centre into Eastside.

The development is scheduled to be completed in 2003. The Martineau Galleries scheme (located in close proximity to the Bull Ring scheme) is comprised of 185,000 square feet of retail space including three large stores and twenty-five smaller retail units. It is estimated by the city that the two developments will lead to the generation of 8,000 new jobs by 2007 (Birmingham Marketing Partnership, 1999b). More importantly, the Birmingham Alliance scheme is seen as vital to the future of the city centre as a quality shopping location for residents and visitors. As Michael Lyons (the then City Council's Chief Executive) noted in 1999:

> The Birmingham Alliance has our total support as it shares our vision of rejuvenating the city centre. We are convinced this is the best way forward, not just for shopping, but also for its future economy and growing role on the world stage.
>
> (Birmingham Marketing Partnership, 1999b: 1)

City Park

This development represents one of the most important elements of the Eastside initiative –

the creation of Birmingham's first city centre park since the nineteenth century. The new nine acre park (located in close proximity to the Millennium Point and Bull Ring developments) is expected to function as ' the magnet to attract visitors just a short walk away from the main core shopping area and will provide Birmingham with a green oasis in the heart of the city' (Birmingham City Council, 1998). The park is scheduled to be completed in 2002. The last park developed in close proximity to the broader central area of Birmingham was opened in 1876 by the (then) Lord Mayor – Joseph Chamberlain.

City Centre Living

The promotion and development of housing within the city centre is a key element of the city council's strategy for making Birmingham an attractive location for business investment. Birmingham city centre is home to more than 16,000 people – the majority of whom live in council-rented properties within deprived municipal housing estates such as Ladywood, Lee Bank and Highgate. At 1991 a total of 71 per cent of accommodation in the city centre was council rented, compared with a city-wide average of 26 per cent. In contrast, only 10 per cent of city centre accommodation in 1991 was owner occupied, compared with a city average of 60 per cent.

Since the first City Centre Strategy emerged in 1987, a key policy of Birmingham City Council has been the promotion of 'City Living' – the encouragement of city centre housing development. Such housing development is seen as a crucial element in the promotion of a wide variety of land uses within the city centre and creating the atmosphere of a '24-hour' central area. Birmingham is not alone in the promotion of new city centre housing – with cities such as Manchester, Glasgow and Nottingham (to name a few) also implementing similar policies.

The promotion of City Living in Birmingham was regarded by the city council as offering a number of benefits and advantages – namely: adding a 'continental' dimension to the city centre; breathing new life into the centre; stimulating life and vitality outside of normal office hours; making use of vacant and underused upper floors; creating a safer environment in the centre; and reducing the need for city centre workers to commute into the centre. The City Living project is also in keeping with national government policy concerning the reversing of out-migration of urban residents to the suburbs and rural areas; promoting housing development on 'brownfield' sites and thus reducing development pressures on greenfield areas; reducing the need for city workers to travel to and from work; and enhancing the vitality of urban centres (Department of the Environment, Transport and the Regions, 2000).

The role and function of City Living is noted in a number of Birmingham City Council's strategic documents. For example, the City Centre Strategy notes that:

> Housing adds another substantial ingredient to the mix of [city centre] activities. It brings life to the City Centre outside of business hours, provides an element of self-policing and would be an ideal way of bringing vacant floor space back into positive use. Residential development should form part of mixed-use schemes, particularly adjacent to waterways and as an integral part of commercial and leisure developments throughout the City Centre. This will add more of a 'continental' dimension to the Centre's attractions. The retention and improvement of the existing housing stock [predominantly rented council properties] is essential, in order to maintain the vitality of the City Centre, along with new residential development which results in a more balanced mix of housing provision, beyond the large concentrations of Local Authority stock.
>
> (Birmingham City Council, 1992: 6)

Also, the Draft City Living Strategy produced by Birmingham City Council notes that:

> More housing will add to the vitality of Birmingham. It will add a continental dimension – by creating activity after normal office hours, Birmingham will become a twenty-four hour,

> seven-day week, city centre. A vibrant city centre will make Birmingham a much more attractive place for business investment.
>
> It must be recognised that more housing will provide life to the city centre after workers have gone home. It will help towards the sustainability of the city centre by providing accommodation for people who do not wish to live too far from their place of work and it will establish a community that will contribute to the quality of environment and facilities within the centre.
>
> (Birmingham City Council, 1997)

Birmingham City Council has established (through a number of studies undertaken during the late 1990s) that there was significant demand across a range of different types of households and different income groups for city living. However, this demand was most prominent amongst the 25- to 35-year-old age group. For this group, City Living was 'increasingly seen as offering a better value and quality of life style with ease of access to social and leisure facilities' (Birmingham City Council, 1997: 11). In response to this demand the city council noted that:

> To meet the needs of this diverse spectrum of people, a wide range of types and tenures of housing will be required: private housing, ranging from low cost flats to the penthouse type accommodation, and to complement this a city council or housing association element.
>
> (Birmingham City Council, 1997: 11)

The impact of the City Council's City Living Strategy has been evidenced by a massive increase in the number of luxury and upmarket private housing developments that have taken place in the city centre over the past four years. During the early years of the City Living project (prior to 1997), most new City Living developments were undertaken by local housing associations. These housing association developments largely focused on new build and refurbishment schemes (one- and two-bedroom flats) for rent – targeted mainly at lower-income groups. During the period 1991 to 1997 there was a 44 per cent increase in the number of housing association properties in the city centre area and a 27 per cent increase in the number of private sector properties (see Table 5.2). However, over the same period there was a 7 per cent decrease in the number of rented council properties – largely due to demolitions. Indeed the decline in the number of rented council properties within the city centre (363 properties) amounted to more than half of the new housing built by housing associations and private sector housing developers for home ownership over the same period (602 properties).

The scale of the transformation of Birmingham city centre's residential stock is reflected in the fact that at December 2000, a total of 222 new-build houses (35 per cent of which were for-sale properties) and 1,145 new-build flats (65 per cent of which were for-sale properties) had been completed within the city centre since 1990. Furthermore, residential property conversion over the same period had produced 257 new flats for sale with a further 347 conversions planned or under construction.

However, the most striking aspect of the City Living project has been the massive growth in 'for sale' high-cost luxury housing developments – geared at high-income groups and professionals. One example of such development is the Symphony Court scheme (which forms part of the Brindleyplace mixed-use development project), completed in December 1996. The Symphony Court development, which is comprised of 143 residential units (one-, two- and three-bedroom properties), is located adjacent to the ICC and NIA. Most of the residential units are luxury apartments – located in six-storey buildings – although ten townhouses are provided alongside a canal frontage. The development includes a central courtyard that is serviced by a single-gated vehicular entrance with 24-hour concierge security, and one of the buildings includes a fitness centre.

By March 1997 all of the homes in the Symphony Court development had been sold, with the 'drawing board prices' (1996) ranging from £70,000 to £250,000. Also, a number of properties were purchased for private renting to meet a demand, which was (circa 1997) seeing

Table 5.2 Changes in tenure (number of properties) in Birmingham city centre 1991–7

	City council	*Housing association*	*Owner occupied*	*Private rented*	*Total no. of properties*
1991	5,202	904	753	426	7,285
1997	4,839	1,304	955	476	7,574
Change (nos.) 1991–7	−363	+400	+202	+50	+289
Change (%) 1991–7	[−6.9%]	[+44.1%]	[+26.8%]	[+11.7%]	[+4.0%]

Source: Information provided by Birmingham City Council

rental values averaging at £1,100 per month. At March/April 2001 housing units within the Symphony Court development were for sale at between £175,000 (2 bedroom) and £370,000 (3 bedroom) and available for rent at levels ranging between £1,000 to £1,500. Furthermore, the Symphony Court development has been followed by large-scale luxury developments adjacent to the ICC and NIA. These developments include The Mailbox – 140 luxury rooftop apartments and 60 canal-side apartments; and King Edward's Wharf – 100 luxury apartments including penthouses. The purchase prices of such developments are currently advertised at a price range between £140,000 and £295,000 (with 'duplex' top-storey flats selling at over £600,000 and some penthouses priced at £1 million) in the case of the Mailbox development, and a price range of between £195,000 and £435,000 (with some penthouses advertised at £1 million) in the case of the King Edward's Wharf development (figures at March 2001). Such property prices are significantly beyond the affordability of most Birmingham residents, including the city's disadvantaged groups – in particular, current city centre residents living in council or housing association accommodation.

BIRMINGHAM'S CITY CENTRE FOCUSED STRATEGY: SOCIAL EQUITY IMPACT

The development of Birmingham's high profile, city centre prestige projects was predicated on three basic assumptions: first, that prestige projects will both directly and indirectly produce substantial economic benefits to the whole of the city of Birmingham; second, that all residents would benefit from the developments; and third, that the public sector costs of its pro-growth strategy would be minimised by private sector and European Community resources and outweighed by the economic benefits to the city as a whole (Loftman and Nevin, 1992). Civic leaders have publicly justified Birmingham City Council's massive financial investment in its prestige projects in the following terms:

> for the past seven years, the city council pursued a strategy for regenerating Birmingham which has involved attracting private sector investment into the City, which has led to a change in the City's image from a provincial manufacturing centre to that of a major European City. It has resulted for example in the development of the International Convention Centre, the Hyatt Hotel and the Birmingham National Indoor Arena as the focal points of inner city regeneration in the western part of the City Centre.
>
> (Birmingham City Council, 1991)

Whilst the achievements of Birmingham's city centre revitalisation have been widely acclaimed, the benefits and costs of the attempted transformation of Birmingham from the 'manufacturing city' to the 'international business and leisure city', have not been equally shared between city residents, and between city residents and outsiders. In many respects the residents of Birmingham appear to understand the current contradictions in urban policy. In

2001 the city council commissioned a residents' survey, conducted by MORI, and between November 2001 and January 2002, a total of 1,026 residents were surveyed. The results showed that 88 per cent of residents believed that Birmingham is 'an international conference and business city' and 78 per cent believe that the area is 'economically successful'. However, only 34 per cent regard the city as being safe and only just over half believe that neighbourhoods contain good quality housing. The MORI survey explored residents' views of the balance of public sector investment in the city and concluded:

> In several qualitative research studies MORI has conducted for Birmingham City Council over recent years, many feel investment has been concentrated in the centre of the City and that the outer areas have been neglected. The 2001/2 survey shows that over half (54%) of residents think investment should be focused equally on both the City centre and local areas. Meanwhile, 39% think it should focus on local areas rather than the city centre as that is where investment is most needed. . . . In many ways this would appear to reflect a perception that work improving the city centre is now well advanced; the key priority for residents now lies with taking forward successful initiatives locally.
>
> (Birmingham City Council, 2002: 16)

Few of the proposed and existing city centre flagship developments, highlighted in this chapter, were formulated and developed with the needs of the urban poor as a priority. Indeed, the needs of such groups are not considered in these public policy initiatives – as public access to such facilities is wholly determined by ability to pay. However, the positive impacts on the city's socially excluded groups and deprived neighbourhoods are yet to materialise. Indeed, it is argued that the city council's massive investment in prestige developments and city centre physical revitalisation has had a negative impact on the delivery of basic services in Birmingham within a context of limited local authority resources.

Between 1981 and 1990 it is estimated that Birmingham lost £599 million in central government Rate Support Grant (Loftman and Nevin, 1992). In such a climate of central government cuts in local authority expenditure, Birmingham City Council's significant financial investment in prestige developments has resulted in the diversion of scarce public resources away from 'basic' services that the city's disadvantaged groups are particularly dependent upon. Over the period 1986/7 to 1991/2 the ICC and NIA prestige developments alone accounted for 18 per cent of total Birmingham City Council capital spending (amounting to £228 million). Invariably the large-scale investment in the prestige projects, civic boosterist projects and general city centre refurbishment over this period, had a significant impact upon services such as housing and education. Over the period 1986/7 to 1991/2 the city council spent £123 million less on housing than the average for local authorities in England.

The relative lack of emphasis on investment in council housing by the city council over the past two decades has generated a backlog of housing improvements and repairs amounting to £3 billion of repairs over the next thirty years. In a bid to meet this challenge the city council proposed to transfer its housing stock to a 'not-for-profit' company, which would be able to refinance the council housing stock to meet the resources required for improvement of its housing stock. The local authority spent £12 million on consultancy fees to advance the project and secure the finance necessary to improve 63,000 council properties and demolish a further 25,000 over a ten-year period. The national framework which governs the transfer of council housing to new Registered Social Landlords specifies that a majority of tenants must vote for the proposal. The ballot of tenants was held over a three-week period in February and April 2002, and the result revealed that two-thirds of tenants who voted were opposed to the transfer. This result of the ballot has completely undermined the city's housing strategy, and the response of the city council has been to announce the creation of a commission to examine the future of housing in Birmingham, this action reflecting the lack of an alternative strategy to secure the necessary housing investment in disadvantaged neighbourhoods.

A similar pattern of local authority under-investment also emerged in relation to the city's education service, with the council's capital spending falling by 60 per cent during the construction of the prestige projects (Loftman and Nevin, 1998). However, it was the city's education revenue budget that was the most adversely affected by the cost of the city council's boosterism initiatives and the operational losses of the ICC and NIA. Between 1991/2 and 2000/1 the ICC and NIA accrued an accumulated operational deficit of £65.7 million. This figure excludes the capital financing costs of the two prestige projects, which over the period 1991/2 to 1994/5 alone amounted to £88.3 million (Loftman and Middleton, 2001). The effect of this revenue expenditure meant that between 1988 and 1993 the city council's revenue expenditure on education was consistently lower than that recommended by central government, and in 1990/1 the city council's £46 million revenue underspend on education accounted for nearly half of the national education budget underspend (Loftman and Nevin, 1998).

Furthermore, Birmingham's boom in city centre retail, leisure and housing developments is likely to exacerbate further the widening divisions between the city's 'haves' and 'have nots'. It is anticipated that the redevelopment of Birmingham's Bull Ring shopping centre will result in the displacement of most of the centre's previous small low-cost traders – as a result of a massive increase in store rental levels after the redevelopment work. Also, many of the shopping centre's low-income shoppers are unlikely to be catered for by the new 'up-market' stores that are expected to take up the retail units once the redevelopment work is completed (Middleton *et al.*, 1990).

Many of the new leisure developments planned for in the city centre's Westside and Eastside are unlikely to cater for the needs of low-income groups, as access to such facilities will be based on ability to pay commercial entrance fees. For example, access to most of the entertainment and leisure facilities included within the Millennium Point development (in particular the Discovery Centre and the Hub) is limited by ability to pay. Indeed, the city's Science Museum (which was available to visitors and residents without charge) was closed in 1999, and the museum's collection has been included in the fee-paying Discovery Centre.[3] Finally, the trend towards the development of private sector, luxury high-cost housing aimed at high-income groups, as part of the City Living Project, is likely to exacerbate divisions between the poor and the rich residents living within Birmingham's city centre.

In summary, whilst the city council's pro-growth strategy and prestige projects have facilitated a dramatic physical and cultural transformation within Birmingham's city centre over the last fifteen years, there is little available evidence which indicates that the wealth and jobs generated from such a strategy have significantly benefited the city's most disadvantaged groups and neighbourhoods. Indeed it is likely that, in the longer term, the city's pro-growth strategy will further exacerbate pre-existing social and economic divisions within Birmingham as a whole. Furthermore, the city council's historic under-investment in basic services (such as education and housing) as a consequence of financing its city centre prestige developments, may in the long term undermine Birmingham's aspirations to enhance its status as an 'international city'. The evidence and arguments made above beg the question '*Who* benefits from the notion of the aspirational "international city", and *for whom* will it exist?' In this context, Birmingham city centre, in the future, may remain 'an island of prosperity and affluence' in 'a sea of poverty and social exclusion'.

Acknowledgement

The authors would like thank Kevin Broughton of the Centre for Public Policy and Urban Change, University of Central England in

Birmingham for his comments on various drafts of this chapter.

NOTES

1 This chapter draws on previous work undertaken by Loftman and Nevin (1992, 1994, 1996, 1998) and by Loftman and Middleton (2001).

2 The inner ring road (a dual carriageway which includes elevated sections and underpasses) was completed in 1971. The inner ring road has been dubbed the 'concrete collar' as it restricts the expansion of the city centre core and pedestrian movement.

3 At January 2002 admission charges only for the Discovery Centre (now known as 'Thinktank') were £4.50 for a child aged 3 to 15 years; £18.00 for a family of two adults and two children; and concession rates of £5.00 were available to senior citizens, unemployed and disabled persons (Source: Thinktank promotional literature).

6 TIM HALL

Car-ceral Cities: Social Geographies of Everyday Urban Mobility

> from the perspective of the individual the car remains the century's most liberating and most desired technological product. It is cheap because it is manufactured in volume and is subsidised; it is practical because cities have not been planned to rely on public transport; and it is an irresistible cultural icon that delivers glamour and status.
>
> (Rogers and Gumuchdjian, 1997: 35)

> The car is probably the chief means by which environmental inequalities are created and sustained.
>
> (Reade, 1997: 98)

> The lives of the poor are blighted by inequitable mobility: the capacity of the rest of the world to move past them, and their own incapacity to move away.
>
> (Monbiot, 2000: 22)

INTRODUCTION: SOCIAL GEOGRAPHY AND URBAN MOBILITY

Mobility, daily movements through, around and between cities on a variety of modes of transport, is a topic that has received little attention within mainstream social geography. It appears little, for example, in social geography textbooks (see for example Ley, 1983; Knox and Pinch, 2000; Valentine, 2001). Despite some notable exceptions, mobility is a topic that has been relatively marginalised within the broader realm of sociology (Hawkins, 1986), urban studies and urban geography, although there is some evidence that this is beginning to change. This omission is extraordinary for a number of reasons.

Everyday life in cities is all about mobility. As Jarvis *et al.* argue:

> We have moved into an era where we are not simply concerned with a trip to work and back but with the multiple journeys that have become not just desirable but necessary in order to sustain our lifestyles each and every day. It is not an exaggeration to suggest that much of our life in cities is bound up with issues of how to get somewhere.
>
> (Jarvis *et al.*, 2001: 2)

Movement in cities is not socially benign. The most popular mode of transport in cities is the private car. Trucks and vans are the normal mode of transport for produce around the city. These modes have negative impacts on the local environment, the global environment, non-renewable resources and urban societies and cultures. These impacts include toxic emissions, noise, vibration, road accidents, the loss of public space and the severing of communities through road development. These impacts are returned to later.

The negative impacts of private transport are socially mediated. Not all groups in the city tend to suffer the negative impacts of private transport equally. They fall most heavily on the more disadvantaged groups in cities. These groups generally correspond to those with the lowest rates of private car ownership and hence those with the most restricted mobility. The more advantaged are able to insulate themselves from at least local impacts, by driving large, safe (from a driver's perspective) cars and living away from areas blighted by heavy traffic and the negative consequences of road provision. Within all groups there are divisions along the lines of age, gender and ability.

Differential mobility is a major dimension of social exclusion in (and beyond) cities. Resources necessary for the sustenance of urban life (housing, jobs, retail outlets and leisure facilities) have become increasingly dispersed around and beyond cities following an assumption of universal private car ownership. Clearly, there are immediate and obvious disadvantages for those groups without access to private cars. Unfortunately, increasingly degraded public transport (mass transit) networks offer no comparative service (either in terms of access, convenience or economy). Transport-deprived groups can be identified geographically, within certain neighbourhoods, or by age, gender and ability. The recent attention to 'food deserts' in British cities (a lack of affordable local food outlets accessible by transport-poor populations) is evidence of the problems restricted mobility poses to transport-poor urban populations.

The car has reconfigured the social life of cities in new ways. While social scientists have focused a great deal of attention on the role of new technologies in social change, the car has received relatively little attention (Sheller and Urry, 2000). This is despite the massive transformative effect it has undoubtedly had on urban social life. Sheller and Urry argue: 'that mobility is as constitutive of modernity as is urbanity, that civil societies of the West are societies of automobility . . . a civil society of quasi-objects, or "car-drivers" and "car-passengers", along with disenfranchised "pedestrians" and others not-in-cars, those that suffer a kind of Lacanian "lack"' (2000: 738–9).

Policies for sustainable transport development need to 'connect' with the multiple social and cultural geographies of mobility if they are to be successful. Recent years have seen plans, policies, models and strategies emanating from a range of advocates, practitioners, academics and government bodies to reduce the volume of private car use and ameliorate the problems associated with private car use (see for example Rogers and Gumuchdjian, 1997; Department of the Environment, Transport and the Regions, 1998; Engwicht, 1999; Crawford, 2000). It is becoming apparent that much car use is not amenable to being affected by such efforts, as it works 'beyond or between' such strategies and thus seems destined for only partial success or outright failure (see Jarvis *et al.*, 2001: 155). It seems imperative that if sustainable transport development strategies are to succeed, they engage directly with the multiple social and cultural geographies of urban mobility or even emerge out of them (see also Sheller and Urry, 2000).

Policies for sustainable transport development are driven primarily by environmental and economic concerns; social concerns have made less impact on the formulation of such policies – indeed environmental policies and social policies might, in some cases, be in conflict. The problems caused by the dominance of cars in cities are complex and multiple. Solutions, while they may address environmental and/or economic concerns, may do little to address other, for example social, concerns. These may require entirely different sets of responses. Sustainable transport and mobility strategies appear to be driven primarily by environmental and economic issues. Such policies are likely to do little to alleviate the social problems caused by differential urban mobility. However, some innovative social strategies do exist that might be expanded to address mobility-based social exclusion in cities.

This chapter aims to explore some of the social geographies of intra-urban mobility. It begins by looking briefly at the historical relationship between the city and mobility. It moves through a review of the problems caused by the dominance of private car use in cities and considers a range of responses to these problems. It then shifts to consider calls to connect transport strategies with the social and cultural geographies of urban mobility. The chapter is concerned primarily, although not exclusively, with cities in the west.

MOBILITY AND URBAN FORM

While far from being the only driving force behind urban morphological change (Hart, 2001: 105), as Peter Muller argues, there is a 'persistently strong relationship between the intra-urban transportation system and the spatial form and organisation of the metropolis' (1995: 26). Broadly similar processes of successive decentralisation, first of residential and later non-residential land uses, have characterised the development of cities in both Europe and the USA since the nineteenth century (and indeed Australia, which is now a highly car-dependent nation – see Short, 1991: 147–51). However, the extent to which cities in the two continents have decentralised differs greatly. While European cities typically display extensive decentralisation within a tighter spatial agglomeration, cities in the USA have been 'turned inside out' to the extent that commentators have begun to employ a new vocabulary to describe these emergent urban forms and the new urban order in the USA. This vocabulary is one of the 'suburban downtown', the 'technopolis' the 'galactic metropolis', the 'edge city' and the 'exopolis' (the city without) (Hartshorn and Muller, 1989; Scott, 1988; Lewis, 1983; Garreau, 1991; Soja, 1989).

Muller (1995) and Hart (2001) have outlined the stages of urban development linked to intra-urban transportation. The main points of their analysis are summarised in Table 6.1. The shifting locations of employment growth in US cities clearly illustrate the most recent and fundamental shift in the form of US cities and their geographies of economic and residential activity. Attracted by benefits such as extensive areas of cheap land, access to large pools of 'green' labour (particularly married women seeking to re-enter the jobs market) and improved transport and communication infrastructures, and deterred by the growing disadvantages of central city locations, the USA has seen successive waves of employment decentralisation, initially back-office functions, but subsequently headquarters functions and occasionally entire companies, to new suburban locations (Cervero, 1995: 330). The major growth period in suburban employment has been after 1976.

> Between 1976 and 1986, employment in the suburbs in the 60 largest US metropolitan areas rose from 16 million to 24 million jobs. By 1990, around two-thirds of all jobs in US metropolitan areas were outside central cities, up from 45 percent a decade earlier (Hughes, 1992).
>
> (Cervero, 1995: 330)

This suburban employment growth has been accompanied by associated growth in suburban residential populations and retailing activity.

The tale of seemingly relentless suburban expansion outlined above, of car-bound populations seeking the good life, is only one tale of twentieth-century urbanisation, however; the other is one of being left behind and abandoned as the city pushes outwards. This tale is of older inner city and inner suburban neighbourhoods and the groups who have come to occupy them. It is a tale of communities who soak up the noise and pollution of urban expressways that plough through or over them, splitting their communities, while they rely on ever more decaying and restricted public transport systems to get around. It is a tale of the emptying out of local economies as opportunities for employment move to the outer suburbs, edge cities or abroad and as retailing follows, of diminishing health, education, welfare and financial services and of restricted mobility as access to resources and opportunities comes to depend increasingly on the cars that they do not own. It is a tale of ghettos, food deserts, dead local economies, financial abandonment and social exclusion, of being immobile in car-cities. As Alan Thein Durning argues:

> Sprawl makes owning a car a necessity of life, which can transform a low income into a poverty income. It also siphons customers away from inner-city groceries, which raises local food prices and again makes poverty more expensive. It draws jobs, investment capital, and tax base from urban to suburban areas. The flight of the successful leaves behind neighbourhoods short on role models, local businesses, volunteers for the

Table 6.1 The relationship between transport and urban form, based on Muller (1995) and Hart (2001)

<table>
<tr><th>Dates</th><th>Hart (2001)</th><th>Muller (1995)</th><th>Europe</th><th>USA</th><th>Modal share (per cent)[1]</th></tr>
<tr><td rowspan="2">1800–1860 –1890</td><td rowspan="2">Walking/ handcart cities</td><td rowspan="2">Walking/ horsecar era</td><td colspan="2">Some dispersal of industry, commerce and residential land uses to sites beyond city boundaries. Concentration of new employment within cities. Working-class housing densities increasing to accommodate growing need for labour.</td><td rowspan="2"></td></tr>
<tr><td></td><td>Some dispersal of residential land use associated with horsecar and later rail transport.</td></tr>
<tr><td rowspan="2">1860–1890 –1920</td><td rowspan="2">Star-shaped mass-transit cities</td><td rowspan="2">Electric streetcar era</td><td>Development of urban trams, elevated and underground railways, some motorised road vehicles and specialised roads. Emergence of numerous examples of high-density middle-class suburbs.</td><td>By the early 1890s the dominant mode of intra-urban transport was the electric streetcar. Middle-class suburbanisation at low densities. Central city growth began to flatten out.</td><td rowspan="2">1870
USA W. Europe
W 87 91
OP 4 4
PT 9 5</td></tr>
<tr><td colspan="2">Cities began to spread more rapidly in spatial extent. Residential dispersal but still city centre dominance in employment. Towards the end of the era, appearance of flows in other directions, beginning to transform city form from star shaped towards web or network patterns.</td></tr>
<tr><td>1920–1945</td><td></td><td>Recreational automobile era</td><td></td><td>Introduction of cars at a relatively slow pace compared with subsequent growth rates. Accelerated the development of suburban areas between mass transit axes. Patchwork of distinct social areas in suburbs began to emerge.</td><td>1930
USA W. Europe
W 18 29
OP 27 10
PT 55 61</td></tr>
<tr><td>1945–</td><td>Multi-centred/ low-density city</td><td>Freeway era</td><td>Centres of the largest cities experienced growth, but otherwise centres suffered decline in their share of employment and retailing. This was associated with urban decline in adjacent areas. There was extensive decentralisation, but suburbs generally retained links to their city rather than develop as independent ‘edge cities’.</td><td>The car became generally affordable following the economic growth of the post-war period. This was associated with extensive highway building. There was a massive urban decentralisation adding extensive suburban zones to cities between 1950 and 1990. The central business district lost the advantages previously associated with its centrality. New independent ‘edge cities’ developed beyond the boundaries of existing urban areas to radically reshape the urban hierarchy.</td><td>1990
USA W. Europe
W 3 10
OP 88 71
PT 9 19</td></tr>
</table>

Note: [1] W = walking; OP = other private transport; PT = public transport

> community center, homeowners, contributors to the PTA, and hope. And these abandoned neighbourhoods are prone to succumb to the concomitants of poverty – welfare dependency, teen pregnancy, violence and school failure. Completing the vicious circle, these in turn speed flight to the suburbs by those who can afford to go.
>
> (Durning, 1996: 26)

THE IMPACTS OF AUTOMOBILITY

Since the 1960s the negative impacts of the car, and of provision for the car, have become well known. The first cries of opposition to automobilty in the USA came from those communities, typically inner city, experiencing high levels of poverty, with high ethnic minority populations and suffering political disenfranchisement, who were being displaced or devastated by the urban highways that were the cornerstone of the post-war 'freeway era' in the USA. The decimation wrought by the construction of the Cross-Bronx Expressway, for example, on the Bronx neighbourhood in which he grew up is powerfully related by Marshall Berman in 'Robert Moses: The Expressway World' from *All That Is Solid Melts Into Air* (1983) (see also Haymann, 1994). We now recognise this as only one of a number of impacts that have reshaped the morphologies, the physical, economic and social geographies of cities in the second half of the twentieth century and continue to do so into the twenty-first century, increasingly and ominously in the cities of the developing world. While we can recognise four sets of impacts of cars on cities (environmental, economic, health and social impacts), it should be emphasised that the first three sets of impacts are cut across sharply with a social dimension. The environmental, economic and health impacts of cars are not felt evenly across different social groups but fall most heavily on non car users, a disproportionate number of whom are from low-income groups (Newman, 1999: 177).

Environmental impacts

The impacts that have received the greatest attention in recent years, and which have pushed traffic reduction strategies and agreements forward, have been the environmental impacts of cars. Cars produce both local and global, short- and long-term environmental impacts. Automobile fuel combustion, and emissions from the industries that produce, maintain and fuel them, are major contributors to the greenhouse gases responsible for increased rates of climate change. For example, automobile fuel combustion accounted for 45 per cent of fossil fuel derived carbon dioxide emitted in the Pacific Northwest region of North America in 1993 (Ryan, 1995 in Durning, 1996: 8). Given subsequent increases in car use in the region the current figure is likely to be nearer 50 per cent. In the UK the Department of the Environment, Transport and the Regions has recognised that road traffic is the most rapidly growing contributor to climate change (Department of the Environment, Transport and the Regions, 1998).

Cars and other road vehicles are the major cause of air pollution in cities. For example, cars release 55 per cent of the air pollution in Washington, DC and considerably more in many other cities (Durning, 1996: 8). Poor air quality in cities is directly responsible for increased levels of respiratory disease and lung cancer. In addition, lead emissions have been linked to poor school performance in heavily polluted areas (Stutz, 1995: 377). Unsurprisingly, air pollution is frequently cited as one of the contributory factors to the suburban flight of middle and upper income populations (Rogers and Gumuchdjian, 1997: 35). This leaves lower income groups in the very areas where air pollution is most concentrated, namely in inner city areas and along transport arteries, especially highways, where it has a disproportionate impact on their lives and their health (Hodge, 1995: 372; Stutz, 1995: 383). Thus air pollution in cities displays a distinctive and familiar unequal social geography. Car travel also has other localised environmental impacts, such as noise generation and vibration that can contribute to sleep-deprivation, annoyance, hearing impairment and stress amongst those affected (Stutz, 1995: 377–8).

In addition the provision of roads has major environmental impacts on local ecosystems during both construction and subsequent use. These impacts include: water pollution from oil leaks and the application of road salts, the disruption of the hydrological cycle through increased run-off and evaporation from paved surfaces leading to increased storm surges, aquifer depletion, pollution, habitat disruption, especially to valuable and sensitive wetlands, soil erosion following vegetation removal and decreases in biodiversity as indigenous fauna are stripped to make way for transport corridors and typically replaced with a narrower range of imported fauna or left to invasive weed species (Stutz, 1995; Durning, 1996).

Economic impacts

Car use in cities generates a number of economic costs. These include the direct and indirect costs associated with congestion. Directly, traffic congestion in the USA has been estimated to cost $150 billion per annum through lost time and fuel. The World Resources Institute estimates the additional social costs associated with congestion at a further $300 billion (Rogers and Gumuchdjian, 1997: 38). In the UK recently the Confederation of British Industry estimated these costs at £18 billion per annum and rising, as congestion in the country's major towns and cities steadily worsens (Walters, 2001: 1). These costs can be hugely significant for individual cities. The economic cost of congestion for the Seattle area, for example, has been put at $740 million per annum (Litman, 1995 in Durning, 1996: 23). Put simply, traffic congestion makes cities uncompetitive. Where cities suffer from congestion and lack an effective public transport alternative, the economic impacts of congestion on low-income households are particularly serious. Low-income households can commonly spend in excess of twenty per cent of their income on transport costs and, if they live in peripheral areas, face three to fours hours per day commuting to and from places of employment opportunity (UNCHS/Habitat, 1994 in Newman, 1999: 377; see also Hodge, 1995). The indirect economic costs of traffic congestion include the loss of investment, and hence jobs, in cities where it is seen as a problem (Newman, 1999: 176). Car crashes also carry with them serious economic as well as health and social costs (Durning, 1996: 23). Furthermore, the subsidisation of motoring in the west means that car drivers do not pay for the full economic (let alone environmental, health and social) costs of their automobility (Durning, 1996; Newman, 1999). These costs are borne by non-car drivers and institutions like health services.

Health impacts and road safety

In addition to the negative health impacts associated with air pollution, noise and vibration, road traffic causes a great deal of death and injury amongst urban populations. This impacts disproportionately on non-car users, especially groups such as children and adolescents. Peter Newman has argued that 'globally, road traffic accidents are thought to be the leading cause of death among adolescents and young adults' (1999: 177). Busy roads are more likely to be occupied by low-income groups, while children on social housing estates are more likely to use streets as play areas because of a lack of access to gardens or other adequate play areas (Lucas *et al.*, 2001: 6).

Curiously, cars are not perceived as dangerous compared with other more highly publicised, yet less serious (in quantitative terms), causes of death and injury in cities.

> Since 1980, motor vehicles have killed almost 31,000 (North American) northwesterners and injured more than 2 million – far more than have

> died or been injured as a result of violent crime. Tragically, people often flee crime-ridden cities for the perceived safety of the suburbs – only to increase the risks they expose themselves to.
>
> (Durning, 1996: 24)

However, the interrelationship between mobility and health is more complex than this. Macintyre *et al.* (2000) explored the relationships between transport, housing and well-being in the Glasgow and Clyde Valley area of Scotland. Their project explored the social and mental health of car and non car users (along with owner-occupiers and people living in social housing). They found that access to a car was associated with a number of positive psychological attributes such as mastery and self-esteem. The perceived benefits of car ownership they found as choice, convenience, privacy, and safety. These benefits were found to a much lesser extent amongst users of public transport (Macintyre *et al.*, 2000: 8–11). Transport mode and housing tenure, they concluded, were associated with differences in physical and to a lesser extent psychological health.

> Social renters and non-car owners were typically one-and-a-half times more likely to report health problems than their ownership counterparts, had higher levels of anxiety and depression and lower levels of mastery and self-esteem.
>
> (Macintyre *et al.*, 2000: 11)

These differences occur because housing tenure and car access are related to socio-economic characteristics, psychosocial benefits and psychological characteristics which are independently related to health and features of the dwelling, local environment and modes of transport that are either health promoting or health damaging (Macintyre *et al.*, 2000: 13–14).

Social impacts

> It is both the deficits of mobility in some places, as well as its excess in others, that are symptomatic of contemporary urban inequalities and problems of social exclusion.
>
> (Sheller and Urry, 2000: 748)

> an inability to access transport can lead to people missing out on jobs, education and other social opportunities.
>
> (Lucas *et al.*, 2001: 9)

Social exclusion can be defined in terms of individuals' lack of access to, and hence participation in, civic resources and opportunities. Mobility is an essential aspect of social inclusion given the dispersed land use patterns that now characterise western cities. These patterns mean that a car is now a virtual necessity for full access to employment, educational, retail, leisure and social facilities (Hay and Trinder, 1991). While it is overstating the case to suggest that transport inequalities cause social exclusion, a lack of mobility is clearly a key dimension that exacerbates individuals' experiences of social exclusion (Grieco, 1995; Huby and Burkitt, 2000; TraC, 2000; Lucas *et al.*, 2001). Selected findings of a Scottish Executive (2000) study of the role of transport in social exclusion in urban Scotland are instructive in this regard:

- Women, the unemployed, elderly, people with health problems and those in low-income groups are more likely to experience transport-related social exclusion.
- Excluded groups are heavily reliant on walking and public transport, and rely on lifts from family friends and neighbours.
- Lower-income groups (gross household incomes below £100 to £149 per week) spend more on fares for public transport than those in higher-income groups.
- Regular car access is strongly associated with a higher income level (gross household incomes above £100 to £149 per week), home ownership and lower levels of public transport use.
- Reliance on lifts and irregular car access is responsible for an increase in average journey times to reach local facilities.
- Elderly people and people with health problems were more likely to find it difficult to use buses and taxis and walk for at least 10 minutes.

They conclude that not owning a car makes a real difference to individuals' abilities to access employment, retail and social opportunities.

> access to transport is a real concern for some people and . . . differential transport access does affect participation in what are considered to be normal activities of citizens . . . . It is clear that transport does affect some individuals' ability to participate in these 'normal' activities, specifically employment and labour-related activities such as job seeking and the attendance of interviews. There is also evidence that leisure activities are influenced by transport considerations.
>
> (Scottish Executive, 2000)

Further, in a study of transport and social exclusion conducted in five mixed urban and rural areas in the UK, Lucas *et al.* (2001) highlighted eight dimensions of the relationship between transport and social exclusion that emerged from focus group interviews with residents of these areas. These were:

- The problem of poor availability and affordability of local services . . .
- the low mobility aspirations of some people . . .
- the cost of public transport and disparities in fare structure and concessionary fares between different areas . . .
- the inadequacy of public transport services in terms of providing reliable, frequent and well-routed services to key destinations . . .
- poor public transport vehicular access and supporting infrastructure,
- the problem of personal safety in the local area spilling onto public transport . . .
- policy ignorance of the car as a basic need for some low-income groups . . .
- the knock-on policy effect of inadequate transport and/or access to activities.

(Lucas *et al.*, 2001: vi)

In aggregate terms, the impact of transport deprivation on the lives of the poor is significant at national levels. Figures for the UK, for example, from the Office for National Statistics Family Spending Survey 1999–2000, indicate that over two-thirds of the poorest twenty per cent of households do not own a car (2000). This accounts for approximately 8.4 million people in the UK. This situation is made worse when one realises that deprived areas are, commonly, poorly served by public transport, doubly disadvantaging groups living in these areas in terms of their mobility and access to civic resources and opportunities. This transport deprivation is even more acute for women, the elderly or mobility-impaired individuals (Hamilton *et al.*, 1999).

The role of transport and mobility received little attention in the early social exclusion literature, and indeed in traditional transport planning (Grieco *et al.*, 2000) or policy (Lucas *et al.*, 2001: v). However, there is evidence that this has begun to change (see Scottish Executive, 2000; Grieco *et al.*, 2000; Lucas *et al.*, 2001). The issue has recently become a focus of central government attention from bodies tasked with finding solutions to social exclusion. In the UK, the government's Social Exclusion Unit launched a major consultation exercise in July 2001. The exercise sought to gather the views of transport-poor individuals and groups with a view to recognising more clearly the relationship between transport, mobility and social exclusion and to find appropriate and effective solutions (see also Department of the Environment, Transport and the Regions, 2000).

In addition to the destruction of communities through urban road building noted earlier, further social impacts of automobility include the degradation of the social lives of urban spaces. This degradation is both quantitative and qualitative. Quantitatively increasing proportions of urban land are being given over exclusively to the car. In addition to the increasing privatisation of the public realm, now widely recognised as characteristic of the contemporary city, we can observe ever-increasing car-only environments in western cities.

> About one-quarter of the land in London and nearly one-half of that in LA is devoted to car-only environments. And they then exert an

> awesome spatial and temporal dominance over surrounding environments, transforming what can be seen, heard, smelt and even tasted.
>
> (Sheller and Urry, 2000: 746)

Qualitatively, urban social life is also degraded by traffic. Evidence from this comes from Don Appleyard's much-cited study of social contacts within streets with different traffic levels (Appleyard, 1981). In a study of streets from different San Francisco neighbourhoods with different levels of traffic, it was found that the level of social interaction between people on the streets declined as the volume of traffic increased. On streets with light traffic people reported having an average of 3.0 friends in the street and 6.3 acquaintances. However, these figures declined on streets with heavy traffic to only 0.9 friends and 3.1 acquaintances.

> This study points the finger at urban traffic as a fundamental cause for the alienation of the urban resident, an effect at the heart of the erosion of modern day citizenship.
>
> (Rogers and Gumuchdjian, 1997: 36)

Finally, Sheller and Urry (2000) argue that automobility has fundamentally reconfigured the nature of urban social life in the twentieth century. While social scientists, and especially sociologists, have been quick to recognise the impacts of technologies such as media technologies and computers on ways of urban living they have been curiously silent on the role of the car. Arguably the impacts of the car have been every bit as great as, or greater than, any of these other technologies, in that they have reshaped the time and space scapes of the modern urban dweller, creating 'distinct ways of dwelling, travelling and socializing in, and through, an automobilized time-space' (Sheller and Urry, 2000: 738). They seek to address this lacuna by mapping the civil society of automobility. They define societies of automobility through six characteristics as a:

> *manufactured object*. . . . The major item of *individual consumption* . . . [a] powerful machinic complex . . . The predominant global form of 'quasi-private' *mobility* that subordinates other 'public' mobilities of walking, cycling, travelling by rail and so on . . . [as] The dominant *culture* that sustains major discourses of what constitutes the good life . . . [and] The single most important cause of *environmental resource use*.
>
> (Sheller and Urry, 2000: 738–9)

Sociologies of automobilised societies are clearly overdue.

RESPONSES – TRAVEL-REDUCTION STRATEGIES

The numerous responses to the growth of car travel have all aimed, by various means, to reduce travel. This aim is pursued either by attempting to reduce the number of trips made by cars, reducing the distances of trips made by car, reducing the numbers of cars making trips and/or reducing the time spent travelling in cars, for example, by encouraging driving at off-peak and hence less congested times (Marshall, 1999a: 90–1). These strategies have been embedded in policies emanating at all levels from the local to the international.

The main motivation behind these travel-reduction strategies has been to reduce the environmental toll of road traffic. However, advocates have been quick to promote the economic benefits of reductions in congestion. Indeed Marshall (1999b) has discerned a discourse of 'sustainable growth' within the travel-reduction strategies of many cities, whereby environmental concerns are pursued within an overall strategy of continued economic growth:

> It would appear that as long as travel-reduction policies are in line with these wider city objectives, they have a chance of promotion and implementation. Where they do not, however, those policies are likely to suffer.
>
> (Marshall, 1999b: 169)

Responses to the growth in car travel appear to be much less concerned with addressing the social impacts of unequal mobility. Initiatives that directly address mobility-based social exclusion are relatively few (though see Garrett and Taylor, 1999; Sawicki and Moody, 2000; Scottish Executive, 2000). Indeed, some authors have argued that policies that aim to restrict travel, for environmental reasons, may actually be in conflict with social aims (Lucas *et al.*, 2001). Furthermore, it is only recently that central governments have begun to ask local authorities to audit their transport policies from a social equity point of view (Lucas *et al.*, 2001: v).

Marshall and others have conducted extensive research into travel-reduction strategies (Marshall *et al.*, 1997; Marshall, 1999b). This research has identified six mechanisms, involving either switching or substitution, which can contribute to travel reduction. These are:

Substitution mechanisms:

- linking trips: rather than making a number of trips each for a single purpose, a single trip is made which addresses a number of purposes;
- technology: physical travel is replaced by electronic communication;
- trip modification: the trip is modified by type, for example, a mobile goods delivery service replaces individual shopping trips.

The effects of each of the above is to reduce the number of car trips made.

Switching mechanisms

- mode switching: whereby drive alone car travel is replaced by car sharing or alternative modes of travel such as public transport;
- destination switching: whereby more local destinations are created or selected rather than more distant destinations;
- time switching: whereby travel takes place at less congested, off-peak times, thus reducing travel time.

The effects of the above are to reduce the number of kilometres travelled or the time spent travelling (Marshall *et al.*, 1997: 295; Marshall, 1999a: 91).

A survey of travel-reduction strategies in seven European countries identified a total of sixty-four specific strategies that have been employed, to varying degrees. These were classified into ten separate categories, including: road capacity management; pricing and taxation measures; various land use planning measures, from encouraging compact mixed-use urban development to provision of infrastructure for alternative modes of travel; developing communications and technology; 'push' and 'pull' policy measures; physical measures such as giving priority on roads to public transport and cycles; location/time and/or user restrictions; public awareness; and various other measures (Marshall *et al.*, 1997: 296–7). Despite considerable cross-European variation, Marshall *et al.* (1997: 301) found that the most implemented of these measures were subsidies and spending, and location/time/user restrictions.

Evidence suggests that, in the face of the relentless growth of car ownership and personal mobility, travel-reduction strategies have had only a limited impact. Maat and Louw, for example, conclude from their survey of case studies from a number of European cities that 'much of the policy implemented in this area does not appear to have been very effective. Reductions in car use have occurred to a limited extent' (1999: 160). This judgement recognises a number of limitations and qualifications that need to be applied to travel-reduction strategies.

Maat and Louw have identified a gap between 'the assumptions underlying policy measures on the one hand and the actual behaviour on the other' (1999: 152). The effect of this 'policy–behaviour gap' is that travel-reduction targets are typically not met. Often travel-reduction strategies have unforeseen outcomes that compromise their aims. Examples of this, outlined by Maat and Louw, include 'pull' policies, for example provision for public transport, generating new trips as people's transport options are enhanced, people switching from one sustainable mode of transport to another, commonly from cycling to public transport, rather than from drive alone car commuting to sustainable modes, and traffic restrictions merely displacing car travel to other non-restricted areas, for example, from city centres to out-of-town locations (1999: 154). Furthermore, Maat and Louw identify a number of barriers to the implementation of travel-reduction strategies including policy and legal barriers, resource barriers and social and cultural barriers (1999: 156–7).

Marshall, in considering the geographies of traffic-reduction strategies, has recognised a particular social and spatial targeting of push and pull measures. Push measures, those aimed at dissuading drivers, tend to be targeted towards inelastic situations, for example at commuters who have to travel to work and are largely unable to change their employment location, and towards inner city areas. By contrast more elastic situations, where travellers have more choice, both in terms of whether or not to make a journey and in terms of potential destinations, tend to be the targets of pull measures. Typically pull measures are directed towards outer city areas. Consequently, because there are few measures in the outer city to dissuade drivers, and despite there being enhanced public transport options following the implementation of pull measures, there is little if any check to the growth of car travel (Marshall, 1999b: 173–5).

> The impression, then, is of city centres becoming carefully controlled 'fortresses' against traffic, safely enclosing ambulant, cash-dispensing shoppers and tourists, while outside, the traffic and development is left to grow wild, and in the 'free market' of convenience, the car is the winner.
>
> (Marshall, 1999b: 174)

Marshall goes on to outline the challenge facing transport policy makers.

> The challenge for policymakers, then, having tackled the inner city, is to turn attention to the outer areas of cities. It is here, on the edges of the city and its hinterland, that most growth is taking place. It is here that transport solutions such as mode switching appear to be least effective. This suggests that a wider range of policies is required, which must offer convincing alternatives to conventional car use in the suburbs.
>
> (ibid.: 178)

In addition to these limitations it is worth briefly rehearsing some further criticisms of specific travel-reduction measures. A significant debate has also grown up around the relationship between urban form and sustainability. This follows widespread advocacy of various high-density compact city models by, for example, the European Commission's *Green Paper on the Urban Environment* (1990), the Department of the Environment in the UK in their *Strategy for Sustainable Development* (1993), and a variety of planning groups and environmental pressure groups (Breheny, 1995: 83). The basis of this advocacy is that the compact city, it is argued, offers a number of environmental benefits over more decentralised urban forms typical of the UK, USA and Australia. These supposed benefits include: reduced fuel consumption through enhanced proximity and accessibility deriving from smaller city size and mixed-use development; a reconciliation of the needs for urban growth and rural protection (Breheny, 1995); accommodation of public transport provision; and the potential for urban regeneration in dense mixed-use neighbourhoods (Blowers and Pain, 1999).

Despite the appeal of the pro-compact city rhetoric, critics have convincingly argued that there is little, if any, hard evidence to back it up and that, consequently, it is based largely around some questionable assertions. The most compelling critiques of the compact city model, and of the urban-containment policies necessary to achieve it, are those advanced by Michael Breheny (1995). It is worth considering these in some detail.

Breheny advanced three main critiques of the advocacy of the compact city model. First is the fundamental point that compact city advocacy is informed by very little hard evidence, and rests primarily on a set of false assumptions about both compact and decentralised urban forms. Breheny's review of the little evidence that does exist suggests that significant savings in fuel consumption following compact city development are highly unlikely to be realised. Even following an extreme policy of extensive re-urbanisation, rates of energy saving are only likely to be a modest 10 to 15 per cent:

> Would advocates of the compact city be so forceful if they knew that the likely gains from their proposals might be a modest 10 or 15 per cent energy saving achieved only after many years and unprecedentedly tough policies? Although proponents of the compact city have not generally specified expected levels of energy saving, it is to be assumed that they have in mind levels considerably higher than this.
>
> (Breheny, 1995: 95)

The wisdom of pursuing such extreme policies is further thrown into doubt when Breheny points out that similar levels of energy saving could be achieved through less painful methods such as incorporating improved technology into vehicle design and raising fuel costs (ibid.: 99). Breheny argues that compact city advocates who almost exclusively attack decentralised urbanisation have failed to take into account that other factors, such as household income and fuel prices, have a greater effect on journey behaviour than urban form or size. In seeking to contain or reverse urban decentralisation, compact city models pursue the wrong quarry and are, therefore, unlikely to result in significant savings (ibid.: 92).

Second, Breheny, among others, has argued that advocates of the, necessarily extreme, policy measures required to achieve compact city development have also failed to take into account the social costs of these measures. Given the prevalence of decentralisation as the dominant process of urban development in much of the developed world, achieving compact city development would involve nothing short of a fundamental reversal of this prevailing process. It is readily apparent that the policies required to achieve this are far from socially benign. The cramming and increased density that seems necessary is likely to result in an increase in overcrowding and homelessness in cities, it has been argued (Breheny and Hall, 1996; Blowers and Pain, 1999). Further, preservation of the rural environment might also come with unforeseen social consequences. Blowers and Pain argue that the lure of preserved rural environments would be in stark contrast to the increased population and building densities of the compact city. This is actually likely to increase flight to the suburbs and beyond for those who can afford it (Blowers and Pain, 1999; Haughton and Hunter, 1994). Taken together, Blowers and Pain argue, these problems suggest that compact city development is likely to heighten inequality within cities, itself a dimension perpetuating unsustainability. In addition Mohan (1999) has pointed out that imposing constraints on development, a further necessity of urban containment policies, in one locality may simply result in the relocation of environmentally harmful activities to areas without such restrictions.

Finally, Breheny argues that it is simply unrealistic to expect compact city models to be implemented given the current nature of urban development. He argues that, in the UK, for example, the forces of decentralisation are too strong, fuelled particularly by employment movement to the suburbs, and planning and political will too weak for any significant degree of urban containment to be realised (1995: 91). Urban containment policies operative in the UK in the recent past have brought only limited success, with development 'boundary jumping' a common problem. The widespread endorsement of city living necessary for the success of compact city models would involve a reversal of deeply held aspects of the cultures of many countries that heavily idealise the rural over the urban (Short, 1992; Colls and Dodd, 1986) and more tangibly goes against the wishes of many house builders and the market (Mohan, 1999). Finally, the commitment to, and investment in, the existing urban realm is likely to create inherent inertia which is likely to seriously mitigate against any significant change in urban form in the foreseeable future (Blowers and Pain, 1999: 280). Neatly echoing this critique, Blowers and Pain (ibid.: 281) argue that compact city models for sustainable development are 'impractical, undesirable and unrealistic'.

It appears that the most practical response to the apparently irreconcilable demands of unfolding urbanisation and sustainability is not to try and reverse decentralisation, but to focus growth around strong sub-centres, rather than allowing uniform low-density sprawl (Brotchie, 1992). Gordon *et al.* (1991), for example, recognised that multi-centred urban areas actually tend to reduce commuting and hence motor fuel consumption per capita. In addition, strong urban sub-centres would allow the development of effective public transport networks within cities. Andrew Blowers (1993) has developed this theme in his model of the *MultipliCity*. In addition to developing strong public transport nodes Blowers advocates urban infilling, some new settlement development and some development in rural areas as a practical path to more sustainable settlement forms within the context of current patterns of development. Crucial, Blowers argues, is provision for sustainability at the regional rather than simply the city scale.

Much has also been made of the potential of electronic technology and communications to alleviate some of the need for physical travel through space. A number of agencies have recently advocated the potential of electronic communication and technology to reduce levels of social exclusion (see Shearman, 1999; Social Exclusion Unit, 2000). One strand of this advocacy has sought to explore the possibilities of reducing the dependency of socially excluded groups on physical journeys through space, to access goods, services and opportunities, with increased electronic exchange (Grieco *et al.*, 2000). While there is undoubtedly potential for making some services and opportunities more readily available to socially deprived communities in electronic form, it is worth sounding a note of caution for those advocating overly technocratic solutions to the problems of *carceral* cities. The straightforward switching relationship from physical movement to electronic communication, that has underpinned much technocratic rhetoric, has not been observable historically. What has characterised the complex relationship between physical movement and electronic technology and communication has been 'interdependence, complementarity and synergy' (Graham and Marvin, 1996: 283) rather than simple switching or substitution. Rather than replacing the need for travel, historically telecommunications links have actually tended to generate physical trips between interconnected nodes (Saloman, 1986; Mokhtarian, 1990; Graham and Marvin, 1996). As communication between places increases, so does the potential to generate extra trips between those places. Although it is too early to tell, given a lack of rigorous research to date, there appears the possibility that the relationship between socially excluded groups, electronic communication and physical travel, within this advocacy, might be conceived of in an unrealistically straightforward manner.

Furthermore, the application of some computer technologies may actually increase the numbers of cars on existing roads, thus, indirectly at least, contributing to greater levels of social exclusion, not to mention worsening environmental, health and road safety impacts. There has been some powerful technocratic advocacy that has stressed the potential of technology to maximise the capacity of existing road networks. For instance, guidance technology and in-car computers could potentially allow far more cars than currently use them to travel on roads in the future. Such technology would allow cars to travel much closer together and significantly cut down journey times as electronic co-ordination regulates traffic speed for maximum efficiency. Such advocacy sees potential increases in road capacity of between three and seven times (Graham and Marvin, 1996: 296) as unproblematic when viewed in terms of potential capacity, management and efficiency. However, this perspective singly fails to take account of any of the negative impacts of road traffic. For example, Whitelegg maps out some of the potential consequences of such a future:

> Telematics can increase capacity on the highway system but what happens when all those cars pour

> into cities and search for somewhere to park? If we increase highway capacity between our cities then we must allocate even more land in cities to vehicles and in so doing transform cities into sterile wastelands dominated by noise, air pollution, danger and lack of people. Alternatively new technology offers the possibility of driving less wealthy people off the streets with pay-as-you-go smart card technology to allocate road space through ability to pay.
>
> (Whitelegg, 1993: x–xi)

CONCLUSIONS

It is clear that for a variety of reasons the numbers of cars in towns and cities need to be reduced. Their growth is not sustainable, equitable nor healthy. However, this path is far from unproblematic. Setting aside the considerable opposition that car-reduction strategies generate and the likelihood of car reduction of a desired volume being achieved, these goals need to be realised in ways that do not compound existing levels of social exclusion through reducing or constraining mobility or making it more expensive for those who most rely on it to access basic civic opportunities and resources. Clearly, the benefits that the car has brought to some groups, such as women (particularly in terms of personal security) and the third of the lowest quintile of the UK population who own cars, cannot be ignored.

To achieve this, four guiding points might be borne in mind. First, it is important that efforts are not directed into policies that are unfeasible and simply unlikely to yield the desired results despite the considerable efforts and resources that might go into them. The most obvious in this respect are compact city models and policies. Second, policies where the environmental aims compromise social welfare and equity should be excluded. An example here is the idea of road pricing or charging a 'congestion fee' for cars to enter city centres. This is simply likely to exclude those unable to afford the charge. The idea of buying access is not socially sustainable despite some sound economic and environmental arguments. Third, resources should be explicitly directed towards policies or initiatives that will alleviate mobility-based social exclusion. These might include information technology initiatives or travel alternatives for welfare-dependent populations moving into employment (Sawicki and Moody, 2000). Finally, all of the policies and initiatives aimed at reducing car use in cities must be linked directly into the multiple social and cultural geographies of mobility if they are to be successful (Jarvis *et al.*, 2001). It is important to understand the many reasons that people use cars and the multiple meanings of mobility before changing those geographies can occur.

PART III

Practising

The informal building of pigeon lofts, Skinningrove, the Cleveland coast, UK (Photo: Malcolm Miles)

INTRODUCTION

Part III, Practising, moves from critical reflection on images in Part I, and the generic concern with movement, access and liberating tactics of Part II, to specific fields in which urban forms are produced and received.

Iain Borden asks what can be considered as radical architecture. Bearing in mind Tafuri's pessimistic assessment (1976: 181, cited in this volume, page 6), it may seem a lost cause to look to architecture itself for new ideas. Yet Borden demonstrates that architectural theory makes eclectic use of the constructs of other disciplines with which it has a synergy. The work of architecture, then, is not confined to designing façades, but takes on more socio-economic aspects.

Borden begins by citing the fusion of architecture with other currents of visual culture and political awareness in the early twentieth century. He gives Futurism as an example, with its images of crowds in restless, energetic movement, its multi-sensory world, and its fantasy cities in the work of Sant'Elia. He notes also the social welfare stance of one aspect of inter-war modernism, and its extension in post-1945 housing policies. But he remarks that modernism has always included a critical strand, leading in the case, for instance, of Dutch architect Aldo van Eyck to a study of intimate urban spaces – the streets of personal use (in which now personal stereos provide a new shell). But his question, having brought the historical sketch up to postmodern contingency and globalisation, is not what new aesthetic forms this gives, but what are its political positions. He sees the old welfare state model of architecture as no longer viable, citing Richard Rogers as foremost among those seeking to adapt its principles for the new, private–public partnership model currently popular with governments in the affluent world. Rogers incorporates a nod to ecology and espouses an imagined social life of the piazza, and Borden emphasises Rogers' commitment to urban density and mixed-use zoning – ideas which can be traced to Jane Jacobs in the 1960s – and the inception in the UK of Architecture Centres aiming to popularise what has hitherto seemed to many an over-privileged and almost mystical profession.

But Borden is uneasy at some of Rogers' omissions. He sees Rogers' notion of the city as both generalising and derived from given constructs of civilisation as the art of city living, rather than the occupation of its spaces. Borden writes that the model of polite society wins through in the city of big Sunday papers, and the leisure to read them. Particularly problematic for Borden is Rogers' conceptualisation of the piazza, mapped onto sites such as Trafalgar Square, London, and its ignorance of difference, though the concept is part of the common parlance of postmodern discourse – and an actuality in urban lives. It is against this elitism that Borden looks to other kinds of radicalism, in a world where everything is on the move and where buildings cannot now be thought of as permanent structures designed for unchanging use. He concludes that there are, in such a situation, many radicalisms inside and outside architecture. Elsewhere (Borden, 2001) he has written at length of skateboarding as embodied architecture in its intricate utilisation of space; here Borden ends by saying that the most radical architecture challenges and denies its own conception, so that architecture in the conventional terms of building design is dead, yet in a permanent revolution in which, of course, the activities of urban mobilities persist.

Patricia Phillips deals with a practice situated between art and architecture – public art. This, as a move out of the gallery into community or liminal spaces such as deserts, was seen by many artists in the 1960s and 1970s as a means to refuse the art market's imposition of commodity status on the art object. By making no objects, artists offered dealers no shows. Happenings, murals and earthworks in remote places could not be transported and thereby were not for sale. Since then, two things have happened: the market has proved itself adept at subsuming into its mainstream the most non-object-based practices, trading instead with reputations; and the administrative world has caught up with public art, inserting, particularly in the UK, a new class of arts managers to professionalise and institutionalise it.

What might once have been radical is, then, made safe, or sanitised, as the ever-present steel sculpture replacing the ubiquitous Henry Moore and before that the cannon in the park. But, as Phillips argues, that radicalism can be salvaged. Public art, she begins, exists at the boundaries, occupying an inchoate space between public and private realms. However much past pieces of public art might now be disliked or removed, as a non-gallery practice public art remains a catalyst to contestation of urban space.

Phillips cites Arendt's *The Human Condition* (1958) to effect that publicity, the condition of being in the public realm, is multiple perception – of all by all, in their differences. This presence of others reassures, re-stating reality in contrast to the privations which may characterise the world of indoor spaces, the house, the sphere of domestic confinement. Others since Arendt have adapted and extended this concept, noting the habitual masculinity of the public streets of cities, but still regarding publicity (in Arendt's sense) as a foundation stone, as it were, of the need – discussed also in Part II of this book – for a realm of free identity formation and maturity of self. But, as Phillips also states, that realm is now labyrinthine in its complexities and shades, in its accommodation of so much in which belief is no longer possible. From that position, Phillips reconsiders a range of recent projects using art in public spaces in New York City, in sites as prominent but contrasting in senses of ownership as Grand Central Station and Union Square. Among the cases reviewed, too, is *Flow City* by Mierle Ukeles, an installation and visitor access project derived from that artist's work, initially unfunded, as artist in residence with the New York Sanitation Department. Here art meets ecology in the question of what a metropolitan city does with its mountains of waste. In another work, *Touch Sanitation*, in the 1970s, Ukeles impacted power by shaking the hands of all New York's garbage collectors – touching the hands that touch the filth, recognising their invisibility to the producers of waste and reasserting their right to the city. All that waste was until 1999 taken in barges to Statten Island, and dumped in a landfill site (incidentally within sight and smell of a mall). The site is now closed, though part of it is currently used to sort the debris from Ground Zero, and, as Phillips writes, Ukeles is engaged in major plans to reconfigure its 3,000 acres as green space. From the cases she cites, and quoting Ukeles, Phillips ends by arguing that art in the public sphere is an instrument of negotiation which does not flee the conflictual side of urban space.

In a different way, but facing an equally vast and post-industrial entity, Tim Collins and Reiko Goto seek to explain the role of restoration ecology in the creation of new post-industrial, urban landscapes. The background to their chapter is the project *Nine Mile Run Greenway* in Pittsburgh (see Miles, 2000: 129–52), a 240-acre site of slag dumped, much of the time illegally, by the steel industry on which that city's wealth is founded. This has recently extended, following completion of a research project drawing together the agendas of public space and bio-diversity in reclamation of a third of the site as green space, to a new proposal – *Three Rivers, Second Nature* –

covering the wider river systems around Pittsburgh and its outlying industrial communities. Hidden in thickly wooded valleys are both sites of outstanding scenic beauty and monster industrial installations of a kind more often associated with non-affluent, or ex-communist bloc, countries. Over the dark trees pour out the vapours of satanic mills, and into the sometimes clear waters (at times good enough for swimming) flow acid mine wastes and other effluents. The task, clearly, is one of reclamation through widespread participation, begun in *Nine Mile Run Greenway* with, amongst other means, a series of charettes, or workshops, in which local people were empowered to speak authoritatively through a dissemination of information and provision of a seat at the same table as planners, developers and scientific experts.

In this chapter, then, Collins and Goto describe the practice of restoration ecology to which they have as artists gravitated. This is differentiated from conservation, with its often nostalgic view, and from preservation of species in isolation from social and cultural change. They argue that restoration ecology – and in Pittsburgh this did not mean re-creation of a pure, pre-settlement landscape – is not only a practical tool for post-industrial cities, but also a new way of thinking. It entails interactivity between disciplines as well as between experts and dwellers, and can be traced, possibly, to the work of Frederick Law Olmsted, or more immediately to that of botanist William Jordan III in the USA in the 1980s. But key to its relevance now is its democratic intention through citizen participation – an element in which Collins and Goto see artists, as floating between levels of power, incidental to but involved and immersed in the situation, as having a particular capacity to be effective as communicators and social as well as ecological researchers. They trace a number of routes into restoration ecology as a cultural practice, from earthwork art, for example, and note, too, the principles of radical ecology. They see struggles within environmentalism between those who see the natural world as having an absolute right to life, even if this places its needs above those of humans, and other, more pragmatic strands of green thinking, and, too, eco-feminism's framing of the question in terms of masculinity. But, again, they see art in its creation of metaphors and signs as offering an interface at which such struggles do not require the kind of resolution which is a closure of the argument, but offering more an open space in which to breed compassion and the unfinished reconstruction of meanings in the natural world and its human habitation.

7 IAIN BORDEN

What is Radical Architecture?

Architecture has been recharged at the beginning of the new century. Infused with all kinds of energy from postmodern forms and theory, social and city challenges, new science and innovative art practices, architecture is once again placed centre stage in urban agendas worldwide. Does the new surge of architectural endeavour also mark a shift in the politics of architecture and the city? Is it the end of a certain twentieth-century tradition of architectural concerns with social change, economic progress, social and, even, revolutionary activity? Yes and no, for what we are witnessing here is at once the end of politics in one form, and the resurrection of a politics of another form; this is death and birth as one of those great dialectics of modernisation, that between production and reproduction, and it seems that the latter is gaining the upper hand.[1]

POLITICS AND ARCHITECTURE

That architectural modernism has always been infused with political concerns, inspirations and associations is undeniable. From the first battle cry of Marinetti, Sant'Elia and the Italian Futurists, singing the song of great crowds, sounds, riots, electric light, power stations and aircraft in the twentieth-century metropolis, modernist architecture has had revolution at its heart. Of course this has not always been revolution of the order of the Bolsheviks and October 1917, and has more often than not sought collusion with the State rather than its overthrow, but nonetheless modernism has always wanted to change things, and to do so in no small manner. The history of architecture between the two world wars is then in effect the history of modernism and social change. On the one hand there were such social democratic, proto-welfarist experiments as those of Ernst May and the housing department of Frankfurt (1925–30), building the *seidlungen* of Römerstadt, Praunheim and Westhausen to accommodate upper-working-class residents. Similar experiments were undertaken at the same time by Bruno Taut, Martin Wagner and Hugo Häring in Berlin, Karl Ehn and Schmid and Aichinger in Vienna, and Michael Brinkman and J.J.P Oud in Rotterdam. Berthold Lubetkin and Tecton provided a British equivalent slightly later with the Finsbury Health Centre in London (1935–8). While down another political avenue, the various manifestations of the Bauhaus school in Germany, together with the state-funded RFG research body, sought in the 1920s not just to reform the education of architects but to alter the social relations of production to include the collaborative endeavours of design, engineering, building construction craft and aesthetics.

Still grander political ambitions were implicated within Le Corbusier's *Ville Contemporaine* (1922), *Ville Radieuse* (1935) and Algiers city projects (1930s), for even though Corbusier would happily work with any political grouping to achieve his architectural ambitions – whether socialist, syndicalist or the Vichy government – the political clout necessary to achieve such plans was necessarily large.

Italy and the USA in the 1930s both undertook State-led programmes to regenerate their Depression economies and political regimes, involving the construction of new towns, dams, roads and other elements of urban infrastructure. Most significantly of all, the continuing revolution of the young USSR in the 1920s produced true avant-garde activity across all the arts and sciences, seeking not only to change modes of production and forms but also the basic conception of what might constitute architecture and other disciplinary boundaries. Workers' clubs, communications centres, collective apartment blocks and linear towns were paralleled with new ways of painting, filming, acting, composing and simply living – this was complete political revolution in the making, addressing every aspect of life at its most fundamental grounding.

If this was the victory modernism won for itself in the inter-war years, it was a victory that had its most important effects in the post-war years. Here, in the magnificent era of the State, modernism became the universal language for schools, housing, government buildings, theatres, airports – for all the buildings of the city, and across all cities in both east and west, socialist world and capitalist world. Corbusier at Chandigarh for the Indian state of Punjab (1951–68), Walter Gropius for Harvard University (1950), Ludwig Mies van der Rohe for Seagram in New York (1954–8), Leslie Martin and the Royal Festival Hall for London County Council (1951), the second competition for the Palace of the Soviets in USSR (1957–9), Auguste Perret's reconstruction of Le Havre in France (1945–54), Oscar Niemeyer at Brasilia for the Brazilian government (1957–79), Arne Jacobsen and the National Bank in Copenhagen (1965–71), Kenzo Tange's Plan for Tokyo (1959–60), Gino Valle in Pordenone for Zanussi (1961) – these were merely some of the more well-known signs of the new and various bonds of architecture, modernism, the State and the corporate institution that infiltrated just about every aspect of the urban realm. Modernism and State politics were now inherently linked, both part of the great post-war project for planned economies, planned cities, planned modernisation, planned progress, planned health and education, planned lives.

And yet modernism has always been a critical enterprise, both of itself and of its relation to politics. It is, then, unsurprising that precisely in these decades of the omnipotent modernist State that the first signs of a critique of this relation should begin to emerge. While by no means wholly dismissive of the State per se, architects like Alison and Peter Smithson in Britain, Shadrach Woods in the USA, Giancarlo di Carlo in Italy, and Aldo van Eyck and Herman Hertzberger in the Netherlands began to explore a more intimate kind of politics, variously stressing such things as urban place, street life, social associations and collective patterns. Projects like the Golden Lane competition entry (Smithsons, 1952), Frankfurt-Römerberg competition entry (Woods, 1963), student dormitories in Urbino (Carlo, 1962), orphanage in Amsterdam (van Eyck, 1957–60) and school in Delft (Hertzberger, 1966–70) demonstrated a concern with structuralist anthropology, society and historical context and thereby marked a shift from the grand projects of the State to a more integrative politics of local democracy, dispersed power and everyday life. In a somewhat different vein, but still following an implicit critique of the dominance of the State, modernist architects explored such areas as pop culture and new technology (Archigram, 1961 onward), a formal investigation of the aesthetic possibilities of modernism (the New York Five of Peter Eisenman, Michael Graves, Charles Gwathmey, John Hejduk and Richard Meier, *c.* 1969–72), semiological meanings (Robert Venturi and Denise Scott Brown) and design participation and ecology (Ralph Erskine and Lucien Kroll).

All of this has been, of course, part of that by now well-known paradigmatic shift from modernism to postmodernism (the latter not necessarily excluding the former) that has accompanied and abetted the corresponding shift from a planned State with a Fordist economy to the much more fractured post-Fordist State and economy. The latter has been characterised by such things as: flexible accumulation; a highly mobile international capital and a global division of labour; the simultaneous development of regionalised, internationalised and hybridised cultures; an apparent diversification of aesthetics and culture, including a 'popularisation' of signs and symbols; the increasing commodification of everyday life, including the privatisation of public space and the development of hyper-real and simulacra-built environments; and a burgeoning bourgeois culture of fear (of crime, of difference, of others). Above all, this has involved a move from an emphasis on an explicitly planned production to an equal emphasis on an apparently free (yet nonetheless highly orchestrated) consumption. We are no longer a society solely of car-makers, medical scientists and agricultural workers but also of movers, carers and restaurant diners, of people who consume as much as they produce, who create attitudes, services and ideas rather than objects, things or facts. And our predominant political agency is correspondingly no longer that of the single controlling State or ruling political party but of the joint venture between self, society, company, institution and government, of a kind of multi-body collective enterprise where New Deal, New Labour unemployment offices are run by Reed International staff dressed in T-shirts and trainers, and where Coca-Cola sponsors everything from schools to football. We are, in short, at the end of the homogenous State of production and the beginning of the fragmented State of consumption.

So what are the politics of this, and how is architectural modernism to respond? Clearly, the old model of the politics of the welfare state, of organised provision, is no longer appropriate. Instead some modernists have sought to radically revise and extend the old model, seeking to transpose the role of State authority into that democratic partnership of private and public sectors now being conducted by governments in the USA and across Europe. Foremost among them is Richard Rogers, whose *Cities for a Small Planet* (Rogers and Gumuchdjian, 1997) has proposed a newly recharged intersection of architecture, ecology and the social life of late twentieth-century cities. If carried out, these ideas would amount to nothing less than a complete overhaul of architectural aesthetics and spatial planning of cities, redefining 'urbanism' in architecture to incorporate ideas about social justice, environmental responsibility, participatory democracy, globalism and new technologies.

Of course these are not in themselves new ideas, particularly when seen in the context of the history of modernism and politics, but it is extremely rare to find such things being voiced by an architect in the 1990s and 2000s, and rarer still linked coherently to specific architectural projects. Rogers undertakes sustained attacks not only on the familiar architectural bugbears of 'single-minded spaces', cloyingly nostalgic aesthetics and conservative planning agencies, but also the increasing inequality of wealth distribution, air pollution, global warming, greed, poverty, unemployment, short-term Thatcherite profitism, social alienation and the over-arching dominance of the automobile.

In their stead Rogers proposes dense, compact cities with 'open-minded' mixed-use neighbourhoods composed around public transport nodes. These cities would incorporate participatory planning, environmental education, pollution taxes, 'creative citizenship' for all and especially the young unemployed, parks, squares, renewed existing buildings and wastelands, a revived architectural profession, contrasting architectural aesthetics, and lightweight buildings exploiting natural air and temperature resources. For London, for example, this would mean new Architecture Centres, greater use of the River Thames, a network of cycleways, river buses, efficient and even free public transport, symbolic public buildings, a new elected authority, re-using shop and office buildings for housing, and a focus on the inner core of deprivation. All this would be set within an overall urban masterplan to regulate growth.

Rogers is here serious yet concerned, global yet local, detached yet impassioned, precise yet (occasionally) poetic. His is a magisterial manifesto in the grand modernist tradition of

Corbusier's *Urbanisme* (1925) and Frank Lloyd Wright's *Living City* (1958), but in its comprehensive social and political motivations it is more like urban geographer David Harvey's *Justice, Nature and the Geography of Difference* (1996). When the politics of architecture is increasingly being distracted into arenas of poststructuralist meaning, professional squabbles and single-issue lobbyism – all of which ultimately only serve to send architecture back into itself – this kind of architectural expression is at once necessary and timely.

Yet Rogers' ideas also lead to a certain sense of unease, for while they feel largely right *in what they say*, it is what they do not say, what they only discursively imply, what they leave out, that causes some concern. At the core of this unease is the model of urban living in Rogers' proposals. These are cities of *civilisation* in the oldest sense of being the art of living in cities: the art of music, theatre, galleries, opera, grand public squares. And while there is the occasional nod to everyday life as not just high culture, a certain model of polite society always permeates through. Above all, it is the city of public squares, gentle wanderings, spoken conversations and square-side cafés. It is the city of mocha, big Sunday papers, designer lamps, fresh pasta and tactile fabrics. It is not, however, the city of all the disparate activities that people do in cities. It is not the city of sex, shouting, loud music, running, pure contemplation, demonstrations, subterranean subterfuges. It is not the city of intensity, of bloody-minded determination, of getting out-of-hand; nor is it the city of cab ranks, boot sales, railway clubs or tatty markets; nor is it the city of monkish seclusion, crystal-clear intellectualism or lonely artistic endeavour.

In physical terms, Rogers' conception of the public city too often seems to mean the urban square. Trafalgar Square, Thames-side walks, tree-lined avenues, Exhibition Road and other London sites are all seen as places for the promenade and chance encounters of *la passeggiata*. Squares are very pleasant places, but they are far from being a universal panacea to the problems of public space – and in Rogers' schema of things even the proposed Hungerford Bridge is reconceptualised as a 'piazza' from which to view London, thus erasing the qualities of direction, exhilaration, transgression, transience, momentariness and in-betweenness that bridges so uniquely offer.

This may be Rogers' idea of the city (in a newspaper interview he once even described his own living room as a piazza) but it is not everyone else's. People of different backgrounds, races, ages, classes, sexuality, gender and general interests all have different ideas of public space, and they subsequently use and make their own places to foster their own identities as individuals and citizens. Beyond the square, piazza and avenue, cities need hidden spaces and brutally exposed spaces, rough spaces and smooth spaces, loud spaces and silent spaces, exciting spaces and calm spaces. Cities need spaces in which people remember, think, experience, contest, struggle, appropriate, get scared, fall in love, make things, lose things and generally become themselves. We as living individuals need spaces in which we encounter otherness and sameness, where we are at once confirmed and challenged. Otherwise we too are erased from view, removed from the square, censured from ourselves, denied the right to the city.

This, then, is where the politics of modernism need to be recharged along with modernism itself. In effect, the project is to modernise the modernist politics. It is no longer in the politics of homogenous consensus, of the universal man, of high and low culture, of the Sunday square, of right and wrong that we will find social salvation, but in the politics of difference, of celebrating and emphasising what Henri Lefebvre has called *differential space* – a space yet to come but which, in contrast to the homogenising powers of the abstract space of capitalism, will be a more mixed, inter-penetrative space where differences are respected rather than buried under sameness (Lefebvre, 1991b: 48–50, 52, 60, 409).

> [Differential space] will put an end to those localizations which shatter the integrity of the

> individual body, the social body, the corpus of human needs, and the corpus of knowledge. By contrast, it will distinguish what abstract space tends to identify – for example, social reproduction and genitality, gratification and biological fertility, social relationships and family relationships.
> (Lefebvre, 1991b: 52)

Proposals such as Rogers' may well go some way towards such a differential space, but in order to do so there must also be a concomitant development of the modernity of life, of rethinking the human subject and their relation to modernism in their everyday life.

One indication of how this might happen occurs – in a somewhat ambivalent manner – on the pages of *Wallpaper* magazine. Promoting 'the stuff that surrounds you', this lifestyle journal is at first sight seemingly about things as objects – clothes, food, designer items, places, furniture and so on – yet at least three things set it apart from similar publications. First, modernist architecture and design (some of it deeply unfashionable) is absolutely embedded into everything it displays: furniture by Florence Knoll, typography by Saul Bass but above all architecture by Paul Rudolph, Albert Frey, RMJM, Oscar Niemeyer, Alvar Aalto, Allies & Morrison, Richard Seifert and so forth are its preferred tastes. Second, this is an embedding of design into the lives of its semi-fictionalised consumers. This is not modernist architecture and design in books, in the art gallery or in the university lecture room, but brought within the realm of the contemporary metropolis. Third, and most importantly, there is a politics here. Now this is also where *Wallpaper* can get more than a little tricky, for it generally promotes a kind of affluent elitism that revels in the relative affluence and youth of its chic, LA–New York–London–Berlin–Milan–Tokyo axis metropolitan readership. But its politics are there nonetheless, not so much in the objects but in the pictorial and textual narratives it weaves in and around these objects – as a politics of sexuality. The models here are men and women of unstated sexuality, probably gay but not explicitly so. The story-lines suggest sex without showing it, with rumpled bed clothes, multiple bodies of indeterminate relationships and suggestive strap lines like 'Hard and Fast', 'Behind the Screens' or 'Some Like It Off'.[2] Above all, in encouraging people to 'Sort out your Space',[3] they therefore suggest the independence of the person to be who they want to be, with modernist design as an intrinsic part of that social construction.

Of course, we would not want all the world to be a *Wallpaper* world. But it does show (albeit when taken with a heavy pinch of salt) a fragment, the merest glimpse of how modernism might be part of someone's personal world, some part of their celebration of difference and life itself. In a way this is a return to some of the earliest moments of modernism (Adolf Loos and Otto Wagner) and to its most far reaching social dimensions (the German *existenzminimum*, Aalto, Corbusier, Ludwig Hilberseimer) where modernism was not just about the modernisation of the world but about *modernity*, the experience of the modern. Similar glimpses may alternatively be seen in the symbolically loaded theatricality, discord and narrative of Branson Coates' post-NATO bars and shops, in the social-consciousness of firms like Koning Eizenberg Architects in Los Angeles who mix expensive residences with non-profit work for affordable housing or single-room occupancy accommodation, and in the theorisations of Bernard Tschumi for a politics of space, pleasure and architecture. Above all, these are all designers and projects that carry within them some implication that ultimately what matters is not the building itself but the experience of it, such that human beings might be characterised less as the users and inhabitants of architecture and more as the *subjects* of architecture, who reproduce it through consuming it, and hence produce a truly lived-in architecture, and hence also a true politics of modernity. Marshall Berman defines modernism as 'any attempt by modern men and women to become subjects as well as objects of modernization, to get a grip on the modern world and make themselves at home in it' (Berman, 1983: 5). This is the real challenge for a politics of modernism in architecture.

WHAT IS RADICAL ARCHITECTURE?

How then to face this challenge? What is 'radical' architecture in a world where everything moves and changes so rapidly? In the contemporary metropolis, we are used to seeing (indeed, we demand as much for our essential urban fix) an ever-accelerating turnover of fashions, signs, buildings, music, tastes and events. According to some sociologists, a 'generation' now covers just three years, as each age group rapidly takes on yet another set of defining values and aesthetics. And as one phenomenon arrives to shout its presence with the audacity of the brazen new, another leaves and quietly dies; constantly and continually, this year's black becomes last year's brown (and vice versa). As Karl Marx (and Marshall Berman more recently) stated, 'All fixed, fast-frozen relations, with their train of ancient and venerable prejudices and opinions are swept away, all new-formed ones become antiquated before they can ossify. All that is solid melts into air . . . ' (Marx and Engels, 1967: 83; Berman, 1983: 21). If the city and its culture is perpetually mutating, one aspect of this changing world is architecture.

So is radical architecture anything that is novel, anything that adds its own unique impetus – however small that might be, just as long as it is new – to the maelstrom of the city? When we try to filter the on-rush of metropolitan data thrust upon us as we go about our daily lives, is the radical architecture we seek necessarily the new architecture, the bold surfaces, intricate spaces and complex technological structures that have most recently appeared? The answer to this, it seems to me, has to be no. And it has to be no on two accounts, one of which relates to temporality, and the other of which has to do with something at once more substantive and more subjective: the content of architecture.

To deal with the first part of this response, the temporality of architecture is quite different to that of much of the rest of the city. Innovations in fast-track design, site management and rapid-construction steel-frame and decking structures, fuelled by the extraordinary cost of interest charges that accrue on the development of an as-yet-unoccupied building, may have led to the paring of production times to 25 per cent or so of their former levels, but the erection of a building is still a pretty slow endeavour, particularly when compared to the turnover of, for example, a fashion designer's new collection or the release of new music recordings. So if the skyline of London is currently filled with the silhouettes of innumerable cranes, and the back-streets are blocked by hordes of bulky HGVs unloading yet another cargo of Swiss cladding, this is not because architecture is a rapid enterprise but precisely because it is, comparatively, a slow one – it takes years to conceive, plan, finance, gain clearance for, design, construct, sell and occupy a building, and no amount of global speed-up is going to reduce this process to a snap of the fingers.

Nor does architecture disappear as quickly as do other aspects of the city. Having concentrated truly enormous amounts of time, money and effort in the construction of a building, no one is going to want to pull it down. In English cities, surrounded as we are by the inherited fabric of buildings 50, 100, 200 or more years old, we tend to think that buildings last forever. Even in Tokyo, where the tear-down-and-rebuild attitude to urban renewal is rife, the city you say goodnight to is still, for the most part, still there in the morning (or the following year). Architecture is a long-term investment in more ways than one, and this investment takes time to pay back.

Perhaps this is why architecture is also comparatively slow to shift in another facet of its make-up: that of ideas and theories. I know of no other practice whose students are still routinely informed about a central theory that is now some two thousand years old. While 'Firmness, Commodity and Delight', the three principles of architecture set out by the Roman architectural theorist Vitruvius, may not be exactly foremost in the minds of architects today as they deal with the complexities of modern architectural construction, their very survival within the theoretical rubric of architecture shows how slowly things can move in this world. Even concepts like 'simulacrum' or 'the body', highly fashionable in recent architectural theory, have origins in philosophic and architectural discourse even older than that of Vitruvius. Compared with diverse contemporary practices like medical science or popular music, architecture is older and more intractable in its millennial manifestations.

This does not mean that architecture is an inherently conservative project. Just because it is not turned on its head every two years does not mean that no substantive changes occur. But it does mean that in thinking about what might be radical in architecture we have to take some care to distinguish between those inventions which are simply novel, and those which are to some degree truly new. This is what Henri Lefebvre sees as the distinction between 'minimal differences' – those which simply repeat extant forms, making changes in terms of quantity and superficial variation – and 'maximal differences' – those projects which make a change in terms of quality and more profound reconfigurations (Lefebvre, 1991b: 372).

To be simply new, then, is not to be radical. To be radical means to make a change to things, to make a difference in terms not only of quantity but also of the concept, essence, quality of architecture and the city. This, then, brings me to the second part of my response to the apparent newness of architecture. If radical architecture is not simply the novel, but is to do with something more substantive and transformative, what is the nature of that content and that action? Where can we identify radical change in architecture?

The complexity of this question was recently made very apparent when three of us – Tom Dyckhoff , Alicia Pivaro and myself – sketched out a conceptual structure for the 'Manifestos' exhibition on radical architecture.[4] In the course of our review of the history of post-war British architecture, two things rapidly became clear. The first was that architecture which was radical in some way might not necessarily be radical in another. To give an example, the much-fêted work of Future Systems may well be 'radical' in its dogged pursuit of aeronautical construction techniques and pure Brancusian forms; the unadorned and shiny curved forms of the new Media Centre hovering over Lord's Cricket Ground do provide a radical shape in the context of that site and that client, and possibly, although more questionably, in the context of architecture in general. Yet in other aspects of their work, Future Systems are most definitely deeply unradical – notably the way in which the imaginative creativity of the Architect and the immutable perfection of the building as Idea are held apart from the bothersome world of finance and from the dirty realities of more commonplace physical construction, or of everyday inhabitation.

Above all, it is the notion implicit in avant-gardist practices such as Future Systems that somehow, surely, *all* architecture ought to be designed and constructed this way that is profoundly conservative. If there is one certainty in the postmodern conception of architecture it is that architecture is no longer about only one kind of architecture – there are many different kinds of architectures, even within one person's or office's output, and even within one building. Consequently (and this is the second realisation of our 'Manifestos' review), there are many different ways in which architecture might be radical.

The death of the architectural and artistic avant-garde, conceived as the elitist group, small in number, somehow apart from yet ahead of the rest of society and prescient of its singular future direction, is then all but complete. We must move on from Manfredo Tafuri's position of despair in which, faced with what he viewed as the historical end of the avant-garde, both incapable of working against capitalism and irrelevant to the workings of capitalism, architecture was seen to be fit for no more than a position of silence (the say-nothing, 'devoid of all meaning' aesthetics of Miesian high modernism and Stirlingesque-formal games) and watchfulness (the vigilance of the architectural critic and historian, ever-hopeful for the onset of the socialist future).

Revolutionary architecture – architecture which in itself brought about revolution – was impossible, thought Tafuri, and instead all that could be done was to await a future architecture of revolution once the latter had already occurred (Tafuri, 1976, 1987).

The new position of radicality comes from a different position. Those who take up this challenge are not those who set themselves apart from society, but those who are knowingly within it, working not wholly oppositionally but ironically and irritatingly against the dominant systems of capitalism, colonialism and patriarchy and their constitutive agents. They are those who, through their work of whatever kind it may be, including architecture, say something that is not only a negative critique but also a proposition – a critical suggestion as to what might be done next. Their purpose is not to enact a total, all-or-nothing revolution but to make a radical difference.

To be radical in architecture is then not to be novel, but to be new in the sense of making a statement through architecture that seeks to change architecture and the city. To be radical in architecture is to adopt a critical position within, and not outside of, the social and cultural milieu. This automatically implies a purposeful sense of direction, for otherwise how are we to make a choice between, for example, the radical rightism and Christianity implicit in the architecture of someone like Quinlan Terry and the radical community orientation discernible in the architecture of those such as muf Architects? And a choice must indeed be made, lest we fall into the trap of inept pluralism. To be radical is to seek, perceive and make a difference of a deliberate kind – an unashamedly utopian position that knowingly considers not only where we are going and where we want to be going (these are not necessarily the same destination), but also for what reasons and with what procedures. To be radical is then to be emancipatory, idealistic and transformative, as well as ephemeral, provisional, questioning and transgressive.

Clearly it is no longer possible to see a simple opposition between radical and non-radical, or between radical and revolutionary architecture. As already seen, the same architecture can be radical in some parts, and non-radical in others. Similarly, old-style clarion calls about the one true path of architectural progress, such as Reyner Banham's exhortation to architects to keep up with the 'fast company' of new technology (Banham, 1960: 329–30), no longer hold sway. Instead the radicality of architecture lies in a more differential field, displaced from the arena of the singular building, the specific architectural practice or even the particular city into the spatially, temporally, culturally and socially fragmented and heterogeneous nexus of urban life. This does not mean, however, that radicality has disappeared, rather that it has to be sought in different ways. In short, the question we are presented with here is less the degree of radicality in architecture today, and more the kind of radicality that we wish to address through architecture.

Where, then, do we look for radicality in architecture? Of course we cannot have any definite answers here, but we can adopt a positioned response wary of a differential field of possibilities. With this caveat in mind, I suggest six potentially fruitful avenues.

Materiality and tectonics

Given that the oldest meaning in English of 'radical' refers to fundamental physical qualities (Williams, 1988: 251), it is perhaps appropriate that one of the most commonly explored arenas of radicality in architecture concerns the technologies and materials available to architectural constructions in any given era. What are these technologies, and, consequently, what are the different technical means by which energy can be managed, structures erected, forms shaped, and services programmed – what of such things as smart materials, pneumatic structures, advanced CAD and GIS software, and responsive architectures? What are the most cutting-edge technologies being developed and how might these transform architecture – what of nano-technology, non-linear systems, hybrids of organics and machines, and of genetic modifications? More radically still, what are the new technologies which might be developed and enacted through architecture and which might be exported to other parts of human endeavour – what can architectural technologies operating at the intersection of the physical, imaginative and the social add to the development of other fields?

Throughout these kinds of enquiry one should also consider what these technologies might do, not just in terms of their technical performance, but also in terms of their 'political' performance. What can they offer in the context of the restless search for social change? What is their integration within capitalism, and what is their potential for integration into everyday life? To ask the same question, but slightly differently, how can we avoid the latent religiosity that pervades so much discussion about all manner of technology, and which presumes that, once organised according to the demands and capabilities of technology, we will be able to conquer everything (Lefebvre, 1991b: 30)? Such technologies are most radical, we should remember, not just for what they can offer in terms of measurable effects, but also for the way these new potentialities are brought out into everyday life, for the way in which they become lived in social space and time.

Form and its experience

Similar questions and qualifications also apply to the form of architecture and to the aesthetic experience that people have of it. On the one hand architecture can be radical in its dispositions of composition and montage, colour and texture, shape and line, harmony and disjuncture. On the other hand, there is also the aesthetic experience of such qualities, using means of engagement ranging from vision and other sensory engagement, taste and connoisseurship, colour psychology and technical appreciation to empathy and movement, narrative and allegory, meaning and symbolism. Few if any of these characteristics are by any means new, but they may nonetheless be re-radicalised when posed in the context of the contemporary urban condition – to give one example, Baroque considerations of movement and allegory may be particularly useful in countering the tendency in the late twentieth century to concentrate on the purely visual aspect of architecture, to the detriment of more imaginative meanings and bodily experiences.

Social relations and culture

One of the longest traditions of twentieth-century architecture is that it might somehow be the agent of social change, bringing about new kinds of living and qualities of life unfamiliar to previous generations. The notion of social determinism – that architecture can directly and necessarily bring about such social changes – is now entirely discredited, either through the denial on the part of more socially minded urbanists that anything as prosaic as bricks and mortar can bring about qualitative change, or through the rather different rejection on the part of many others who, while implicitly accepting the power of *architecture* to make a difference, nonetheless repudiate the right of *architects* to repeat the perceived 'errors' of the modernist past.

While few entirely adopt the extremity of either of these positions, they remain the particular Scylla and Charybdis between which any socially minded architecture must currently navigate. Rather than the safety of suburban and countrified domestic houses, of heritage-inspired shopping malls and leisure complexes, of vaguely classical office buildings and call centres, all of which imprison social possibilities within reactionary nostalgic forms, we need radical architectures that call into question not only the clothes in which we drape our private and public bodies, but also the very social relations by which they are constructed. We need, therefore, radical architectures which promote hybrids and differences in social qualities, that celebrate all manner of human activities and beliefs.

Relations of production

Architecture is not just a thing or an effect but is also a process. Radical architecture can then challenge not only its constituent technologies, forms and social qualities but also the relations of production – the way in which architecture is designed, constructed and managed as an activity shared across a number of different people. Radical architecture here might be composed from the relations of contractors, project managers, engineers and designers, through contracts, working methods and responsibilities, and also, and here rather more radically, through the recognition given (or not) to the distinction between these different spheres of operation. Community architecture, self-build architecture, design-and-build, construction management contracts and other such process-oriented entities are all 'radical' in the way that they address these kinds of concerns, as much blurring as distinguishing particular spheres of architectural production.

Yet another relation of production concerns that of architectural education. Who gets taught, how and by whom? It is worth remembering here that one of the most radical centres of architectural activity, the Bauhaus of inter-war Germany, was above all else a school of architecture and industrial design, intent on changing both the relations of production and the way in which producers were trained in these methods.

Discourse

If architecture is a process, then part of that process is the *reproduction* as well as the production of architecture. Given that architecture is, in the main, an essentially static entity, grounded as specific buildings in a fixed place and hence immediately accessible to a comparatively limited number of people, then to have its greatest effect such architecture must be seen and thought about by other audiences and, hence, through other means. The most radical architecture is, after all, that which is eventually adopted by others; to follow here the terminology of technical innovation, its *invention* as a particular building or project must be followed by its *innovation* as a more generalised architectural idea or practice.

Indeed, the discourse of architecture – as texts, photographs, debates, exhibitions – can even be viewed less as the reflection or dissemination of the most radical or canonic architecture, and more as operating in such a way as to actually create that canon. As Victor Burgin says, 'the canon is what gets written about, collected, and taught . . . the canon is the discourse made flesh; the discourse is the spirit of the canon' (Burgin, 1986: 159). The radical discourse is then that which identifies new kinds of architecture – new kinds of buildings, technologies, forms, social effects and production relations – and which in doing so fundamentally defines and produces that architecture.

It is also (for discourse is, in itself, beyond transparent communication, a creative and technical practice) involving of the means by which architectural discourse expresses and communicates itself. Here the radical architectural discourse is that which not only identifies new architectures but which also finds new ways of doing so. My own discipline, that of architectural history and criticism, might then become most radical whenever I set aside my keyboard and the subjects of which most is known and turn instead to making filmic or text-image constructions about that which, for the moment at least, very little is known.

The concept of architecture

The five areas identified above all contribute their own particular version of architectural radicality. Each is a field in which the radical can be pursued. But taken together, these five all add up to a sixth area, something far more tenuous but also, potentially, far more substantial. I am aware that I am treading here in the murky waters of utopia and speculation, yet it must be said. The most radical of all architectures is that which challenges and denies its own conception, that which suggests that architecture is not just a building, not just what architects do, not just what people experience or write; it is that which asserts that architecture as we presently know and conceive it is dead. The death of architecture – as with the death of all institutional forms, be they philosophy, the State, nations, art, labour, the family, politics, history, the everyday and so forth – is to be invoked in a complete and permanent revolution involving new forms of education, production, creativity, desires, self-management, territory, etc.

Most importantly, this is a profound revolution of human beings themselves in the way in which they create the world and themselves. Revolution is, after all, nothing if we do not change ourselves. In short, the question with which I started, 'What is radical architecture?', must be answered through its inverse, and through a supplement; the most radical architecture is that which dares to say, 'What is architecture?' and 'Who are we?' in the same breath.

NOTES

1. Sections of this chapter were first published in earlier versions as Iain Borden (1999) 'Resurrection Politics: Modernism and Architecture in the Twentieth Century and Beyond', in Helen Castle (ed.) *Modernism and Modernization in Architecture*, London: Academy, 22–9, and as Iain Borden (1999) 'Revolution: Radicalism, protest and subversion in architecture and design', *Blueprint*, 163, July/August, 36–48.
2. *Wallpaper* (May–June 1998), *passim.*
3. *Wallpaper* (March–April 1998), cover.
4. 'Manifesto: Radical British Architecture in the Post-War Years', RIBA, London (July–August 1999). Co-curator with Alicia Pivaro and Tom Dyckhoff. Exhibits included contributions from Nigel Coates/NATO, Peter Cook/Archigram, Rod Hackney, Matrix, muf, Cedric Price, Richard Rogers, Walter Segal, Alison and Peter Smithson and Neil Spiller.

8 PATRICIA PHILLIPS

Public Art: A Renewable Resource

> When we have forgotten that we participate in the shaping of the world and become enslaved to shaping left us . . . then a cunning arts-worker may appear, sometimes erasing the old boundaries so fully that only no-way remains and creation must start as if from scratch, and sometimes just loosening up the old divisions, greasing the joints so they may shift in respect to one another, or opening them so that commerce will spring up where 'the rules' forbid it. In short, when the shape of culture itself becomes a trap, the spirit of the trickster will lead us into deep shape-shifting.
>
> (Lewis Hyde, 1998: 279–80)

Public art balances at the boundaries, occupying the inchoate spaces between public and private, architecture and art, object and environment, process and production, performance and installation. In both reality and rhetoric, it operates in the seams and margins. If not entirely unique, its vantage points offer discursive angles of vision. Public art inhabits contemporary civic life unpredictably; its saga has been varied – triumphant and tumultuous. It frequently is time-consuming to plan and treacherous to bring into existence. Then, it is often difficult and demanding to support and maintain. The stories increase of works contested, removed, abandoned, and neglected. Public art is owned by everyone and no one. People may share a stake in it, but few feel any responsibility for it. Perhaps that it exists at all, that some people choose to support, produce, and write about it, confirms the efficacy of fierce beliefs or imponderable miracles.

Generally, public art offers new opportunities for artists to work within cities and communities, for or with citizens/audiences who are identifiable, occasionally collaborative, and frequently challenging. Sometimes, the opportunities or limitations of public sites encourage artists to pursue inventive processes and strategies. But there are many other dispiriting examples where concepts are crushed by unbearable compromises and pointless regulations. All methodologies are imperfect; each situation has extenuating circumstances. The tribulations of public art are prodigious and well documented.

It is shockingly commonplace for public art to fail because of the heterogeneity of civic life it attempts to recognise and reflect. The consensual model of public art often leaves citizens and communities dissatisfied with, bewildered, or unmoved by, the aesthetic outcome. Paradoxically, public art falters when it attempts to be for everyone. As an aesthetic field, it is rarely taken seriously, if not routinely ignored, by a contemporary art world whose criteria for success and failure have only partial application for public art. What makes artists continue to work in this field where success is often thwarted, interrupted, or deferred? And why do public arts administrators persist and persevere to support and develop new work? With all of the obstacles, what signs of promise and passionate convictions drive new ideas?

In his exploration of the interactions of pictures and texts, W.J.T. Mitchell suggests that 'all media are mixed media, and all representations are heterogeneous; there are no purely visual or verbal arts, though the impulse to purify media is one of the central utopian gestures of modernism' (1994: 5). Although the large abstract (or figurative) sculpture in the plaza generally conforms to modernist conventions of an established, confident aesthetic order, it is difficult to imagine anything more impure than contemporary public art. Processes and intentions, materials and media, content and contexts. There is a magnanimity of contrasting conditions and projects tenuously held by the diaphanous threads of a network called public art. It is a challenge to determine how to define and defend something that seems to be constantly changing, dispersing, and disappearing – seemingly moving from a solid to fluid state.

Public art's vast variety of intentions and outcomes makes any theoretical work a tentative and speculative enterprise. What is public space? Its many different contexts make topological analyses intriguing, but multiple and approximate. Although it is assumed that public art is made with the presumption that there is someone in mind (a public), response and reception offer fascinating but undependable sources for critical analysis. Of greater significance is the volatile and quixotic nature of public itself.

If modernism sought to separate and contain independent media, a view of public was once more confidently single-minded. Operating from the premise of an attainable consensus of a cohesive civic life, 'public' possessed a syntax of dependable social and physical characteristics that made it possible to identify and situate. But just as all media are recognised as mixed and often errant, public has become multiple, molecular, and plural. In *The Human Condition*, Hannah Arendt describes a phenomenon of the public realm:

> It means, first, that everything that appears can be seen and heard by everybody and has the widest possible publicity. For us, appearance – something that is being seen and heard by others as well as ourselves – constitutes reality. . . . The presence of others who see what we see and hear what we hear assures us of the reality of the world and ourselves.
>
> (Arendt, 1958: 50–1)

In fact, the public now is the realm where the relationship between seeing, believing, and knowing has become most labyrinthine. There are many aspects of the public realm that cannot be seen – that are not witnessed by people together. We believe certain things that can never be seen. We encounter things in which we do not believe. We know things to be real that are too small or too far away to apprehend. The problematic project of public art may offer some illumination on this entanglement. Seeing, believing, knowing, making, and performing are the structures for the formation of public values and civic cultures. What must we experience in order to know something? Can public art offer experiences that shape insights on public life – the human condition?

A selection of recent public art proposals and projects in New York (all developed within the past few years) offer constructive case studies for the questions raised with increasing frequency about public art – its vexing problems and unrealised potential, as well as its passages of incisive insight and unerring vision. At very least, these new works offer concrete and explicit evidence of a range of approaches to public art in a major United States metropolitan area at the juncture of two centuries. Does this current work elucidate important themes and issues? Do these examples of recent work, considered together at a particular time and place, highlight some of public art's most prevalent and pressing issues? Can meaningful observations and plausible predictions emerge from this circumstantial collection of recent projects?

The recent restoration and renovation of historic Grand Central Terminal by architects Beyer, Blinder, Belle has brought justifiable attention to a legendary urban landmark. Once obscured by accumulated grime and over-sized advertising, the main concourse has been restored to a spare and elegant space that features the movement of thousands of people. North of the concourse's great expanse of civic space, people descend deep beneath the city to catch trains or wander through an extended network of underground pedestrian tunnels.

In two of these subterranean passages, visitors encounter a series of visual metaphors embedded in the walls.

There are twenty dazzling chapters in a visual anthology of worldwide mythologies of the earth's cycles and the inestimable mysteries of the night sky. Made of mosaics, glass tiles, bronze, and digitised images, Ellen Driscoll's *As Above, So Below* (1992–9) establishes a connection with the constellations painted by Whitney Warren with Paul Helleu and Charles Basing prior to the building's opening in 1913, now brightly illuminated by fibre optic lights on the main concourse's vaulted ceiling.

Driscoll explored and compared the cosmologies of different world cultures, including scientific and apocryphal explanations for cyclical, predictable phenomena – the diurnal pattern of night and day, seasonal and annual cycles. Weaving together myth, conjecture, and scientific observation, Driscoll incorporated Lenape/Iroquois creation stories, Albert Einstein and the Theory of Relativity, the myth of Sisyphus, as well as Mayan, Aboriginal, and Celtic tales and stories. Within this active nucleus of transit that attracts and expels daily commuters and other travellers, Driscoll's project invites attentive viewers to seek metaphorical connections between their daily commute to outlying boroughs and communities and an expanded theme of global time travel.

Strategically placed on the walls to offer full and indirect views in a cinematic cadence of constructed images and the errant movements of pedestrians, the artist's visual narratives have the capacity to transport individual imaginations beyond specific conditions and circumstances. The characters in these ancient stories are our contemporaries – people from different backgrounds who Driscoll selected to represent these timeless accounts. In this and other ways, she unerringly orchestrates historical narratives with contemporary experiences. The past becomes an active presence; viewers can explore the instrumental role that imaginative mythology and empirical science have played in people's efforts to make the world comprehensible.

The project was done in conjunction with the Metropolitan Transit Authority Art for Transit Program and Metro North Railroad. It is safe to say that these hallways would be spartan and unremarkable without *As Above, So Below*. It remains to be seen, given the scattered, subterranean, and marginal location, if the project will acquire the legendary proportions of the main concourse's celestial ceiling. Will it? Should it? Calculation and necessity led to a cohesive, yet multiple strategy. Rather than a single, grand narrative, Driscoll presents an anthology of multiple views and voices. Anthologies are never authoritative. The confederation of many visions may encourage citizens to honour different stories – to let public spaces be both symbols and repositories of transformative civic values.

Sited almost thirty blocks to the south in Union Square, establishing a distant conversation with the great south-facing clock on the exterior of Grand Central Terminal, is *Metronome* (1996–9). Long-time collaborators Kristin Jones and Andrew Ginzel's immense project is a vivid evocation of different ideas and experiences of time in the city. Installed on the north elevation of One Union Square South, a new multiple-use building designed by Davis Brody Bond Architects, the genesis of the project is pertinent.

The Related Companies, private developers of the project, enlisted the Public Art Fund, Inc. in New York to organise a national competition to define and articulate a 98x200 foot expanse designated by the architects as an 'art wall'. Sponsored by a private organisation, aspects of the project were organised and administered by a not-for-profit arts organisation. Particularly in new buildings, the spaces designated for public art demonstrate a narrowness, if not hostility, towards art in architecture. Perhaps dutifully, generously, and conventionally, the architects provided (or left) a single space with a frame on the building elevation for public art. Even with the most high-minded intentions, the project was shaped by unusual liaisons, difficult circumstances, a convulsive site, and an unexceptional building.

Several years in development and completed with a legion of collaborators, Jones and Ginzel created an enduring (accepting the shrinking lifespan of most buildings) image of the temporal. Deploying the scale and systems of architecture to render a poetic testament to time, *Metronome* appeals to different dimensions of a collective civic imagination. The central vertical panel consists of courses of bricks laid in a pattern of concentric rings that radiate from a dark cavity near the top of the building. In contrast to this active fluidity, a casting of an immense boulder from the northern edge of Manhattan island invokes the enigmatic dimensions of deep time.

Like a living organism, *Metronome* embodies the immanence of change. To the west of the central tower, the slow, exacting rotation of a golden sphere is calibrated to the moon's movement around the sun. On the eastern wall, a ceaseless sprint of electronic illuminated numbers converging in the centre concurrently add and subtract the remaining and elapsed time each day. At the blur in the centre, the unintelligibility of time acquires a visceral presence. Constructing palpable and intermittent metaphors of the temporal and intangible, at noon and midnight each day, an audible summons and plume of steam are released from the dark cavity.

At the top of the wall, a golden hand in a gesture of inquiry or invitation makes an explicit connection to the equestrian statue of George Washington by Henry Kirke Brown in Union Square (below and north of the wall). A sole concession to the complex urban site it overlooks, the connection paradoxically is obvious yet remote. *Metronome* transcends or side-steps the long, turbulent history of political activism in Union Square. The theatrical meditation on time conforms to the specifications of the 'art wall', but it generally avoids the historical and social intricacies of the situation. There is no evocation of the personal in the public. Members of the public remain removed spectators who never become participants in the meaning of the work. The artists' experience and intelligence could not overcome the conditions and limitations of the project prospectus.

Shifting from grand theatricality to the vivid intimacy of the human body, Jones and Ginzel created *Oculus* (1998) in the Chambers Street Subway Station of the World Trade Center. (The artists report that the work survived the devastating destruction at this site on 11 September 2001.) A Percent for Art Project for the City of New York, the vast, meandering project is a riveting elicitation of the gaze in public space. Using a large format Polaroid camera, the artists photographed the eyes of hundreds of schoolchildren in New York.

These 'ocular portraits' were transformed into stunning mosaics installed throughout the network of underground passages. As people pass through the corridors of the station, pair after pair of gazing eyes form a hauntingly beautiful eye-level horizon. Drawn to the shimmering tesserae, rich colours, and exquisite details, people pause to study individual eyes or glance at frame after frame as they hurry through the subway passages. Pedestrians' focused or furtive looks are met by the hundreds of searching, silent stares that return their gazes.

Historically, eyes have been endowed with symbolic significance. They are windows to the soul, the centre of individual identity. The eyes – in fact, hundreds of pairs of individuals' eyes – offer compelling information about gender, race, and ethnicity as physical attributes and social constructions. The eyes of the city's children are poignant representations of its vigorous diversity. Clearly, the gaze can be intrusive and aggressive, but *Oculus* sensitively demonstrates that it also can be compassionately connective. The project is a moving and generous image of the multiple dimensions of contemporary public life.

Oculus shares an affinity with another project Jones and Ginzel installed earlier in the new home for Stuyvesant High School north of Battery Park City. *Mnemonics* (1992) consists of 400 small glass vitrines embedded in the walls of hallways, rooms, stairways, and other spaces of this large urban public high school.

The glass cases are diminutive cabinets of curiosity, filled with objects and artefacts from natural, cultural, regional, and personal collections. Some cases remain empty, awaiting the contributions of future graduating classes. The choreography of students' and staff members' movements creates a dynamic collection of many experiences and stories. As Tom Finkelpearl (Director of the NYC Percent for Art Program during the development of the project) suggested, a studied consideration of all of the elements of *Mnemonics* constitutes a reasonable equivalency of a challenging high school education.

It is commonplace to think of buildings, subways, railroad stations, public schools, and plazas as public sites. In spite of increased security systems and regulated access, tradition and convention claim these kinds of places as open and available to one and all. Concurrently, it is generally acknowledged that the home has acquired public attributes through new media and communications technology. The radio, television, Internet, and other instruments make the home a live circuit to global phenomena.

What about the human body? This final frontier of privacy and intimacy is also a site of public interrogation and transformation. Public is cellular, as well as spatial. A number of artists have explored issues of publicity and the body. The publication of Paco Cao's *Rent-a-Body* (1999) documents a provocative performance/investigation of public space sponsored by Creative Time, Inc. in New York. Having advertised on the Internet and distributed information materials on *Rent-a-Body*, Cao engaged in a number of transactions with people who chose to lease his body for various activities. Mimicking the different levels of car rentals, the artist established three categories: Basic (Body as Prop), Premium (The Active Body), and Deluxe (Total Mind Function). While the artist's rental of service did not discriminate on the grounds of sex, race, religion, culture, health, or social class, there were some restrictions. The agreement included:

> The renter may not, under any circumstances, inflict irreversible physical or psychological damage on the rented body, or administer to the rented body or take from it any substances whatsoever without a previous arrangement. The renter may not use the rented body for criminal and/or illegal purposes. The customer must respect the physiological needs and limitations of the rented body. SEXUAL CONTACT WITH THE RENTED BODY IS NOT ALLOWED.
>
> (Cao, 1999: 92–3)

The genesis of Cao's project was a desire to be direct about the idea that the audience should be an active part of art. Prior to *Rent-a-Body*, Cao felt that the connection was distant, brief, and imprecise. The project attempts to overcome these restrictions. 'If only we were able to come to a consensus, make this conversation productive, and avoid the constant problems of miscommunication. For a long time I've been trying to bring you closer, make you more complicit, and earn your encouragement. But I always come up against the harsh reality of your flattering response. . . . It's time that this body become an extension of yours' (Cao, 1999: 94). *Rent-a-Body* confirms that private and public have ceased to be entirely reliable or meaningful distinctions. The project inhabits an ambiguous, fluid space; the shifting ground shares characteristics that might simultaneously be defined as private and public. Cao has made no lasting amendments to the city, but like many other ephemeral projects, its inevitable disappearance has caused a spectre of inquiry to radiate.

Several years ago, Bradley McCallum completed *The Manhole Cover Project: A Gun Legacy* (1997) in Hartford, Connecticut. Concurrent with the large exhibition *Sam and Elizabeth: The Legend and Legacy of Colt's Empire* at the Wadsworth Atheneum, McCallum presented his project on an exterior terrace of the museum. Colt's lectures on nitrous oxide (laughing gas) led to the unanticipated invention of anaesthesia. Undoubtedly, he is better known – in fact, renowned – for the large munitions factory he built in Hartford in 1847, and as the inventor of the Colt 45 gun which inaugurated the solemn history of portable firearms in the United States. The Colts were generous cultural benefactors and philanthropists (hence the blockbuster exhibition featuring their many gifts to the Wadworth's collection), but McCallum believed that this was an important occasion to reflect on other persistent and catastrophic dimensions of the Colt legacy.

In 1992, Connecticut Governor Lowell Weicker mandated the destruction of all weapons confiscated by the state police. Prior to his executive order, impounded guns were typically sold in auction and often ended back on the streets. McCallum learned that in compliance with Weicker's executive order, the collected illegal handguns were being melted down at a Massachusetts foundry that produced utility covers. This engendered a compelling metaphor of malevolent instruments being transformed into ubiquitous street utilities. McCallum commissioned 228 specially designed 172 pound manhole covers representing the volume of weapons melted down since 1992.

The inscriptions on the manholes covers included: 'MADE FROM 172 LBS OF YOUR CONFISCATED GUNS' as well as the Colt family motto, '*VINCIT QUI PATITUR*: He Who Perseveres is Victorious. He Who Suffers Conquers.' Following the temporary exhibition of the utility covers stacked on pallets on the outdoor terrace of the Wadsworth Atheneum, many of them have been installed throughout the city. The lifespan of a manhole cover is 40 to 100 years. Perhaps these objects placed in neighbourhoods devoured by gang and gun violence will serve as memorials of loss and agents of change.

Continuing an investigation of the consequences of urban violence in the lives of individuals, McCallum and artist Jacqueline Tarry installed *Witness: Police Violence in New York City* (1999–2000) in the dark interstitial spaces of the immense Gothic Cathedral Church of St John the Divine. Police violence is not a new subject, but two atrocities recently placed the New York City Police Department under severe scrutiny. The torture of Abner Louima while in police custody for questioning (1997) and the shooting of Amadou Dialio by police officers as he stood unarmed on the front steps of his Bronx apartment building (1999) characterise the chequered and politicised history of police work in this major city.

With first-hand testimony as the nucleus of the work (the testimonies of trauma unit/health care workers, families, and friends were an aural dimension of the Hartford project), McCallum and Tarry gathered audiotaped stories of individuals affected by police violence. The accounts included recollections of parents, siblings, and friends of victims, as well as officers who have witnessed police violence and misconduct. Developed in two parts, the first instalment in St John the Divine was a visual and visceral anthology. Different incidents and stories were portrayed in the dark, narrow passages perpendicular to the central nave. Including audiotapes, videos, back-lit transparencies, and objects, viewers witnessed the hushed testimonies of Nicholas Heyward, Sr whose 13-year-old son was fatally shot in the Gowanus Houses by a police officer. Or Iris Baez, whose adult son was choked to death by a police officer enraged that a misguided football had inadvertently hit his parked car. Or Graham Weatherspoon, a retired New York Transit Police Officer, who has encountered police misconduct and now does advocacy work for minority officers.

Witness combined the immaterial sounds of spoken stories with the tangible encounter of architecture, illuminated images, and artefacts. Drawn into discomfiting marginal spaces of the large gothic building, people experienced the confining weight of trauma. Standing alone or in small groups in the darkened spaces of *Witness*, viewers heard searing accounts of the horror of violence and loss. The excruciating stories of shattered lives become part of each viewer's life. Tragedy and injustice were palpable; these concepts resonated in the words of those who continue to bear witness. The experience of witnessing had an exponential effect;

viewers became witnesses. Testimonies of loss became insistent passages of a shared public conscience. As the project is circulated, changed, and amended, the New York-based stories of despair, disbelief, and grief may reverberate nationally. Perhaps the most urgent specificity always has general significance.

Mierle Laderman Ukeles has had a twenty-year appointment as the (unsalaried) artist-in-residence for the New York City Department of Sanitation. In an early project entitled 'Touch Sanitation' (1979–84), Ukeles spent an entire year shaking hands with more than 8,000 municipal sanitation workers. A daunting project, the artist developed an intricate cartographic and time-based system of 'sweeps' that brought her into five boroughs of the city during the three daily eight-hour sanitation shifts. For the past decade, Ukeles has been working on *Flow City*, a visitor access art project in the Marine Transfer Facility at 59th Street and the Hudson River.

At a site in operation twenty-four hours every day (with the exception of Christmas), Ukeles has designed a pedestrian corridor constructed of recyclables that leads to an observation platform. The view to the east features the majestic New York City skyline. With the Hudson River as backdrop, the view to the west reveals the transfer facility's tipping floor, where hundreds of sanitation vehicles dump trash each day into barges destined for the Fresh Kills Landfill in Staten Island. The observation area is a joint of the civic body, where idealised and functional perceptions of the city are negotiated. Images of the glamorous and visceral converge and overlap.

In 1999, New York Mayor Rudolph Guiliani announced the closing of the 3,000 acre Fresh Kills Landfill no later than December 2000. (The site did close, but was reopened in September 2001 to receive the mountains of debris – and undoubtedly body parts – from the World Trade Center site following terrorist attacks and the collapse of the twin towers.) Fresh Kills is the final destination of virtually all of New York's solid waste. In an appropriate denouement of her many years dedicated to the culture of labour, maintenance work, municipal waste, and an art process based on an ecological and feminist model of recursive tasks, Ukeles is involved in the reconnaissance, recovery, and conceptual development (including issues of access, transparency, public safety, and of course public art) at this fantastic, almost surreal site. A major national competition to select a design team to shape the future of this site is under way. Presumably, Ukeles will serve as the artist collaborator with the winning team.

Ukeles envisions this magnificent wasteland as a place to understand the entire culture, as an environmental laboratory and cultural repository of unusual proportions and possibilities. Fresh Kills is the definitive social sculpture; it is a site where the artist and many other collaborators will face incredible circumstances for public art (this, of course, is compounded following the events of 11 September 2001). They will attempt to make art in a public space of nowhere – within the great mountain of everything that citizens have pushed away, rejected, and refused. She will excavate public meanings in utter dispossession. Ukeles will challenge New Yorkers and others to reclaim Fresh Kills – and other forgotten and forbidden sites – as growing phenomena of public space.

Subway platforms, train terminals, building elevations, churches, the human body, performances, interventions, installations, and testimonials. Public art in New York. A microcosm? An idiosyncrasy? An aberration? These recent case studies suggest the scope of the philosophical dilemma to define public art. Without hesitation, all of these projects are identified as public art. In spite of contrasting methods, intentions, contexts, and conditions, all of the artists involved in these different propositions and projects believe they are making public art. The sponsoring organisations (city or federal agencies and not-for-profit arts organisations) believe they are supporting public art. Is everyone doing what they think they are? Is there a creative, constructive way to investigate and analyse public art that provides a clear but flexible intellectual and aesthetic framework to accommodate such contrasting ideas?

A major obstacle to defining contemporary public art is the shifting borders of public. Public has become an errant idea. At moments, everything seems public. Other times, everywhere appears private. Increasingly, new technologies and changing civic cultures have rendered both private and public ambiguous and volatile concepts. There remains a physical and territorial dimension to public space, but there is a growing recognition of its more quixotic, intangible characteristics. Public space has become cellular and molecular, dynamic and granular. Paradoxically, it often seems shockingly fragile and infinitely replicable.

Public moves back and forth through boundaries that once dependably distinguished private from public. It collects and congeals temporarily, then shifts to another time, place, or set of circumstances. Public space has become immaterial and invisible. It is both virtuous and vile; it is widely honoured and loudly admonished. It is our highest ideal and lowest common denominator. Access and quality, rights and ownership. It has become the contested territory (National Endowment for the Arts, public money, public education) of the culture wars. It is no wonder that ideas and experiences of it are vertiginous.

The architectural, aural, tactile, and ocular dimensions of contemporary public space (and privacy) are increasingly, often unexpectedly, explored, expounded, and redefined. Last year while riding a crowded bus to New York City, I witnessed the eruption of an angry dispute. A man was engaged in a noisy, unpleasant conversation on his cell phone. Following fifteen minutes of loud, pointless chit-chat, he made what appeared to be a racist comment. He was verbally attacked by several exasperated and offended passengers. In high-pitched fury and frustration, the following ideas emerged. The bus was a public space. Others were trying to sleep, read, or talk quietly with their neighbours. What right did he have to saturate a space of public transit with his private conversation? With neither eloquence nor sophistication, the physical, ethical, and perhaps legal lineaments of public space were debated and drawn as insults (and insights) reverberated off the interior wall of the bus. Public space has not disappeared, but it often appears in new disguises.

My current employment at a state (public) university has raised other questions about public. Is public education something to be honoured or endured? Does it represent the highest level of civic commitment, or has it simply become a compromised, often contested site? Who controls curriculum in a public university? What is the public contract? Certainly, at the primary and secondary level, the debates concerning charter schools have deformed and eroded the once clear division between private and public education in the United States. Is public transportation a marvellous convenience, or an affordable, typically disappointing, form of transit? Is the public realm something that we strive to create, or something that we settle for? Is public life an invaluable collective experience, or is it an impoverished, if not abandoned, historical relic?

There is a profound, if perplexing, ambivalence about public institutions and values. Firsthand understanding of public is almost always episodic and often distracted. Frank Kermode writes: 'We interpret always as transients' (Frank Kermode quoted in Hyde, 1998: 289). While he refers to reading and literary interpretation, the experience of public space is also migratory. Public is a threshold; people constantly move in, out, and through it, but it is rarely inhabited (except by disenfranchised or homeless individuals who have no private living space). As Lewis Hyde observes about our place in the world: 'We can orient ourselves, but we cannot arrive' (Hyde, 1998: 289).

In the past two decades, emerging ideas in public art have raised other difficult questions about access and art. Public art is frequently praised, practised, and defended as a way to bring art into cities and communities. Its democratic, if often unrealised and imperfect, potential is to insinuate art, aesthetics, questions, and ideas in the daily lives of ordinary citizens. People choose to visit galleries or museums with an expectation to encounter some form of art. People rarely end up in museums by accident. Presumably, public art accommodates and acquires deliberate, as well as fortuitous, encounters. Some people choose to go out of their way to see public art (just as they might to go to an exhibition in a gallery), but often public art is witnessed as an unintended consequence of transit through the city – during a routine commute, errand, or new route that brings people into art's vicinity. Access to and expectations of public art are varied.

If all art is mixed media, public art, in particular, presents numerous case studies of impurity. Not only are media and methodologies scrambled, but questions of context and community introduce puzzling variables. Why is it art? This inquiry implies the determination of distinguishing aesthetic features. The objective is not to isolate public art, but to discover and define what it shares with other art. Then it may be possible to identify additional characteristics.

When is it art? This query implies a temporal component – a dimension of public art that acknowledges time and duration as significant factors. There are well-documented, contrasting requirements, circumstances, and justifications for both long-term (permanent) and short-term (temporary) public art. More enduring projects require a vision of the future. How will time alter, amend, confirm, or extricate a work of art? How will people's expectations and responses transform or stabilise it over the years? In contrast, short-term public art also requires another kind of commitment that is timely, intense, incisive, and urgent. For those involved in temporary public art, confrontation and implication frequently are dynamics in projects and performances.

Questions of time introduce issues other than qualities of ephemerality. If something or some place is first produced, acknowledged, accepted, and criticised as art, does it always remain art? Is art steadfast? If something begins as art, can it also be (or become) something else? How do these transformations occur? And if something comes into existence, establishes a presence or resonance, that was not produced or intended to be art, can it become art? If we embrace these dynamic, contextual, and circumstantial factors, what are the implications of unanticipated metamorphosis – the inherent instability and impurity – of art? Public art inhabits boundaries; it is perennially in a condition of suspension. This is its most perplexing internal conflict, the source of its greatest creative potential, and a daunting challenge for public art criticism.

Where is it art? Of course, this asks what constitutes an appropriate context for art. Are there particular situations where public art can or cannot be found? Site-specificity is a perplexing and contested subject, especially as ideas of site have become so malleable, multiple, mutant, and intangible. Rather than possessing particular characteristics or a location in a specific site, public art can happen anywhere. Just as public has drifted and changed, public art has the capacity to be everywhere or nowhere. It can be some thing or object, a performance, event, or activity, a place or environment, or a story or proposition. If the prophecy of site-specific art has been spent, and neither context nor formal characteristics nor media offer clear perspective on meaning and significance, perhaps the performance – the place of exchange and passage – of public art offers a more promising environment for critical work.

There has been a great deal of theoretical speculation about boundaries and edges. Public art inhabits boundaries – everywhere and nowhere. It embraces the undefinable; it occupies the immaterial, metaphorical spaces that hover between public and private, sharing qualities of both. Boundaries are always provisional. People honour, ignore – respond with indifference or hostility to them. Some people will go out of their way to respect and confirm them. Others only begrudgingly acknowledge them. So it is with public art. Some people will actively seek it out, but generally it is encountered – engaged with or overlooked – by people involved in some activity of their daily life.

People generally experience public art when they are doing something else. Rather than a distraction or dilution of an art experience, this coupling of art with daily life can produce an enrichment of a pure, idealised, and increasingly rare concept of aesthetic contemplation. An encounter of public art is more complicated and dynamic. Public art is a joint or hinge. Like the space between bones, it hardly exists. But the space is indispensable to the activity and agency of the body. It ensures flexibility, agility, methodical movements, as well as spontaneous actions. It transforms a part of the body in response to different neural messages. Public art can be this active hinge between private and public, unthinking habit and civic consciousness, and past and future.

To live together in the world means essentially that a world of things is between those who have it in common, as a table is located between those who sit around it; the world, like every in-between, relates and separates (people) at the same time. . . . The weirdness of this situation resembles a spiritualistic seance where a number of people gathered around a table might suddenly, through some magic trick, see the table vanish from this midst, so that two persons sitting opposite each other were no longer separated but also would be entirely unrelated to each other by anything tangible (Arendt, 1958: 52–3).

How is it art? This question raises difficult issues of instrumentality. What does public art do for us? What is the nature of the transaction between public art, communities, and members of the public? Like the table – the human artefact – that Arendt describes, borders connect and divide. Ephemeral public art offers some insights on this paradox of the public realm. There is a fascination with the sporadic or episodic possibilities of a temporary public art that circulates through cities and communities, addresses and ignites issues, responds to and defines timely, often pressing issues, and sometimes subversively defies the normative requirements and expectations for public art. The time of public art involves questions of intimacy and immediacy – those transactions when art becomes part of an individual's life, thoughts, and ideas. These moments of connectivity, if elusive, imprecise, and incalculable, enable people to envision their lives within a community, to see and seek some equation between private interest and common good.

Public art needs to reveal something to us. It is the 'world of things' that is shared in common. Of course, revelation takes many forms, but it generally allows people to move over, through, or around some boundary. Public art can assist individuals passionately and strategically to navigate in and connect to the world and with each other. Given an alarming proliferation of 'content-free' public art, it is important to confirm a tangible meaning. Without being single-minded about what is acceptable or critical content, there has to be some poetic urgency of ideas or issues.

Public art is art when it encourages and expedites connections between the private and public, the intimate place and the municipal space, the body and the community. There are moments of reflection when an image, or sound, or space allows individuals to embody, in a unique and often specific way, the vast and various issues of public life, perhaps including the ecology of cities, gun violence, police brutality, the relation of global and local cultures. As Paco Cao suggests, perhaps public art can help to close the distance – to bridge the gap. Perhaps it can enable us to experience concepts and sensations – including the horror of violence – in a palpable way.

McCallum and Tarry's *Witness* has been a provocative catalyst, as have other artists who explicitly and implicitly deal with ideas of memory, exchange, and the very solemn, imaginative transformation that bearing witness entails and inscribes. Without trivialising or diluting the searing history of the witness – the historical testimony of trauma or crisis – public art can encourage people to become active witnesses of an animating idea, a profound aesthetic experience, or an urgent issue of contemporary public life. They are challenged to literally embody and internalise this experience in order to share it in some impassioned and constructive way with other individuals.

Reconsidering these recent New York projects, some of them seek simultaneously to critique and enhance public life. If not fundamentally different from other art, public art suggests its own particular model for thinking about the possible ways art exists in the world. Public art is invariably situational. It seeps into its environment, and the setting inevitably, almost imperceptibly, invades it and changes its meaning. Art is not a thing; instead, it is a dynamic exchange of invention, production, delivery, reception, and action rather than a stable collection of formal characteristics. In its diverse manifestations, it questions what occurs when people encounter and experience it.

In subtle and profound ways, public art insists that critical analysis consider the responses and reactions of viewers who shape, modify, perpetuate, and complete (at least provisionally) its meaning. Public art acknowledges and accommodates the transactions that sustain the transformative potential of all art. As we are changed, so too is art.

It is important to clarify that this connection of response and contingent meaning is not a defence for the effectiveness of public art. A consideration of what art does in the lives of individuals does not imply what it will – or will not – produce. If public art aims to do something to or for people, it is never entirely predictable what this might be. Art is thrillingly imprecise. It can shape a public space of imagination and suggests a sense of consequence of individual desires and actions within a community. But public art is neither user-friendly, risk-free, nor precisely programmed.

As the idea of public art is expanded and effaced, a significant feature is the response and reception of the audience. Public art criticism has often dwelled on environment or context. Other times, an interrogation of the idea of public has provided the subject matter for art and its criticism. And, of course, there is always the reliable review based on formal qualities and concepts. While the idea of public remains a compelling intellectual idea, perhaps critical attention should also be focused on the audience that insistently encounters and defines the work. Public art, in name and its many forms, claims that it has been made with someone – some place – in mind. But it is never clear who will find meaning in it. It accepts the centrality, if undefinability, of the public.

But the idea of audience is evasive. If the relationship to a public is too rigidly defined, criticism becomes a quantitative, instrumental application. And yet, if audience is too narrowly envisioned, some of the most risky, robust, and resonant ideas about the experience of art may never stimulate criticism as an open space for ideas. For all of the problems, public art's deliberate alignment of the creative process and the concept of audience has the potential to provide new insights into the relationship of aesthetic ideas to the renewal of public life. More specifically, what can public art reveal about the art/human dimension, public life and democratic culture, and the spaces of transaction and transformation?

Even as artists embrace the centrality of the audience in public art, they must reconfirm their commitment to image-making. A great deal of public art may share some characteristics and concerns with forms of social activism, but art distinguishes its objectives through the influence of its aesthetic dimensions – the incalculable potential of its images to evoke response and ignite ideas. The role of the audience will be clarified through the insistence and integrity of the forms and images of public art. By broadening and deepening an understanding of what images are and how they circulate in public life, artists and audiences can sustain the ongoing, critical process of forming a civic vision of deep conviction.

Are there some non-negotiable, inviolable characteristics of public art? Is public art always free? Is it always accessible? What does availability mean? Are particular methods of distribution more 'public' or democratic than others? If we expect public art to be 'engaging', what are some of the ways that it can engage? How do we look at this, not from the standpoint of a quality – to be engaging – but with the implication of activity – to engage? Engaging art allows viewers to remain passive, indifferent, and dispassionate. A public art that engages insists on a level of involvement or scepticism.

In *Back to the Front: Tourisms of War*, architects Elizabeth Diller and Ricardo Scofidio write: 'The construction of the sanitized past is called living history. It is a past in which the tourist can go back in time as a passive observer without any effect on the outcome of the future – a classic dilemma of science-fiction time travel . . . the tourist unproblematically accepts the role of voyeur' (Diller and Scofidio, 1994: 36). Of course, this is an easy trap for public art and there are countless examples of it that simply require experience to be – or deliberately limit it to – a form of uninvolved voyeurism.

What is the relationship between public art and urban livability? What makes cities and communities more livable, and how does art assist this process? Certainly, examples abound of murals, sculptures, and art-enhanced amenities that unquestionably make cities more attractive and interesting. Presumably, as a consequence, people feel better about their communities and the physical environment. Who can argue that this is not a positive development? But a more aggressive idea of urban livability embraces ideas of sustainability. Sustainability is not simply the maintenance of an acceptable environmental balance. Sustainability encompasses the vast and intricate web of social, political, economic, and ecological systems. It concerns the travesty of gun violence, the future of waste management in a capitalistic, consumptive society, and the rights of private citizens in the public realm of new information technologies.

What are the some of the ways that public art can inform, inspire, or incite individuals to envision and enact an ongoing experience of democracy? Generally, public art can produce a transaction that confronts and implicates observers as participants – that projects a vision of audience that enriches experience beyond the narrow preoccupations of voyeurism to the radical transformations of the witness.

Public art will become more borderless as definitions of private and public become increasingly migratory and mutant. Public art and its criticism is perpetually at a threshold – inherently indeterminate, never occupying a stable place. Thresholds are fraught with anticipation and fear. Neither in nor out, interior or exterior, they are critical junctures, psychological sites, places of perpetual unrest. Nowhere and anywhere, bounded and boundariless, empty and inhabited. Traditionally they have served as hinges between public and private worlds. Thresholds are physical, symbolic, and imaginative spaces.

Mierle Laderman Ukeles describes public art as 'an instrument of negotiation'. Negotiation is about navigation, connection – moving to a fruitful, animating idea of common purpose rather than a chilling or dulling compromise. Negotiation occurs on the threshold. Transition is an unerring condition. The genesis and goal of all public art must be clear and compelling ideas and issues that lead to open and unforced responses from viewers and participants. Public art may present challenging and difficult situations or, as Ukeles has described, 'zones of conflict'. But these zones of conflict are not unproductive places. If we accept the dynamic, transitional nature of democratic cultures, these junctures for different views and positions are the renewable resources of public life.

9 TIM COLLINS AND REIKO GOTO[1]

Landscape, Ecology, Art and Change

INTRODUCTION

Restoration ecology is the emerging paradigmatic relationship of humanity to nature. It has occurred in response to the industrial revolution and its massive programme utilising nature as both raw material and sink for wastes. Restoration ecology establishes a new relationship to nature by addressing a range of damaged land and water systems in both urban and rural settings. Restoration ecology is a community of disciplines, which acts upon natural systems through the sciences and engineering; and upon cultural values through the arts and humanities.

If we are going to restore ecosystems, practitioners must engage nature through primary experience. Restoration ecology redirects the concepts of agriculture and gardening, to a programme of healing nature. Restoration ecology emerges from a culture that is beginning to recognise its detrimental effect on nature, and is seeking new ways to understand and act upon a range of post-industrial problems. In the following pages, We will explore ideas of applied and cultural ecologies. We will conclude with a manifesto, which places art and ecology (eco-art) in the midst of the radical human ecologies and social-environmental change.

APPLIED ECOLOGIES

Preservation, conservation, and restoration ecology

Restoration ecology is a logical outcome of the projects of conservation and preservation. Preservation and conservation emerged in the years around the turn of the twentieth century in response to growth and development in the American West. Preservation began with the aesthetic/scientific interests of botanisers and gardeners in the subject of trees. Organised groups at that time helped to establish Arbor Day and promoted a plan for national forest preserves. This interest in nature was fuelled by the writings of the naturalist/authors Emerson, Thoreau and Muir. A popular movement, preservation was soon balanced by a more practical and scientific voice. The project of conservation has been described by Samuel Hays (1959: 123) as 'efficiency in the development and use of all natural resources'. Established during Theodore Roosevelt's presidency, conservation was defined by Gifford Pinchot and others as a rational approach to land management.

Conservation theory was rooted in an engineering approach to applied knowledge. The ultimate goal was properly to inventory all natural resources prior to a planned development intended to achieve efficient use and minimise waste. This is still the focus of conservation biologists worldwide who inventory natural communities and their movements, then manage habitat so that select species (either migratory or indigenous) will prosper. Conservation and preservation are programmes that are driven by a reaction to human disturbance of natural systems. Conservation projects today involve large habitat areas, nesting areas, and numerous migration areas where landscapes are managed to the best advantage of a single species or groups of similar species.

Preservation and conservation were forged as a binary pair. Preservation was primarily a citizen-based cultural programme inspired by naturalist authors and artists, while conservation was primarily an expert-based scientific programme. Preservationists supplied the social and nature-aesthetic argument for protection of national forests as public land while the conservationists inventoried and planned for the rational development of public lands. This binary pair was alternately oppositional and collegial in pursuit of their goals. The preservationist rhetoric was about keeping wild areas and wildlife from commercial development and use. The conservation policy goals of the era included protecting large tracts of land, watersheds and forests, sustained forestry, dams and water-power, hunting, grazing and mineral-mining rights. The Hetch Hetchy and Yosemite Valleys in California are probably the best examples of the oppositional outcomes of the preservation and conservation movement.

Preservation and conservation were a reaction to the perception of encroaching physical limits within the United States. Preservationists believed that wilderness was in a state of grace, beyond the limits of human habitation. Nature was something to be preserved and contained for future generations. The conservationist believed that wilderness was a resource bounty to be managed and controlled for long-term economic benefits. Both of these philosophical and political positions placed nature (in the form of wilderness and land resource) well beyond the limits of cities or towns.

Restoration ecology is a new way of thinking. It links citizens and experts, as well as cities and wilderness, in a broad programme of ecological awareness and action. It is a community of disciplines synthesising a continuum of diverse cultural practices. On one end lie the arts and humanities, in the middle are the design professions, at the other end science and engineering. Restoration ecology has been touted as a new relationship to nature, one in which the old reductionist paradigm is reversed and scientists are charged with re-assembling a working nature from the pieces discovered over the last 200 years, taking it apart. While the machine metaphor was useful in the disassembly and analysis of nature, it is less useful when re-assembling nature. The aesthetic roots of restoration ecology can be found in the urban-nature design projects of Frederick Law Olmsted (particularly the Fens of Boston, 1881). The roots of its science lie in Aldo Leopold's work restoring the lands of the University of Wisconsin–Madison Arboretum (1934).

It can be argued that this 'discipline' was established in the early 1980s, in part by the efforts of William Jordan III (1984), a botanist and journalist who was employed at the Madison Arboretum and saw the potential of Leopold's ideas in a contemporary setting. During the fiftieth anniversary of the Madison Arboretum, he published a seminal text declaring the import of this area. This was followed by a symposium on restoration ecology, which brought together some of the key thinkers worldwide. This resulted in an edited text, *Restoration Ecology: A Synthetic Approach to Ecological Research* (Jordan, 1987). In Jordan's original document, he was interpreting restoration ecology as a mixture of cultural and scientific efforts, 'active as a shaper of the landscape, yet attentive to nature and receptive to its subtlest secrets and most intricate relationships. The restorationist is in this sense like an artist and a scientist, impelled to look closer, drawn into lively curiosity and the most intricate relationships' (Jordan, 1987: 24). After Leopold, Jordan is clear that restoration is about restoring a 'whole natural community, not taking nature apart and simplifying it, but putting it back together again, bit by bit, plant by plant', concluding, in his own words, 'the ecologist version of healing' (Jordan, 1987: 23). Jordan commented on the import of restoring whole communities in this text, but he also recognised the import of restoring (reclaiming) industrial sites (referencing the noted biologist Anthony Bradshaw's pioneering work on coal mining sites in England). Jordan sees the Madison Arboretum as a research laboratory for work that will be in increasing demand in the future, due to the fact that the industrial revolution has provided humanity with the tools to affect nature on a grand scale. The work that first found its symbolic and intellectual focus as a result of the anniversary of the Madison Arboretum occurs around the world today. Today there are academic, private industry, non-profit and federal government models of restoration practices. In 1988 the Society for Ecological Restoration was announced at the Restore the Earth Conference in Berkeley, California through the efforts of John P. Rieger, John T. Stanley on the west coast and William Jordan in Wisconsin. There are now two journals attending the area, *Restoration Ecology: The Journal of the Society for Ecological Restoration*, published by Blackwell Science, and *Ecological Restoration*, published by the University of Wisconsin–Madison Arboretum. Each year, the disciplines of anthropology, art, biology, botany, ecology, engineering, philosophy and poetry participate with government regulators, first peoples, citizen activists, policy makers and spiritual leaders at the annual Society for Ecological Restoration Conferences (http://www.ser.org).

Restoration ecology attempts both to define and reconstruct nature while staying aware (and respectful) of the complexities of the process, its ethical context and the social potential of its performative aspects. Restoration ecology is an important new area of thinking and acting. It provides us with experiences and knowledges that can transform the human relationship to nature.

RESTORATION ECOLOGY – CONCEPTS AND PRACTICES

Nature can be restored to its 'historical form' if there is a record of the ecosystems' component parts, or a likely reference ecosystem can be identified, and the soils and hydrology remain untouched. In urban settings there is little left untouched, soils and hydrology are almost always affected. The historical model and its expert-defined parameters are the exception rather than the norm. More likely the restorationist is faced with conditions which are challenging: in other words, disturbed soils and a radically changed hydrology. Furthermore, the work occurs in a social setting where the perception of nature and its physical manifestation is quite diverse. In the words of Anthony Bradshaw (1995: 105), 'The primary goal of restoration is an aesthetic one – to restore the visible environmental quality of the area.' This means that the decision to restore is both cultural and discursive, a mix of science, art and democracy. Once the site has lost its soils and hydrology, the expert is seeking to define an aesthetic target that is relevant to a local community.

Philosopher Andrew Light (2000) has written extensively about the import of the social element in restoration ecology. 'Restoration is not inherently democratic, but it does have an inherent democratic potential. The problem is not just to identify this potential, but to make a case for why it is part of the criteria for what identifies restoration as a good environmental practice' (Light, 2000: 165). The democratic potential of restoration programmes is revealed at three steps in the process:

1. In terms of the goal of restoration, which must be set in such a way that it reflects local aesthetics.
2. In terms of the act of restoration, when additional bodies, hands, hearts and minds can invest the project with diverse integrity and stewardship.
3. During the monitoring period after planting, where citizens can gauge the efficacy of their actions.

This democratic potential and its social benefits are not inconsequential. The concept was the basis of the $85 million Changing Places programme of the Groundwork Trust in the United Kingdom (Groundwork Trust, 1999–2001). Changing Places invested money to recover the aesthetic quality of local environments, in communities ravaged by industry. It integrated restoration ecology, social recovery and community arts in the programme.

The definition, goals and ethics of restoration are a consistent point of discussion from a range of points of view. In 'Sunflower Forest: Ecological Restoration as the Basis for a New Environmental Paradigm' (Jordan, 1994: 205–20) and 'A Field Guide to the Synthetic Landscape: Toward a New Environmental Ethic' (Turner, 1998: 195–204), the botanist William Jordan and poet and philosopher Frederick Turner describe restoration as a new relationship to gardening or agriculture. They see restoration as a culture–nature activity in which benefit goes to the ecosystem rather than the ecosystem providing benefit in terms of products for the gardener or farmer. In this emergent relationship, the concept of nature as an instrument of humanity is expanded to nature as a relational entity worthy of human investment. This expanded concept of nature takes us away from a human use value and towards the recognition of the value of ecosystems and their component organisms.

Other philosophers and theorists with important contributions to the philosophy of restoration ecology include Robert Elliot, Cheryl Foster, Eric Katz and William Throop. The interest of these 'cultural ecologists' can be defined by the pursuit of ethical positions in an arena where the foundation of knowledge is in constant flux, and the context for action is a moving target. Much of the discussion has revolved around the idea of restoration (and some of our 'natural' monuments) as 'counterfeit nature' or nothing more than another human artefact (Throop, 2000: 71–134). Other discussions (primarily amongst the scientists) revolve around the various valances of meaning that surround the descriptive terms of practice and the means of evaluating projects. There has been an ongoing dialogue about the shaded meanings of restoration, reclamation, mitigation or revegetation. Restoration can refer to the recovery of biological communities, or it can refer to the recovery at the landscape ecology scale. Reclamation usually refers to the recovery of strip mining sites, or areas of industrial soil. Mitigation is currently specific to wetland recovery; usually replacements are created to replace those lost to development. Revegetation might involve the recovery of a forest, on farmland. Joan G. Ehrenfeld (2000) has identified at least four approaches to restoration ecology (Figure 9.1). These four strands of practice demand separate goals and evaluation methods. The items on the left describe less disturbed sites, the items on the right describe highly disturbed or, in the case of mitigation wetlands, lost sites. This range of goals (bottom text in each box) demands diverse theory and analysis. (It is assumed that the first step of restoration is to preserve land and conserve species wherever possible.)

The community of disciplines working in the area of restoration ecology necessitates respect for a diversity of knowledge, practice and pedagogy. The immediate future of restoration ecology will be defined by a struggle between discipline specific knowledge standards. The arts and humanities relish the idea of thinking about the foundation of knowledge, seeking new metaphors, narratives and shifts in human value. The sciences relish the idea of building on accepted foundation knowledge, seeking the empirical tools to understand, preserve, conserve and restore natural systems. This dynamic tension is essential if we are to achieve a new community of disciplines responsible for both a theoretical and applied knowledge, capable of addressing both rational and speculative approaches to a changing nature/culture relationship.

Restoration ecology with its attendant citizen/expert programmes and integration of nature/culture goes to the heart of fundamental environmental questions of the twenty-first century. Restoration ecology is an emergent community of disciplines. The form and theory of its social–aesthetic interface is as yet unrefined. There is both intellectual and aesthetic

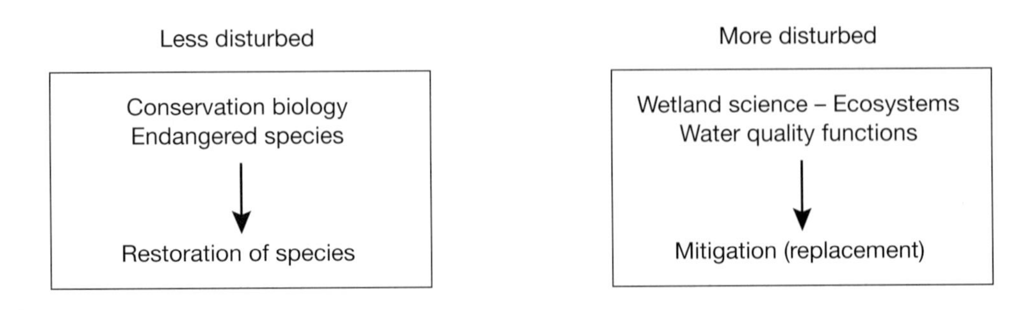

Restoration ecology

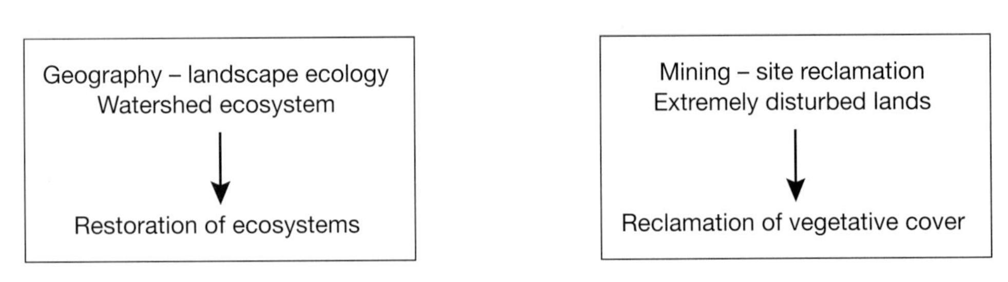

Figure 9.1 Restoration ecology: systems, disciplines, focus and goals (after Ehrenfeld, 2000)

space for artists to participate in this paradigmatic shift in nature/culture relationships.

CULTURAL ECOLOGIES

Background, art and landscape

Artists have worked with the nature/culture dichotomy for generations. Europe has a rich landscape painting tradition. In the seventeenth century Claude Lorrain painted Italian landscapes specifically chosen for their contextual references to Roman history. Intertwining images of nature and nostalgic culture, his work provided a primary source of ideas about pictorial framing. Nature was the formal setting to consider the ruins of Rome as picturesque elements of an experientially rich landscape. In the eighteenth century Thomas Cole, Frederick Edwin Church and John James Audubon were working in America, painting the wilderness and wildlife of the United States. The American painters illustrated pristine landscapes that were being newly tamed by the technology of the pioneers. With the advent of the modernist period of the twentieth century, the relevance of landscape as a subject began to fade. Painters were disinterested in subjects external to the media and its means of expression. The painter moved away from landscapes and other subjects external to the medium. The painter's effect on the meaning and perception of landscapes and nature diminished accordingly.

Earth-art

Sculpture's historical relationship to landscape is different from that of painting. It is only with the advent of the minimalist era of modernist sculpture in the 1960s that landscape began to play a sustained and primary role in sculptors' thinking.

John Beardsley, Thomas Hobbs and Lucy Lippard are the primary authors on the subject of earth-art. Beardsley's (1977) *Probing the Earth: Contemporary Land Projects* and later (1984) *Earthworks and Beyond* provide the authoritative view on the originators of earth-art. Hobbs' (1981) *Robert Smithson: Sculpture* is the best reference on that artist's works and writing. Lucy Lippard's (1983) *Overlay* is a classic. It transcends the hierarchy of the art world as well as the ranks of the earth-artists, providing her readers with a comprehensive overview of diverse archaeological, historic and contemporary practices. Beardsely and Lippard both describe a post-studio inquiry and practice that integrates place, form and materials. The earth-artists engaged landscape directly, earth was the material, the form oriented the viewer to the place of the work. Earth-art challenged the purpose of art as a collectable object. In some ways it was the first artwork to go public, to establish art as an interface to the world.

Herbert Bayer, Walter De Maria, Michael Heizer, Nancy Holt, Mary Miss, Isamuu Noguchi and Dennis Oppenheimer were just some of the original practitioners that began working as earth-artists, or environmental sculptors. They experimented with simple geometric forms that integrated place, space, time and materials. The work ranged from pristine natural environments to post-industrial environments. Theorist-practitioners Robert Smithson, Nancy Holt and Robert Morris expressed a more integrated relationship to nature as system. Smithson was acutely aware of nature's entropic and eutrophic cycles, and embraced mining areas and quarries as the content and context for his work. Smithson's partner and colleague Nancy Holt was particularly interested in earth/sky relationships, creating works that updated ancient techniques with a modern sculptural vocabulary. Morris addressed post-industrial landscapes, in both form and theory. Writing about his own work in Kent, Washington, his articles about the ethical responsibility of artists working in post-industrial landscapes telegraph issues in restoration ecology, a decade later (Morris, 1979, 1993).

Ecosystem approaches

At the same time, another group of artists emerged with a focused interest in systems theory and ecology. Hans Haacke, Helen Mayer Harrison and Newton Harrison, Alan Sonfist, and Agnes Denes were the original (and continuing) practitioners. They differed from the earth-artists by their interest in dynamic living systems. Where the earth-artists expressed themselves in the landscape, these ecological artists were interested in collaborating with nature and ecology to develop integrated concepts, images and metaphors. While earth-art was the first to go public, these ecological artists were the first to act in the greater interest of nature and the commons. In 1974 Jack Burnham wrote *Great Western Salt Works*, an important book that developed an initial approach to systems aesthetics.

Alan Sonfist (1983) edited *Art in the Land*, a selection of texts which addressed the scope and range of artists working in relationship to environment at that time. In 1992 Barbara Matilsky curated *Fragile Ecologies*. Matilsky's catalogue provides an excellent overview of the historic precedents for this work, as well as some of the most important work of the first and second generation of ecological artists. A text edited by Bylai Oakes (1995), *Sculpting with the Environment*, is unique, and quite valuable as a reference in that he asked each artist to write about their own work. *Land and Environmental Art*, an international survey of both types of artists' projects, was edited by Jeffery Kastner with a survey of writing on the subject by Brian Wallis (Kastner and Wallis, 1998). The text goes into the first, second and third generations of earth and ecological artists, providing an overview of works and accompanying articles. In 1999 Heike Strelow curated *Natural Reality*, an international overview which expanded the concept of ecological art and its range of effort to include the human body as a site of 'natural' inquiry. The accompanying exhibition catalogue provides arguments for the three areas of the exhibition, the unity of humanity and nature, artists as natural and cultural scientists and nature in a social context.

ART AND RADICAL ECOLOGY

Ecological art, or eco-art, is a creative process that results in interface between natural systems and human culture. It is manifest in form, concept and focused experience. It recognises the historic dichotomy between nature and culture and works towards healing the human relationship to the natural world and its ecosystems.

A set of goals (broad intentions):

- eco-art is a compassionate advocacy for natural communities;
- eco-art addresses damage to diversity and dynamics of ecosystems;
- eco-art is an expression of value, in such it is transformative by intent;
- eco-art seeks to balance the technical with the biological and ecological;
- eco-art seeks legal rights, and appropriate advocacy for natural systems.

Objectives (metrics for project analysis):

- interdisciplinary knowledge or participation from other disciplines;
- strategic actions targeting specific communities (human and natural);
- aesthetic and scientific components that are open to supplementary response;
- an integrated, conceptually informed aesthetic experience of complex systems.

Theory and interdiscipline

In *Fragile Ecologies* (1992) Barbara Matilsky provides a robust historical context for ecological artworks. As a contemporary curator, she was interested in form and content, how the product of ecological art presented itself to the world, specifically the art world. As eco-artists and theorist, we are interested in form and content, but I am also interested in the concepts and theories that inform and sustain the practice. We would argue after Harrison and Sonfist (Auping, 1983: 99) that eco-art is fundamentally interdisciplinary, in that we can't rely on the art world as the only point of engagement and interpretation. Furthermore, the artists involved in this practice can't confine their learning or production to art. We must reach out across disciplines to build a platform of knowledge and practice. In the interdisciplinary model, artists find critical social space to expand their practice by moving outside their discipline and its institutionalised relationship to society. In this way, we find opportunities both intellectual and creative that we cannot find within our own discipline (which like other disciplines has turned inward upon itself). Interdisciplinary practice breaks the form of discipline-specific institutions. It expands the combined disciplines and provides the artist with a new path to social engagement. Inherent in that path is the responsibility for the artist to educate him/herself in both disciplines. In turn, the work needs to be received and evaluated for the totality of its intention.

Dualities

The environmental movement can be characterised by a struggle between the oppositional ideas of nature as an autonomous and intrinsically valuable entity unto itself versus nature as both a concept and focus of human exploitation for economic value. The social-ecologist Murray Bookchin (1974: xv) sets up a simple duality, to help us understand these ideas better: 'Ecologism refers to a broad, philosophical, almost spiritual, outlook toward humanity's relationship to the natural world . . . environmentalism [which is] a form of natural engineering seeks to manipulate nature as a mere "natural resource" with minimal pollution and public outcry.' Bookchin's philosophical position on ecologism and environmentalism reflects the earlier discussions of preservation and conservation. But what does this mean for artists? Ecologism provides artists a pathway into a new area of knowledge. In that broad philosophical/spiritual outlook there is plenty of room for artists to experiment with interface, perception and human values. We want to think of interface as a common boundary or interconnection between systems, equipment, concepts or human beings. Interface is the art, the physical manifestation of that 'relationship between humanity and the natural world'. The concept of interface is appropriately open, its form undetermined but its intention explicit, it defines the art of ecology without closing out its options. Perception is the awareness of interface or awareness through interface. Human values are the target, or goal, of cultural agents (the active role of agency is assumed under the interdisciplinary model). Eco-artists manipulate the attendant metaphors, symbols and narratives of the nature/culture interface to shift human perceptions around the dual subjects of their inquiry, research and production – affecting valuation. These are the strategic points of engagement for the eco-artist – interface, perception and human values.

The philosophies

It is important to understand the philosophies that can inform our actions. Where do the eco-artists stand in relation to nature? We can broadly situate the eco-artist in either the wilderness or the garden. This simple duality allows us to consider wilderness as the condition of nature without human impact, and garden as the human condition (or city condition) of nature. Our value systems can flow in either direction. If value is centred in the garden, then it is the use of nature that drives our actions. The garden relationship assumes that we are above nature and capable of some charitable (and not so charitable) contributions to nature. If value is centred in the wilderness then it is the maintenance of that boundary separating humanity from nature that drives our actions. Wilderness (by strict definition) is a condition that can be defended, defined or interpreted but never improved upon by human action.

Human actions now affect global climate; this is an essential awareness of the post-industrial condition. Carbon-based industrial by-products and a lack of effective human advocacy threaten nature. The human effect on the planet has become so great that we have begun to examine carefully the human condition and its intellectual (and psychological) relationship to nature. Three philosophies have emerged which inform a continuum of human thought and action in relationship to garden and wilderness ideas – social ecology, eco-feminism and deep ecology. These three ecologies share a common thread – the negative effect of human civilisation upon natural systems has instigated the need for various radical communities to seek a path to action.

Social ecology calls for a grand decentralisation scheme, with a move to smaller cities, and appropriate technologies. It seeks a balance with nature and an equitable 'human footprint' on the planet. This is the broadest of our three eco-philosophies, firmly rooted in the garden but respectful of wilderness. Social ecology claims a spatial allegiance to city, town and country, believing that nature is a fundamentally human concept that must be resolved in a social setting. Industrial economies, politics and urban forms all must change if we are to achieve an ethical and sustainable relationship to nature and our remaining wilderness lands. The approach is built upon a powerful critique of the oppressive nature of industrial-capitalist society.

Eco feminism examines the historical relationship between women and nature. Eco-feminism claims no spatial allegiance; the focus is upon the biological relationship between humanity and nature. The eco-feminist reveres wilderness, but the eco-feminist project is still garden, it is focused upon resolving humanity's inequitable relationship to nature. The argument is that dominant (male) culture, in its role of master with dominion over all, is unable to acknowledge its dependency upon nature, which has culminated in a threat to survival. 'We must find a form which encourages sensitivity to the conditions under which we exist on the earth, which recognizes and accommodates the denied relationship of dependency and enables us to acknowledge our debt to the sustaining others of the earth' (Plumwood, 1993: 198). The eco-feminist approach is built upon a critique of the oppressive nature of an inherently masculine society.

Finally, deep ecology recognises the intrinsic value of non-human nature. In terms of our garden/wilderness metaphor, it is explicitly wilderness. It declares a lack of critical knowledge in the context of increasingly complex worldwide ecological catastrophes. In this vacuum of rational eco-knowledge it demands an advocacy and activism based on philosophical principles. Deep ecology is bio-centric; it recognises the value of all living things and emphasises non-interference in natural processes: 'The flourishing of human and non-human life on Earth has intrinsic value. The value of non-human life forms is independent of the usefulness these may have for narrow human purposes' (Naess, 1989: 29). The deep ecology approach privileges nature and places humanity on a co-dependent footing. It declares the human interference in nature excessive and calls for a substantial decrease in human population. The deep ecology approach is the most radical of the three ecologies; it is built upon a critique of materialism and technological progress. It obligates action through a dynamic platform first developed by Arne Naess (Naess, 1989) and George Sessions (Sessions, 1995).

These three philosophies, with their spatial commitment to city, town, country or wilderness, and their political commitments to humanity, post-dominion humanity and the intrinsic rights of nature itself, provide a broad intellectual foundation for the eco-artist. This is a foundation of human values that project ways to understand and act in relationship to nature. This foundation provides room for a range of practitioners with a shared interest in the roles that art can take in the changing human relationship to natural systems. The founda-tion can accommodate the artist as witness, advocate or activist, but always as an agent of change in the shifting values of nature and culture. As I have said earlier, the strategic points of engagement for the eco-artist are interface, perception and human values. It is through human values that we know nature, it is through human values that artists act upon culture to change the perception of nature.

Eco-art and paradigmatic change

Restoration ecology is the only applied ecological programme that integrates city, town, country and wilderness. It is the only applied ecology that is equipped to address all three political programmes, of humanity in nature, post-dominion humanity and the intrinsic rights of nature (although most deep ecologists would disavow the restoration ecology programme as too manipulative). Restoration ecology integrates preservation and conservation in its scope of activity. The all-too-obvious challenges (and costs) of restoring complex ecosystems and landscapes dictate the use of stabilising methodologies (preserve/conserve) before any attempt is made to restore.

Restoration ecology is the emerging paradigmatic relationship of humanity to nature. It addresses the full range of philosophical, spatial and political constructs outlined so far. More importantly, restoration ecology is a community of disciplines, which recognises its role in the context of cultural values. If we are going to restore nature (the underlying goal of all three of our philosophies), we must engage nature as a primary human value which we feel responsibility for. Responsibility flows from intimacy. Jordan speaks of it as an ecological version of healing. Light declares its democratic potential. Restoration ecology redirects the age-old concepts of agriculture and gardening, which reaped benefit from nature, to a programme of caring experimentation with/for nature (healing). Restoration ecology emerges from a culture that recognises its effect on nature, and is just now beginning to understand and act upon the problem. While there are conflicting opinions about how best to take responsibility, we believe that intimate knowledge is the primary place to start.

We have argued in the past that we must seek the aesthetic properties (the interface) which define the perception of complex ecosystems. With intimate knowledge we can characterise those properties (after Eaton, 1997: 85–105) as health. Ecological restoration begins the intimate dialogue with nature that will allow us to internalise and define that aesthetic perception. The ecological humanists and artists are faced with the challenge of understanding this aesthetic property, and communicating it to the broadest segment of society. Imagine being so intimately aware of your bio-region that you can ascertain its health by an aesthetic understanding of its complex form. An odd idea, until you think about family, friends and pets – you can 'see' illness in them before you know they are sick. Or your car, computer or some other appliance that you use daily, and in some strange way, you know it is unhealthy, breaking down before it stops working. The perception of health is a multi-sensual mix of signals, which have an appropriate rhythm or pattern. When that pattern or rhythm is amiss, it is clear that the health of that perceived system is compromised. The recognition and definition of aesthetic health within a perceived ecosystem goes directly to our questions of interface, perception and human value.

We would argue that the most important culmination of the project of eco-art, of our three philosophies and applied ecologies, will be legal rights for nature. Within our capitalist democracy the courts provide the ultimate notice of change in human value. Guaranteeing the rights of nature would extend the essential rights of political and economic equality. Recognising nature as an important third leg of democratic society is a logical and essential next step. This is an idea which has been considered in the past.

In 1972, Supreme Court Justice William O. Douglas (Stone, 1972: 74–5) offered the dissenting opinion on a legal case brought against Disney Inc., in support of a grove of trees in California: 'The ordinary corporation is a "person" for purposes of adjudicatory processes, whether it represents proprietary, spiritual, aesthetic, or charitable causes. So it should be as respects valleys, alpine meadows, rivers, lakes, estuaries, beaches, ridges, groves of trees, swampland or even air that feels the destructive pressures of modern technology and modern life.' While we may initially consider the rights of nature an outrageous idea this question is a little less shocking to think about when we remember that 150 years ago the legal status of women and various cultural groups were considered radical and upsetting thoughts. Corporations, merchant marine vessels, comatose humans, and babies all have rights, based on advocacy by invested interest. Why couldn't we extend these basic legal rights to natural ecosystems? A legal interface for nature fundamentally changes perception and value within a specific society and culture. We would argue that without the ability to advocate legally for natural systems, the political and economic systems of democratic nations will never be moved towards a sustainable model. If we can empower this advocacy, we guarantee an era of creative change and renewed democracy.

CONCLUSION

We have attempted to outline a manifesto of eco-art that recognises the emergent restoration ecology paradigm. We have developed a programme with definition, direction, intention and interdisciplinary goals in mind. We have argued that the bridge between areas of knowledge and the focus on issues of the public realm can be fertile ground for both academic and applied research. We have argued the importance of engaging issues, from the rights of nature to the properties, or aesthetic, of perceived health. The arts is the discipline that most consistently questions the foundation knowledge of society. In a culture dominated by science (which defends foundation knowledge), the arts have to develop new critical and strategic tools to act upon society. We need to create a supportive interdisciplinary community of creative individuals that are committed to, and take responsibility for, positive shifts in culture.

- Eco-art is a compassionate advocate for natural communities.
- Eco-art addresses damages to diversity and the dynamics of ecosystems.
- Eco-art is an expression of value; in such it is transformative by intent.
- Eco-art seeks to balance the technical with the biological and ecological.
- Eco-art seeks constitutional rights, and appropriate advocacy, for natural systems.

> *the world that everyone sees is not the world,*
> *but a world which we bring forth with others.*
> (Marturana and Varela, 1987: 245)

RECOMMENDED TEXTS

Bookchin, M. (1990) *The Ecology of Freedom*, Montreal, Canada: Black Rose Books.

Botkin, Daniel (1990) *Discordant Harmonies: A New Ecology for the Twenty-First Century*, New York, NY: Oxford University Press.

Marturana, H.R. and Varela, F.J. (1987) *The Tree of Knowledge: The Biological Roots of Human Undersanding*, Boston, Mass.: Shambala Publications, Inc.

Merchant, C. (1980) *The Death of Nature, Women, Ecology and the Scientific Revolution*, San Francisco, CA: Harper.

Sessions, G. (1995) *Deep Ecology for the 21st Century*, Boston, Mass.: Shambala Publications.

Throop, W. (2000) *Environmental Restoration: Ethics, Theory and Practice*, Amherst, NY: Humanity Books, an imprint of Prometheus Books.

PART IV

Shaping

Reading newspapers, Lisbon, Portugal (Photo: Malcolm Miles)

INTRODUCTION

Part IV introduces material from two areas of social research, one with homeless people's organisations in the San Francisco Bay area, the other from cultural projects in context of urban regeneration in east London; and ends with an analysis from an ecological viewpoint of turbulence as the planetary condition. The first two chapters here are longer than others in the book, justifiably so given the inclusion of hitherto unpublished empirical research and, in the first of these, extensive quotation from individuals consulted in its course. To give space to these voices, which are marginalised in most other contexts, seems important and more than a gesture. The thread which runs through all three chapters is their concern with how social processes are shaped. It could be asserted that all the chapters in the book deal with this, and that its title frames them accordingly. But in this part, questions of how the frameworks, mobilities and practices discussed in the previous three parts feed through in highly specific situations, localised by constituency as well as geography, are addressed more directly.

Anne Roschelle and Talmadge Wright, drawing together material from a much larger, here selectively edited, body of research in the Bay area, note the causes of homelessness there as an expansion of new technologies – a post-industrial, silicon city of high-income levels – allied to tourist promotion and other forms of urban development which simply price out significant sections of the population. They note, too, that policing strategies drive the already evicted out of public spaces, as Mike Davis (1990) sees happening in the parallel if more extreme case of Los Angeles, while emerging labour markets require increasing numbers of semi- or unskilled low-wage earners, many of whom are recent immigrants. Part of the attraction of the San Francisco Bay area for such research is that each of its counties has an autonomous policy for housing, homelessness and policing, so that comparison can be made within a common set of broad economic, social, and cultural conditions. The two counties researched in depth are San Francisco and Sonoma. Over fifty homeless activist leaders were interviewed in the research, and considerable variations found in tactics. Local news accounts and reports from government or non-profit organisations were also scrutinised. Roschelle and Wright then deduce from their material differential success levels in advocacy as well as its divergent aims, then propose regional collaboration to strengthen the voices of such groups and go beyond, while not setting aside, their local struggles. If the context for this is gentrification, as a globalised phenomenon, then the potential for mobilisation among homeless people, and their capacity for organisation – ignored, it seems, in Davis' more melodramatic accounts of Los Angeles – is part of a localised resistance which nonetheless has global implications. The authors, however, are not unaware of the barriers to such collaboration, and see its prospects as dim unless there are other changes in the political climate.

If the Bay area is one laboratory for social research that engages those whose futures it ponders, the East End of London is another. In the neighbourhood of Hoxton, for example, culturation denoted by artists' lofts has led rapidly to gentrification. Now most of the artists have been displaced by executives in the financial services, marketing, or media industries, who live in a new affluent bohemia close by tower blocks whose inhabitant suffer alarming levels of multiple deprivation (that is, in housing, health, education, transport and cultural provision). Hoxton, however, is on an edge between the City (financial district) and the East End. Further east, in Whitechapel or Stepney, for instance, gentrification is more patchy, and co-present with other social scapes of multi-ethnicity, or, to be more accurate, multiple pockets of different ethnicities. Graeme Evans and Jo Foord begin by juxtaposing rival definitions of culture, as high art or everyday life, as economic asset or expression of values, and see the contentions produced from rival definitions as unresolved in much cultural policy and planning in areas of urban development or regeneration. Amongst other factors they note the failure of the trickle-down effect promoted in the UK and USA since the 1980s, so that investment in flagship projects tends not to bring new vitality to all the neighbourhoods around it, nor to bring new prosperity to more than a few of the publics who live or work in proximity to it. This does not stop developers and civic authorities looking for cultural value as an add-on to redevelopment schemes, as the powerful image of symbolic economies, to use Zukin's concept (1995), is projected without the critical analysis Zukin offers of it.

Like Borden (in Part III) , Evans and Foord cite Rogers, seeing him as an influential figure, and go further than Borden in linking his views to a new elitism of the inner city, no longer a powder keg but a site of reclamation as work-and-pleasure zone beside a sparkling Thames. They see this vision of a newly civilised inner city as underpinning much of urban policy for 'problem' areas, and its ethos as one of colonisation. Cultural intervention, thus, becomes an avant-garde not of a revolutionary society but of a new middle-class urban landscape. The case through which Evans and Foord test their theses is Stepney, for which the cultural industries, small-scale offshoots of what Adorno calls the culture industry, are seen as key to regeneration. One result, they claim, of such policies – which bring in culture from outside, in context of those older histories of culture as high art, or means for public education as it was in the founding of public art museums in the nineteenth century – is that local culture, that produced locally by members of diverse publics, is rendered invisible. Hence a Bangladeshi Festival seems to have imported one kind of ethnic culture for one kind of metropolitan audience, leaving British Bangladeshis in east London standing as usual on the side, not empowered but, it could be summarised, marginalised by cultural charity.

Finally, Nigel Clark reconsiders urban life as a set of physical processes the footprint of which runs far outside urban space. But he begins by citing the same novel, Zadie Smith's *White Teeth*, as Rowe in Part I – a neat and unplanned symmetry in the editing of the book. And while much of the critical commentary included in this book is from a cultural or social science perspective, Clark draws his metaphors, and observations, from the natural sciences, from how biological and geoclimatic forces impact city spaces.

Clark revisits Louis Wirth of the Chicago School, and Lewis Mumford, in discussion of the relation, or edge, between city and countryside, and recognises the urban quality of at least some ecological discourse. He links urban sustainability to the concept, from development studies and global economics, of sustainable development (for an introduction to this see Elliott, 1994); then returns to the idea of footprinting, noting that London's requirement of land to service its energy and food consumption, and its excretion of waste, is 125 times the city's area, or about the size of the productive land area of the UK – which does not look very sustainable. In face of such evident disequilibrium, then, Clark sees sustainability as aiming for a redress of the balance, from which comes the interest of an ecological view of cities.

While Collins and Goto in Part III utilise ecological knowledges practically in the recreation of the conditions for bio-diversity, Clark sees ecology as a model for a reconceptualisation of the city, in part deconstructing the rather anti-urban view of cites as – in their parks or leafy suburbs – new arcadias. But now turbulence enters the frame, as an aspect of the natural world generally excluded from its polite representations in blissful landscapes. Destructive forces, such as lightning and fires, play a significant role in shaping natural landscapes as yet outside human engineering. Characterisation of these forces and their impacts as disasters is shaped by social and cultural narratives into which they fit only as threats or apocalyptic others. But, Clark muses, suppose the norms to which these forces pose such danger are not norms at all but constructions which disregard some of the major factors in the situation. Suppose there are no norms, only punctuations. And suppose again that these are exacerbated by human intervention, as in policies which cluster poor people in areas of extreme environmental vulnerability. Beyond that are the consequences of urbanisation for non-human species and other deliberate simplifications of eco-systems to suit what are often economic scenarios. If this seems as gloomy as the disaster scenario of Davis' *Ecology of Fear* (1998), Clark introduces towards the conclusion of his chapter a proposal that improved grassroots participation in decision-making is a vital step towards a more appropriate, and in Darwinian terms perhaps sustainable, urban ecology. At this point, where research in natural disasters, or turbulence, and urban sustainability converges, his chapter seems to converge, too, with some of the findings of the previous two in this part of the book. If there is a lesson to be drawn, then, it is that the knowledges of dwellers, long sidelined in matters of power, money, social formation, and culture, might yet be those most needed if cities are to remain the primary form of human settlement on Earth.

10 ANNE R. ROSCHELLE AND TALMADGE WRIGHT

Gentrification and Social Exclusion: Spatial Policing and Homeless Activist Responses in the San Francisco Bay Area

INTRODUCTION

City redevelopment policies promoting tourism, shopping, sports, and entertainment have facilitated the social exclusion of the poor and have promoted harsh policing strategies in response to declining downtown revenues. Urban policies that lead to rapid gentrification of the city have displaced poor and working people for years. However, recent increases in disparities of wealth combined with a reduction in the social wage, inadequate health care, and the decline of affordable housing have forced the poor and homeless out of desirable public spaces, isolating them in peripheral neighbourhoods and in shelters. This new 'Revanchist' (Smith, 1996a) city reflects uneven development and the intensification of such development through the rapid movement of global capital. The response by homeless activists and poverty workers to this social exclusion has ebbed and flowed throughout the past twenty years. Policing and legislative strategies practised by cities have also shifted, reflecting the dynamic political nature of housing and homeless struggles.

Since the 1980s homeless activism has shaped the debate over poverty, housing, health care, and city redevelopment policies. Homeless people and their advocates organised 'social movement organizations' (SMOs) around the country. Chapters of the National Coalition on Homelessness, Union of the Homeless, Food Not Bombs, and many others organised to resist police practices, urban displacement, and shelter containment. Over fifty cities saw protest actions throughout the 1980s (Cress and Snow, 1988). However, the 1990s witnessed a decline in overall homeless activism nationally and an increase in the institutionalisation of homeless activism within mainstream advocate organisations and government agencies.

The staging of the *Housing Now!* March on Washington, DC in the late 1980s signified the high point of the homeless activist movement. However, as Cress and Snow (2000) have demonstrated, the ability for homeless movement organisations to secure resources and successfully advocate for their constituents is an outcome of having well-developed, viable organisational structures, not merely an ability to disrupt the local political scene (Piven and Cloward, 1992). We found that the inability to sustain homeless activism is a result of the internal and external fragmentation of homeless SMOs, increasingly punitive policing strategies, the criminalisation of the homeless, a lack of movement organisation resources, and the overwhelming nature of sustained activism. We have attempted to understand this decline through a comparative examination of different homeless advocate/activist groups operating in the San Francisco Bay Area.

The success and failure of homeless activist groups at the local and regional level, and their effectiveness in ending homelessness, is instructive in directing us towards future organising and research efforts. A regional analysis explicates the possibilities of coalition formation. It also demonstrates the relationship between advocacy groups and local city officials as they attempt to isolate and exclude homeless people from public areas. The San Francisco Bay Area, in Northern California, provides a natural laboratory to examine homeless struggles and city development on a regional basis. Each city and county has its own particular form of homeless and housing struggles that are affected by the specific history and politics of the region.

This chapter is part of a larger project in which we conducted in-depth interviews with fifty-one homeless advocate/activist leaders from eight cities around the Bay Area during the Summer and Fall of 1998. While we have data on activist perceptions in Alameda County, Marin County, San Francisco County, Santa Clara County, and Sonoma County, this chapter focuses on Sonoma and San Francisco Counties. Each county is unique, San Francisco a heavily urbanised county and Sonoma a mainly rural county. Advocates from the Sonoma County Task Force, the Homeless Service Center, Burbank Housing, the Committee on Shelters (COTS), and the Sonoma County Homeless Union were interviewed in Sonoma County. In San Francisco County homeless advocates from the Coalition on Homelessness, Religious Witness With Homeless People, Food Not Bombs, Homes Not Jails/San Francisco Tenants Union, and Home Base: Centre for Common Concerns Inc. were interviewed as well as independent activists. In addition, interviews were conducted with representatives of People Organized to Win Employment Rights (POWER) and the Northern California Coalition on Immigrant Rights. In addition to interview data, we examined primary and secondary documents of homeless movement organisations, as well as local news stories related to homeless protests and attitudes towards the homeless in the Bay Area. We also used data from government documents and non-profit agencies to situate the research within the local political-economic context.

THE POLITICAL ECONOMIC CONTEXT OF HOMELESS ACTIVISM

Sonoma County, located one hour north of San Francisco, grew from a population of 388,222 in 1990 to 458,614 by 2000 as commuters moved north to escape San Francisco and Marin County's high housing costs.[1] Composed of nine cities within a rural area, Sonoma County witnessed extensive growth along the 101 freeway corridor for the past twenty years. The Association of Bay Area Governments (ABAG) estimated that more than 20,000 jobs were created in the county between 1995 and 2000. The bulk of these jobs were for business services and manufacturing, outpacing agricultural and construction employment. This has meant a corresponding jump in the median incomes of the area, a requirement for higher levels of job skills, and a rapid increase in home prices and rents (Association of Bay Area Governments, 1999).

As the fastest growing mid-size city within the San Francisco Bay Area and the largest city in Sonoma County, Santa Rosa expanded from a population of 134,228 in 1990 to 147,595 in 2000. It is expected to add another 36,800 people by 2020 according to projections by ABAG[2] (Association of Bay Area Governments, 1999). Cities within Sonoma County, especially Petaluma and Santa Rosa, have undergone major transformations as a result of the influx of the telecommunications industry (Agilent Technology, American Communications Technologies International, Microsource, Fibex Systems, and others), generating new levels of local social, economic and political inequality. Export-led growth is increasingly connecting Sonoma County to global markets, further stimulating regional growth. For example, 20 to 30 per cent of local companies involved in the wine industry depend on international exports. Manufacturing exports alone totalled $862.7 million in 1998, doubling since 1993. Other local firms including software developers, consultants, and architects sold up to $430 million abroad. Sonoma County high-tech companies now account for $1.1 billion of the County's $10.9 billion gross regional product. They also employ over 15,000 workers, most of them professionals. The number of high tech companies has doubled, from 219 in 1996 to 471 in 1999. The average salary of this high-technology sector in 1998 was $71,040 compared with average overall salaries of $26,797. As a result of these macroeconomic shifts over the last two years, the median family income in Sonoma County has risen to $63,200 (Twombly *et al.*, 2001).

San Francisco County has followed a similar trajectory. Recent estimates illustrate that since 1990, San Francisco's population has increased from 723,959 in 1980 to 742,613 in 1998, an increase of 2.6 per cent. Demographic projections estimate that the population of San Francisco will increase to 771,048 by 2003, an increase of 3.8 per cent (Northern California Council for the Community, 1998). San Francisco has also experienced the largest influx of immigrants and refugee residents among all California counties and ranks fifth among US cities with the largest number of foreign residents (White *et al.*, 1995). According to the 1990 Census, 246,034 or 34 per cent of San Francisco's residents were born in other countries with nearly half of these new residents (118,292) arriving throughout the 1980s.[3] Although there has been a large influx of Latinos into Sonoma County, San Francisco has one of the most ethnically diverse populations in the world.[4] Even though San Francisco is growing at a slower rate than Sonoma County, it has also experienced rapid increases in population, particularly among Asian–Pacific Islanders and Latinos. Similarly, San Francisco has also experienced shifts in industry that have impacted housing prices and the general cost of living.

By 1994, the nine-county San Francisco Bay Area pulled itself out of one of its longest recessions. What emerged was a different kind of regional economy from what was in place prior to the recession. The Bay Area became a world leader in the new knowledge-based economy. High-tech companies requiring a relatively highly skilled and highly educated workforce were creating most of the new jobs. The number of jobs for people with a high-school degree or less was growing at a slower rate than jobs requiring a college education. In San Francisco business services, publishing and entertainment became increasingly important sectors of the economy (Northern California Council for the Community, 1998). During the emergence of the knowledge-based economy there was a simultaneous shift away from the region's previous manufacturing base. Between 1981 and 1986 the Bay Area lost 12,000 durable manufacturing jobs while at the same time manufacturing jobs in other parts of the state increased. Prior to its decline, manufacturing in the region provided employment with living wages and benefits to workers who had only high school degrees or specialised vocational training. The increase in knowledge-based industries such as computers and electronics, telecommunications, multimedia, biosciences, banking and finance, environmental technology, tourism, business services and wholesale provide fewer jobs for less educated workers. In addition, the jobs available to employees without college or advanced degrees within this industry do not always provide a living wage.

In San Francisco the fastest growing segment of the knowledge-based industry is the multi-media industry. The bulk of the multi-media industry in San Francisco is located in a neighbourhood now called Multi-Media Gulch.[5] Mutli-Media Gulch houses about 200 to 300 multi-media companies that directly employ up to 2,000 people. In addition, the multi-media industry is indirectly responsible for an estimated 35,000 jobs in the South of Market area. In San Francisco, the multi-media industry is growing in terms of the number of companies at an estimated rate of 100 to 300 per cent per year and the annual job rate has grown around 65 per cent (Northern California Council for the Community, 1998). Although not all jobs in this industry require a college education or advanced degree, these jobs do require more education relative to jobs in Sonoma County as well as other parts of the region. As industries in San Francisco become more technologically sophisticated, a larger share of jobs in the county relative to the region requires a Bachelor's degree at minimum. Data also suggest that people with little education, immigrants with limited English language skills, and current and former welfare recipients will face unique challenges as there are fewer and fewer jobs for them in San Francisco.

The increase in knowledge-based industries in San Francisco and the dependence on global exports in Sonoma County has resulted in the influx of professionals to this region. The rapid entrance of large numbers of well-paid, educated professionals into both Sonoma and San Francisco Counties has had profound effects on incomes and on housing costs. In Sonoma County it is projected that between 2000 and 2020 the average household income will increase from $64,100 to $79,500 and in San Francisco will increase from $68,600 to $95,400 (Association of Bay Area Governments, 2001). Throughout San Francisco and Sonoma County housing prices have also become extremely inflated. In Sonoma County, demand for high-priced homes entices developers to build for the luxury market and not the middle-income or working-class markets. At the high end of the market, million-dollar homes are becoming frequent real estate items, with forty-seven such homes sold during the first ten months of 1999 alone, double the number from the previous year. The City of Santa Rosa now ranks fourth in housing price increases out of all Metropolitan Statistical Areas (MSAs) within the United States, experiencing a 78.6 per cent increase between 1995 and 2000 (Office of Federal Housing Enterprise Oversight, 2001). In October 2000 a mid-priced single-family three-bedroom home cost $330,000 in Sonoma County. According to the Association of Realtors, only 19 per cent of households in Sonoma County could afford to purchase a home (Chorneau, 2000). With housing costs increasing rapidly, and affordable housing in short supply, scarce rental units are taken by those who cannot afford to buy homes, locking out the poorest of the poor and decreasing rental vacancy rates to close to 2 per cent. Between 1989 and 1997 rents on two-bedroom units increased 19 per cent to $806 a month. By 2000, fair market rents for two-bedroom units had increased to $1,020 per month. During the same period three-bedroom rental units increased 30 per cent to $1,418 in 2001. In order to afford a three-bedroom rental unit in Sonoma County, a household would have to earn $56,720 a year (Twombly *et al.*, 2001). The rapid increase in housing costs coupled with a decline in the number of affordable units has resulted in severe housing shortages for residents on a moderate, low, and very low income.

While the overall median income continued to climb, those in the lower segments of the working class experienced the greatest housing stress. Extremely low-income households make up to 20 per cent of all households in the City of Santa Rosa. While these people are most at risk of becoming homeless, the crisis in affordable housing is now affecting the fortunes of middle-income people, such as teachers and city workers. As a result of the housing crisis many of the estimated 8,215 homeless people in Sonoma County are finding it difficult to get back on their feet even if they have full employment. With only 434 shelter beds, and an additional 120 beds in the winter, the City of Santa Rosa has had difficulty meeting even the partial needs of Sonoma County's homeless population. Fourteen per cent of those who sleep in Santa Rosa's National Guard armoury are working full time, and of the 150 people who stop at the Homeless Service Center in a day, twenty of them are often working, but not at sufficient wages to end their homelessness (Swartz, 1999).

Similarly, residents in San Francisco are in the midst of a severe housing shortage. As a result prices for owner- and renter-occupied units throughout the city are skyrocketing, placing a tremendous burden on families on very low, low, and moderate incomes who are already devoting a large share of their income to rent. According to the United States Department of Housing and Urban Development (HUD), a very low-income San Francisco County household consists of two persons with a combined income of $27,450 and a low-income household consists of two persons with a combined income of $36,800. According to HUD's definition of unaffordable housing (a household that pays more than 33 per cent of its annual income on housing), the maximum amount of rent that two-person very low-income and low-income households should pay in rent is $650 and $920 respectively (United States Department of Housing and Urban Development, 1998). However, average rents for two-bedroom units in San Francisco range from approximately $1,550 (South Bayshore, South Central, Mission District, Ingelside) to $2,250 (Buena Vista, Inner Sunset) to a high of $2,750 (Marina District, Russian Hill, Nob Hill, North Beach). Clearly, even in the most affordable neighbourhoods, very low and low-income renters seeking two-bedroom units on the open market cannot afford to pay the rent (Northern California Council for the Community, 1998).

Throughout the 1980s San Francisco lost 9,000 of its cheapest housing units to demolition or conversion (Coalition on Homelessness, 1999). Between 1990 and 1995 6,859 housing units were constructed. Of these newly constructed units, 34 per cent were constructed by non-profit housing development corporations which build permanent affordable housing. However, the need for quality affordable housing far outweighs the number of units constructed. For example, on average 76 applicants move into San Francisco public housing units on a monthly basis. However, this figure represents only 1.2 per cent of the total stock of San Francisco's public housing. The vacancy rate in public housing is as low as the current 1 per cent rate on the open market. The number of people waiting to move into public housing units in 1995 was 9,085 (24 per cent seniors, 22 per cent disabled, 54 per cent families). Currently there are approximately 15,000 people on the waiting list for public housing and it is estimated that between 1996 and 1998, approximately 4,000 units of public housing have been destroyed with only a fraction of those units being replaced. In addition, the number of households on the waiting list for Section 8 units (housing found on the open market in which the rent is federally subsidised) also remains high. The number of people waiting for Section 8 housing rose from 1,395 in 1995 to a current figure of 2,000 (Coalition on Homelessness, 1999). Many of these households wait an average of five years before finding Section 8 housing (Mayor's Office of Community Development, 1995). Further exacerbating the housing shortage is the loss of 380 out of 1,189 units on the private market whose occupancies are limited to very low and low-income residents. These private sector units were initially financed with public sector issued mortgage bonds and their respective use restrictions expired in 2000 (Northern California Council for the Community, 1998). Clearly, housing shortages in San Francisco occur in both the public and private sector. A shift in the economy, the growing shortage of affordable housing, and a simultaneous increase in poverty have resulted in a tremendous increase in homelessness in San Francisco over the last twenty years.

Although homeless advocates, service providers, city officials, and policy makers agree that it is extremely difficult to determine the exact number of homeless people in San Francisco, there are various estimates. According to Home Base there are between 6,000 and 8,000 homeless people in San Francisco on any given night and between 11,000 and 16,000 individuals experience at least one episode of homelessness annually (Home Base, 1997). While there is no precise count of homeless families, estimates suggest that families make up 25 to 30 per cent of the total local homeless population. As previously stated, the lack of affordable housing is particularly acute in San Francisco.

For those on very low incomes the situation has been exacerbated by the gentrification of 'skid row' which has resulted in the loss of large numbers of residential hotel and boarding house rooms. In addition, rents skyrocketed in the remaining Single Room Occupancy (SRO) hotels. During the 1980s rents increased 166 per cent for SRO hotel rooms and 183 per cent for studio apartments in San Francisco. For the average General Assistance (GA) recipient receiving $223.45 per month, securing housing, even in an SRO, is difficult. On average, rent for an SRO hotel room without a kitchen or a bathroom is $330 per month. In order to secure housing a GA recipient would need an additional $100 in benefits, which leaves no extra money for food. The problem is similarly bleak for impoverished families. In 1997 a family of three on Temporary Assistance to Needy Families (TANF) received $565 a month, not nearly enough to cover the cost of an apartment. This problem has been exacerbated by a new CalWorks regulation (a statewide component of the 1996 Federal Welfare Reform legislation) which eliminates the automatic cost of living adjustment previously applied to Aid to Families With Dependent Children (AFDC) grants. Thus, grant amounts under TANF will not increase to offset rent and other cost of living increases. Even the 14 to 22 per cent of the homeless population in the Bay Area who are employed, cannot afford a place to live.

Along with SROs, shelters are a part of the emergency services that are available to the homeless. Unfortunately, shelter space for individuals and for families is extremely scarce. In San Francisco there is a total of approximately 1,339 shelter beds available for families, victims of domestic violence, youth and single adults combined. These shelter beds are only enough for 15 per cent of the homeless population on any given night. In fact, all shelters in San Francisco are at or beyond capacity nightly. Two of the largest shelters in San Francisco operate on a lottery system for allotting the nightly shelter beds. For the 85 per cent of those not able to secure shelter space, the alternative is to sleep in squats, on the street, in cars, or to rely on family or friends (Coalition on Homelessness, 1998). Despite the influx of different industries and unique demographic shifts in population in Sonoma and San Francisco Counties, the result is the same: an increase in the cost of living, a lack of affordable housing, a widening gap between rich and poor, a rise in homelessness and a more mean-spirited approach to dealing with the homeless.

CRIMINALISING THE HOMELESS AND PUNITIVE POLICING STRATEGIES

The forming of activist strategic and tactical perceptions needs to be understood in the legacy of policing actions undertaken by city and county government in order to contain and/ or repress homeless populations that become politically active. Emphasis is now placed on urban development oriented towards the new 'fantasy cities' where shopping, finance, real estate and general services shape a public space of consumption over production, of private over public spaces. The new 'riskless risk' downtown entertainment complexes privilege the affluent over those who do not fit, such as the street homeless who must be systematically removed (Hannigan, 1998). In global cities such as New York and London (Sassen, 1991), and indeed in most other cities, urban poverty concentrations and income polarisations (Logan *et al.*, 1992) as well as homelessness (Marcuse, 1996; Talmadge Wright, 1997) have increased over the past twenty years. Initially, affluent class segments moved out of inner cities into the outer suburbs, increasing the emotional and geographic distances between themselves and citizens of different class backgrounds. More recently, as the affluent have moved back into the central city to take advantage of new urban redevelopment policies, poverty and homelessness are increasingly relegated to the margins of the city. Homeless persons in these reconfigured cities are often seen as dangerous threats that need to be eliminated from the sight of the newly affluent. It is this perception which has characterised public officials' preoccupation with removing homeless people from visible areas and containing them in distant industrial areas. A variety of tactics have been used to construct these 'defensive spaces' which exclude undesirable people from public view.

The passing of anti-camping and anti-panhandling regulations are not conspiratorial actions, rather they reflect a community's discomfort with having to visibly confront poverty on a daily basis. Given the power of the business community, homeless people often suffer from a lack of adequate representation in the contest between downtown businesses, consumers, and the general public. In both Sonoma and San Francisco Counties it is often downtown business merchants and their allies in neighbourhood homeowners' associations and city councils who raise objections to homeless people panhandling, camping, living on downtown streets, in city parks, and in local neighbourhoods. Whether a city's response to this pressure is benign or repressive has a great deal to do with who is the mayor, the political make-up of city councils, their constituencies, and the power of the dispossessed to exert effective political pressure. Responses to political advocacy by the homeless have a great deal to do with whether or not a city or county administrative body is under pressure to expand or restrict growth. In Sonoma County, for example, the County Board of Supervisors and local city officials are faced with a conflict between supporting pro-growth strategies that appeal to many incoming wealthy professionals, maintaining high property values, and a safe and orderly county, and developing sustainable growth plans which can support all of Sonoma County's population. In addition to sharing many of these concerns, San Franciscans are also preoccupied with maintaining their massive tourism industry. As a result many merchants and neighbourhood associations have advocated for more punitive uses of law enforcement against the homeless.

For most homeless advocates/activists in Sonoma County the Board of Supervisors were perceived as the main stumbling block to desired changes approved by the cities regarding homelessness. Throughout the decade Supervisors allocated less county money for non-profits, particularly those working with the homeless. In addition, the consolidation of the county government resulted in the neutralisation of advocates working within the administration by centralising power, reducing the autonomy of departments, and increasing the power of the county administrator. Ultimately, this consolidation combined with population shifts resulted in a more bureaucratic form of government in which local politicians invested in the community had less power. In addition, community groups and homeless activists within county government became disempowered. The inability of community groups to continue to effect social change was reflected in the failure of Sonoma County cities, politicians, and homeless advocates/activists to convince the County government to convert an abandoned Holiday Inn hotel into a homeless shelter in 1997 after a long and arduous struggle.

In San Francisco a hard-line attitude towards street homeless has increased over the last two decades despite changing mayoral administrations. The police often criminalise the homeless by taking law enforcement action against people for sleeping, eating, loitering, and eliminating themselves in public. For example, in 1988 there were massive sweeps of Golden Gate Park, Civic Center, and Cole Valley by the San Francisco Police Department. The following year Mayor Agnos arrested between 60 and 100 people living outside in the Civic Center area of downtown. As a result of these arrests a massive vigil (called 'Camp Agnos') was organised which ultimately resulted in even more arrests. Despite these arrests, Agnos, who also got federal funding to build supportive housing units, was depicted as soft on homelessness. He lost the re-election to Frank Jordan, his former Police Chief, who promised to rid the city of its homeless population.

Mayor Jordan began his term by starting an intense anti-panhandling campaign, culminating in the passage of an aggressive anti-panhandling law. In 1993 the Matrix Program began, which was presumably based on a continuum of care philosophy. According to the

Coalition on Homelessness, the Matrix Program was merely an institutionalised way to harass the homeless. In the first months of the Matrix Program more citations were given out for sleeping and camping in parks, drinking in public, obstructing the sidewalk, and sleeping in doorways, than in the previous five years combined. Hundreds of people's property was thrown away and Mayor Jordan tried to pass a law making it illegal to sit on the sidewalk. The Coalition on Homelessness helped defeat the passage of this law and helped individuals sue the City in Small Claims Court for stealing their shopping carts.

Disgusted by Jordan's vitriol towards the homeless and activism by the Coalition and Religious Witness With Homeless People, a sympathetic Board of Supervisors passed a resolution condemning Matrix. In August 1995 as part of his re-election strategy, Jordan orchestrated a multi-departmental sweep of Golden Gate Park. Partially as a result of police abuses against the homeless and Jordan's Matrix Program he lost the mayoral election to Willie Brown, who promised a more humane approach to ending homelessness.

In 1996 Mayor Brown took office promising to end Matrix. Unfortunately, Mayor Brown continued the anti-homeless discourse now pervasive in the City and continued criminalising the homeless by giving out citations for 'Quality of Life' infractions. In 1996 a total of 17,532 citations were given out, almost 3,700 more than in 1995, and 6,100 more than in 1994. Each citation carried a $76 fine and any unresolved or unpaid ticket went to warrant within twenty-one days. In addition, failure to pay the fine within the twenty-one-day period resulted in it being doubled. Under Mayor Brown's administration, the police were also instructed to confiscate and destroy homeless people's personal property if it was in public. Throughout this period one of the authors lived in the Haight Ashbury section of San Francisco, a neighbourhood inundated with homeless street people. Day after day she witnessed homeless people being harassed, ticketed and arrested for panhandling, public drunkenness and, most commonly, for simply being homeless. More often than not, homeless individuals were harassed by police officers for sitting on the sidewalks and offending people's sensibilities.

As Mayor Brown's anti-homeless rhetoric and policies increased, there was a concomitant shift in the way they were depicted in the local newspapers. Throughout the end of the 1990s the *San Francisco Chronicle* and the *San Francisco Examiner* had numerous articles disparaging the homeless. Periodic sweeps of Golden Gate Park and other neighbourhoods continued to occur whereby the homeless were demonised and physically removed from their makeshift homes. For example, on 4 November 1997 the headlines of the *San Francisco Chronicle* screamed 'Crackdown on Riffraff At Golden Gate Park'. Three days later there was an article in which Mayor Willie Brown said that he had arranged for the police to conduct night-time helicopter surveillance flights over Golden Gate Park using special heat-seeking equipment on loan from the Oakland Police Department to rid the park of homeless encampments. Despite outrage from homeless individuals and homeless advocates, the next day's headline read 'Police Chief defends Helicopter Searches'. Ultimately these searches never took place but only because the Oakland Police Department refused to lend the San Francisco Police Department their helicopter.

In Sonoma County, the most frequent trigger of homeless advocacy occurred when there were police actions taken against local encampments or in reaction to city council resolutions opposing panhandling. In 1990 the newly formed Sonoma County Homeless Union protested the US Census count, and in the Fall protested police sweeps of Santa Rosa Creek encampments. In actions taken against the City of Santa Rosa, the Homeless Union forced the city to give camp dwellers advance notice, include the presence of a social worker, and to settle with four homeless persons who had their property unfairly taken by the city.

With the onset of the 1991 Gulf War, homeless protesters began to make alliances with anti-war protesters, increasing the pressure homeless people could exert on Santa Rosa's city council. The election of two progressive supporters on Santa Rosa's city council allowed for a short moratorium on police sweeps until the publication of a news article detailing homeless teenagers living under a bridge in Santa Rosa. As a result of this article in the *Press Democrat* police sweeps resumed. As stories about aggressive panhandling and criminality among the homeless continued, community members became increasingly angry. Community outrage culminated in a series of protests which led to pickets in front of the Mayor of Santa Rosa's house and a demand for the city to protect citizens from a 'threat to their security'. Reporting by the *Press Democrat* in which homeless people were represented as victims or as predators, combined with a consolidation of county government, resulted in punitive actions against the homeless. In 1999 the City of Santa Rosa passed an anti-panhandling ordinance which received little opposition from the activist community. Police sweeps of homeless encampments continue unabated, producing various levels of dissent among homeless persons and within the wider activist community.

In San Francisco, homeless activism also took place in response to the rousting of homeless encampments throughout the city. In addition, activism centred around the need for permanent housing. Religious Witness With Homeless People, the Coalition on Homelessness, and Food Not Bombs united in an attempt to secure housing for the homeless in the Presidio, a former military base that had recently been closed. After many years of struggle they ultimately lost the battle, with many of the housing units going to local professors and students who were also struggling to keep up with exorbitant rents. More recently, Lucasfilm was given permission to move into the Presidio, effectively ensuring that the homeless will never be housed there. Although the Coalition on Homelessness continues to advocate for the homeless in San Francisco, other groups have become less activist in their approach. Unfortunately, in both Sonoma County and San Francisco the quietude of the homeless activist community marks a radical difference between the late 1980s and early 1990s which seemed to contain a promise of building a homeless social movement among a coalition of disparate organisations.

ADVOCATE/ACTIVIST PERCEPTIONS OF STRATEGIES, TACTICS AND EFFECTIVE ORGANISING

Internal fragmentation

In the midst of increasing social and economic polarisation throughout the San Francisco Bay region and punitive policing strategies, homeless activist/advocate groups have experienced both internal fragmentation in struggles over securing funding, and external fragmentation produced by difficulties in crossing organisational boundaries. Homeless advocacy is difficult when frequent changes are made in funding priorities, city policing strategies, or declines in housing availability. In addition, collaboration between groups is often hindered by ideological differences underpinning particular types of organising strategies.

Internal fragmentation common to many homeless advocacy groups was exacerbated by an endless preoccupation with securing funding. Advocates frequently mentioned how competition over funding created divisions within and between groups throughout the Bay Area. Advocates within organisations did not always have a unified strategy for increasing funding and spent inordinate amounts of time strategising on how to get more. In addition, the constant shifting of priorities among funding agencies often caused internal fragmentation among advocates who had to tailor their services around ever-changing funding priorities. Many advocates expressed concern over having to spend the bulk of their time writing grants and reconfiguring their services as a means of obtaining those grants. The constant pressure to secure funding and the political manoeuvring necessary to keep it often resulted in tension among co-workers and caused burnout among some activists. According to one advocate:

> You have to spend the first years making goddamn sure that you let all the funders know that you don't need their money. You need to say this is what we are going to do, do you want to support it? Otherwise funders will tell you what to do, usually in response to something that their board of directors said or responding to a mission statement that they came up with at some fancy ass fuckin' retreat in Marin about what they want to see happen in the homeless programs. It's a real challenge not to shift your agenda when someone is waving these ten thousand dollar checks in your face. You have to come up with a structure that says we don't want to be like this, we don't want to be poverty pimps, otherwise you spend all your time arguing over how much compromise is acceptable and how important the money is to your organization's survival.

The competition for funding and the ability of funders to set the advocates' agenda is extremely frustrating and can sometimes internally fragment a group. Indeed, Lyon-Callo (1998) demonstrates how funding patterns by institutional lenders, foundations, and federal and state agencies work to suppress activism on the part of providers and homeless persons through the indirect regulation of professional behaviour. The concern over losing funding caused some advocates in our sample to limit their militancy and alter their mobilisation strategies. For example when asked about the impact of funding agencies on their political activism and how they run their organisation, one respondent stated:

> We are very aware of how funders see us. We constantly ask ourselves will I look good? Will this agency look good? Will a particular strategy help our public relations and our fund raising? Will an action affect our government sources of funding? How will this affect the program I'm running, the program I believe in? Does this action seem too threatening? Should we be less antagonistic and more accommodating in our activism? We all get sick of asking ourselves these questions and constantly arguing over what the best answer is. It is very draining on all of us.

Clearly, the constant drive to secure funding and to maintain the integrity of the social movement organisation is exhausting, demoralising, and can cause intense friction among group members.

External fragmentation

Unfortunately, the struggle to obtain funding also contributes to external fragmentation. Advocacy groups often feel they are vying for a limited amount of resources and are put in direct competition with other homeless social movement organisations. One activist put it succinctly when she commented on the unwillingness among homeless advocacy groups to share economic resources:

> They're not going to discuss how to increase the pie. They're so busy getting their small piece of the pie, and wheeling and dealing with each other.

External fragmentation resulting from differing philosophical orientations about the causes and consequences of homelessness and appropriate organising tactics also severely hinders the ability of groups to work together. Homeless activists conceptualise homelessness from a variety of perspectives, some of which are in direct opposition to one another. How a homeless SMO 'frames' the issues directly impacts their goals and the types of organising they engage in (Cress and Snow, 2000).

In our research, how an advocacy group 'framed' a particular issue directly impacted their social movement activity. We found that in both San Francisco and Sonoma Counties, styles of organising were a direct result of how homelessness was 'framed', In fact, disagreements between various Bay Area advocacy groups were often a result of different organising techniques which in turn reflected the groups' ideological underpinnings. Whether or not homeless people were included in the organisational structure of an SMO, and how much input they had into setting the collectivity's agenda, was a major source of tension between various advocacy groups, particularly in San Francisco. Similarly, coalition building was only possible among groups who shared similar approaches to dealing with homelessness. For example, in San Francisco, groups like The Coalition on Homelessness, Food Not Bombs, and Homes Not Jails often organised together because they shared a common bottom-up approach to organising, in which homeless individuals set the political agenda. As one homeless advocate stated:

> I think that the different methodologies of the organizers determine a lot about what the issues are that are being organized around . . . how does a small group of people decide for a community what the priority issues are? By having a couple of people from that community? That's not good enough. . . . We don't sit in a fuckin' room and decide what the issue is that we are going to organize around. We do outreach, and once a week for eleven years we've gone into a fuckin' shelter and talked to people and at minimum five to ten times a week we do street outreaches and we talk to people. And then a minimum of twice a month each of our projects has work group meetings and gets input from people. That input is what drives our priorities. Not our staff meetings, not our Board of Directors, not something I came up with. That input is what determines what the priority issues are.

However, even among more radical groups that shared a bottom-up approach and engaged in direct action there is dissonance when homeless individuals are not actively involved in the organising activity. One homeless activist who had worked extensively with Food Not Bombs commented:

> Food Not Bombs has used homeless people far more often than they have served homeless people in terms of a political agenda or a change to the system. Primarily, because Food Not Bombs wanted to have a police response to whatever they did and they did not actively do things that would serve homeless people just so they could have a police response against them which they could publicize.

Nonetheless, there was collaboration between this advocate and Food Not Bombs. Ultimately, organisations that shared a similar political philosophy and utilised similar organising tactics frequently overcame their differences to mobilise for the homeless. The importance of overcoming differences and working together was made clear by an advocate who said:

> If you're going to work together you're going to fight at some point or you're full of shit. If you're passionate about an issue and you're working together on an issue and you have two different organizations with two different philosophies, two different styles, you're going to have to fight at some point. But you should just accept that and then you go on. Some people are more sensitive about that shit, and maybe I should be, but conflict often brings consensus.

Although this activist recognises the inevitability of conflict, particularly among groups with different approaches to homelessness, collaboration was extremely difficult and less common between groups who did not include the homeless in their organising strategies than groups that did.

Clearly, not all homeless advocacy groups in the San Francisco Bay Area include the homeless in their decision-making process. In fact, in San Francisco only a few groups encouraged active, ongoing participation among the homeless. In Sonoma County only the Homeless Union systematically involves homeless people in their advocacy. Whether or not to include the homeless in movement organisations was a source of intense conflict between various groups. Some advocacy groups, particularly the ones that were started by homeless individuals, believed that the homeless must be an integral part of their organisation. For example, one San Francisco advocate stated:

> Direct action only works if it comes from the bottom up. It only works if the leadership is coming from that group, and is not led by a housed person who has never been homeless. It doesn't work if the leadership is ego driven and makes decisions for the group without coming

back to the group and asking, hey look, this is what we wanted, this is what we laid out, here is the response, we need to respond to that, what do you think?

Another advocate from the same SMO also spoke of the importance of incorporating homeless individuals into the organisational structure of the group:

Our Board is set up much differently than most other Boards. In general we are a bottom-up organization, where we don't have the Board coming down and setting the policies. We are accountable to the homeless folks of San Francisco. That is why every project has an outreach component where people are going out and checking out what's going on and then including homeless people in creating change.

Some activists were extremely angered by the marginalisation of the homeless by particular advocacy groups. When asked about the lack of participation of the homeless at Home Base one advocate crystallised this anger when he said:

They wouldn't know a homeless person unless they ran over them in their Lexus!

Many activists questioned the legitimacy of defining an advocacy group as a direct action organisation without the active participation of the homeless. One advocate reflected this sentiment in his comments about Religious Witness when he said:

They call themselves a direct action group and they do direct action stuff except they don't have homeless people there. They are all religious people and they are all housed. So, it sort of takes away their claim of being a direct action group.

Finally, one activist put it succinctly when he said:

The success of Homes Not Jails, in my mind, has always been because of the active participation of homeless people.

Not all advocates share this philosophy. Some groups were adamantly against including the homeless in their advocacy work. Some advocates believed that the homeless were too difficult to organise because of their transience. For example one Sonoma County advocate said:

The trouble with homeless leadership is it is very transient. Somebody will get an incredible idea and get people all interested in doing it and then they disappear.

Other advocates simply felt they could fight for the needs of the homeless without working directly with them. When asked why they didn't include homeless people in their organisational structure one San Francisco advocate replied:

We do not include the homeless in our planning. We try to include the homeless in our events. We see what is happening to homeless people and we try to respond to that. We do not claim to be a group of homeless people. We are an advocacy group working on the concerns of the homeless. But we don't try and claim that we are a group that represents the homeless. The Coalition on Homelessness does so their constituency really are the homeless. We have a different approach.

Another advocate from the same organisation similarly stated:

From the beginning we created a steering committee which consists of interfaith leaders as well as representatives from the service providing community, from housing, from other organizations that are concerned with issues that affect the homeless. We have tried on a very limited basis to include the homeless at the planning level. It's not always worked out. It is very difficult. So we don't have homeless people on our steering committee, they are not involved in our planning, but they are involved in our events.

The decision of whether or not to fully integrate homeless people into an SMO caused a lot of friction between different advocacy groups, particularly in San Francisco.[6] In general, members of bottom-up organisations that did include the homeless in their organisational structure had significant ideological differences with their counterparts who did not.

Nonetheless, many of these disparate movement organisations did in fact work together on several occasions. Cooperation between bottom-up SMOs and more mainstream groups was more likely when the mainstream organisations included homeless people in their political actions. More important than whether or not an SMO included the homeless was the strategies and tactics they utilised to promote social change. Groups using confrontational direct action tactics were often at odds with groups trying to work within the system.

The amount of external fragmentation that occurred over the use of different mobilisation techniques was profound. Members of radical direct action organisations often clashed with folks from groups with more mainstream approaches. Individuals working for charitable organisations were often defined by both direct action activists and mainstream advocates as poverty pimps. Ironically, most of the respondents in our sample articulated the necessity of using a variety of movement strategies to achieve social justice. The following comment by one activist reflects the sentiment of a majority of our sample:

> I believe that everything works. We need direct action, legislation, lobbying, negotiation, and confrontation. I think that everything is important in trying to create social change.

Nonetheless, most advocates in San Francisco were extremely critical of movement strategies that differed from their own. Oftentimes discussions with advocates over the effectiveness of different movement activity became highly charged. The following quote reflects the difficulty some advocates have with confrontational organising tactics:

> I am sympathetic to Food Not Bombs, but I have seen a tone which I think is not effective. You can be nonviolent and strong without being verbally abusive to the police and so forth. I saw so much of that in Food Not Bombs.

Another advocate expressed similar concerns about the aggressive style of the Coalition on Homelessness:

> The Coalition wants to impress everybody that they are really angry. You know, so they yell at the Board of Supervisors and tell them how their houses are going to get blown up and everything else, you know right at City Hall. They shout at people and threaten them. They alienate people in a lot of ways. We don't do that. We might be equally angry but we don't feel that's our way of communicating.

One advocate articulated the limitations of direct action when he said:

> Well I think direct action polarizes communities and tends to make it more difficult to have bridge discussions. They are usually less willing to compromise, they are not very flexible, there is an unwillingness to brainstorm a solution. I think that is a liability of direct action. Direct action opens up windows so that people are forced to think about things they don't want to look at. But unless you get to negotiations all of the direct actions are useless. Negotiations have to happen, I mean that is the goal.

However, for direct action advocates the use of negotiation and compromise to effect social change is counterproductive and paternalistic. One advocate captured this view when he stated:

> The biggest disadvantage of negotiation, is the art of compromise. And for many people who are homeless compromise doesn't cut it because life on the street is very cut and dry. When you start negotiating with others who are deciding your fate and they expect you to give a little bit, it is very difficult to negotiate. It is difficult to negotiate access to health care. When a person is sick, the last thing they want to hear is compromise. It is very difficult to negotiate access to drug treatment. 'Well, gee, yeah, I can't get into substance abuse, well I'm going to have to wait three months to get in.' Is that person supposed to compromise, why go through this bullshit?

Furthermore, direct action advocates who work directly with the homeless believe their social movement tactics are more effective. When asked about alternative strategies, they also often respond with criticism:

> We do direct action by reclaiming public spaces, whether it's in plazas or parks or city squares, and bringing the issues of homelessness, poverty, and economic inequality into public arenas that are extremely visible. And also by confronting what we believe are the reasons for homelessness, namely capitalism and violence in society. Negotiating with the City doesn't work. Look what happened with Matrix? There was lots of negotiating with City Hall and even though the Board of Supervisors passed a resolution banning Matrix it continued under Jordan. When Willie Brown came to power he didn't call it Matrix but he continued to harass and criminalize the homeless. All negotiating does is put a band-aid on the problem and make the do gooders feel better. Radical social change is what we need, not a bunch of middle-class people trying to reform a corrupt system.

When asked about his frustration over how more mainstream advocacy groups are trying to eradicate homelessness, one direct action activist stated:

> The last fuckin' thing poor people need is another goddamn leader. They got more leaders and more lawyers and more academics talking on their behalf and talking about them and nobody is listening to a goddamn fuckin' thing they have to say.

Despite the vitriolic discourse that exists among San Francisco homeless advocates,[7] we found that more often than not, they transcended their ideological differences and frequently supported each other's movement activity in their desire to eliminate homelessness. This was particularly true for protest activity which required a large number of participants. So, despite the lack of daily cooperation between particular SMOs, advocates did mobilise together in crisis situations. The presence of direct action groups, service providers, religious advocates, and other more mainstream movement organisations in San Francisco resulted in more diverse types of political action than in Sonoma County, where the majority of advocates engaged in conciliatory forms of social movement activity. As a result, San Francisco has a more vibrant activist community and is fighting against homelessness on more fronts.

DEVELOPING REGIONAL ACTIVIST NETWORKS

Given the problems with internal and external fragmentation and the philosophical differences that underlie particular types of activism within counties, is it possible to develop activist networks across geographical boundaries? Is it realistic for advocates to reach out to activists in other counties and engage in political activity beyond the local level that could potentially promote regional homeless movement coalitions in the Bay Area?

A majority of network participation among the homeless advocates we interviewed was locally based. Many activists were unaware of the political activity occurring in other parts of the Bay Area. Advocates told us that regional networking was difficult given their preoccupation with the concrete struggles of obtaining funding, maintaining services, providing food and shelter, and impacting local policy. Most advocacy groups were constantly engaged in the struggle to obtain funding and stay afloat.[8] In addition, advocates were totally overworked. Direct action advocates were overwhelmed with preventing the criminalisation and harassment of the street homeless while simultaneously finding them suitable housing. Service provider/advocates were overburdened with trying to provide food, shelter, substance abuse treatment, health care, and mental health services for their clients. Policy oriented organisations were busy lobbying local, state, and federal agencies to implement substantive changes in the law. The overwhelming nature of homeless advocacy was a common theme that emerged during our research and is evident in the following statement made by a Sonoma County provider/advocate:

> Folks have probably not heard of us, and it's OK. We do what we do very well in our little community. We have some connections elsewhere, but

> we're so busy doing our work that we don't take a long time to network with folks elsewhere.

This sentiment was echoed by a direct action activist from San Francisco who said:

> In reality people are so busy and they have so much to do in their own city that it's really hard to think beyond your own city.

According to Carroll and Ratner (1996) in their study of cross-movement activism, activists who are in constant contact with other social justice oriented groups are more likely to 'frame' their issues in the context of social structural inequality and the globalisation of capital. The critique of social injustice, rather than a focus on particular issues, unites these groups in their struggle against inequality. Although many advocates/activists in our study understand the macroeconomic context of homelessness and the structural causes of poverty, their immediate practical interests and geographic location work against forming any regional coalitions. In addition, the same ideological differences that give rise to different movement activity and cause external fragmentation within regional boundaries, are exacerbated across counties. When asked about the possibility of regional cooperation, one direct action advocate summarised his scepticism when he said:

> In order to be really effective they would have to organize and be able to bring to the table, and to hearings, and to regional meetings, a hell of a lot of homeless people. And I don't think there is any organizer who is willing to do that. And if you don't do that then you're just jerkin' off yourself and other homeless people.

For direct action advocates, who insist on including the homeless in their movement activity, transportation was also a hindrance to creating regional coalitions. Many advocates mentioned the lack of available and affordable public transportation between counties.

In addition, the ability to overcome class and race differences across geographic boundaries is also difficult. For SMOs that do include the economically disenfranchised in their movement activity there are concerns over how their constituents will be treated by advocates in other cities. For example:

> I think that social economic status and racial separation determine almost everything in our society. So I think that there is a difficulty when you're crossing people who are professional people who have professional jobs with people at the bottom of the economic ladder who are homeless people, who have no status, who have no money, who have little education, who probably use drugs. You know they just talk a whole different language.

Although these racial ethnic and class differences exist within counties, they are exacerbated when homeless and non-homeless activists who are unfamiliar with each other come together. In cities like San Francisco or Santa Rosa the homeless activist community is connected enough so most of the players know each other. Although there are intense disagreements and class and race cleavages, activists who are familiar with one another are better able to see past these cleavages and work collectively. For strangers, these types of differences are often extremely difficult to overcome.

In addition to demographic differences, another major impediment to regional activism is that the forms of local government and the structural causes of homelessness vary dramatically in each city. As one activist succinctly put it:

> I just don't think it is a viable approach because the political situation, the demographics, the structural underpinnings for homelessness, all differ from city to city. The form of government differs. Like in Oakland you've got a city manager and Jerry Brown[9] is going to be changing that. So, you can't compare it to San Francisco which has a totally different type of government. The same with other cities in the region. Because San Francisco, for instance, is a city and a county. You've got other cities which are members of counties. Over in the East Bay part of the major cities are part of one county. Services are different. The

> fragmented service delivery system, regardless of whether it's housing, substance abuse, or jobs, is fragmented because you got different cities with different constituencies and different politics and demographics all trying to respond to a county-wide effort and it just doesn't work. You're not going to have a homeless movement. You're not going to have a large homeless collaborative movement. What you're going to have is very local groups trying to establish their own local needs and trying to effect the system in such a way as to have those needs met. And to collaborate with others outside that jurisdiction is not going to influence that.

What seems clear is that regional networks within the Bay Area are not very well developed. Groups that have collaborated across geographical boundaries are often primarily concerned with national, not local, causes. Sometimes affordable housing people network with other affordable housing people. Community development people network with other community development people and service providers with other service providers. Overall, activists linking with other activists outside of their particular county is rare, although there has been some collaboration between activists in Berkeley and Oakland (Alameda County) and San Francisco.

In addition, service providers and shelter advocates are limited in their impact on refocusing the debate away from human services to such issues as land use, distorted housing and wage markets, and overall social inequality. Provider advocates who have attempted to do this often become isolated and are threatened with the loss of their funding. As a result, many homeless advocates in the Bay Area lack sustained resources, remain scattered, and have been unsuccessful at building broad social movement coalitions

Unfortunately, the possibility of working collectively, on a regional basis, to end homelessness is dim. However, changes in the social and political climate could precipitate an end to geographic isolation. Activists do come together on occasion to protest specific city or county policies. More collective action is possible in the future if activists can unite over such seemingly diverse issues as affordable housing and gentrification, welfare reform, lack of public transportation, workers' rights, inadequate health care, unimpeded regional growth, immigrants' rights, and environmental degradation. Uniting all of the SMOs fighting for these various causes is their underlying commitment to social justice and an end to inequality. The embrace of a broader political agenda within the context global capital and its local impact might allow for coalition development with disparate groups throughout the Bay Area.

CONCLUSION

Over the last decade, communities in the San Francisco Bay Area have all experienced accelerated rates of gentrification, social exclusion, and policing strategies that expel the urban poor from desirable public spaces. Since the 1990s homeless activism has declined throughout the Bay Area as a result of internal and external fragmentation, an increase in punitive policing strategies, and the criminalisation of the homeless. Despite rapid economic growth since the mid 1980s poverty still persists and is worsening in selected regions of the Bay Area. Unfortunately, the influx of wealthy patrons who have made their fortunes in newly burgeoning industries has produced not only higher housing costs and a lack of affordable rental space but also support for corporate retail shops. These macroeconomic shifts have created a wider gap between the rich and poor and have permanently altered the cultural fabric of the Bay Area. In Sonoma County the inflation of housing costs brought on by the rapid influx of well-paid professionals, able to bid up the costs of homes, has created a crisis not only for homeless persons but increasingly for the middle class. For decades San Francisco was considered a hotbed of political activism, artistic expression, and diversity. As rents and salaries have skyrocketed, many political activists, artists, and people of colour have been forced to leave the city. Similarly, as the cost of retail space continues to escalate local business people have been priced out of business and cannibalised by chain stores. Previously interesting neighbourhoods like Haight Ashbury, the Inner Sunset, and the Mission District are becoming bland reflections of corporatist culture. Sadly, the increased cost of living in San Francisco has meant the inability to support the rich diversity of protest activity, artistic development, and immigrant culture that once made the city famous.

The good news is that local political and economic policies are not uncontested. While the homeless activist community has been brought to a defensive standstill throughout the Bay Area, other segments of the community are mobilising. In San Francisco the fight over live–work lofts and over gentrification and redevelopment has led to the creation of new SMOs. For example, the Mission Anti-Displacement Coalition turned out over a thousand people for a neighbourhood rally to protest the displacement of Latino families and store owners by wealthier non-Latino residents. Immigrants' rights groups, tenants' rights groups, welfare reform groups and others have also tried to stem the tide of displacement. The recent overturning of the pro-development San Francisco County Board of Supervisors signified a new day for low-income housing and homeless advocates in the city. In Oakland a new coalition called 'Jerry Watch' has sprung up to defend the interests of the poor, and is making links with other groups across the Bay. In Sonoma County, the newly formed Sonoma County Housing Advocacy Group (SCHAG) has mobilised on several fronts. Rather than focusing exclusively on housing and homelessness, SCHAG has made alliances with church and labour activists, environmental activists and anti-development coalitions to promote sustainable growth and low-income housing throughout the county.

Our research reveals the necessity for a multi-dimensional, multi-coalition approach to eradicating poverty and homelessness. Knowing when to employ particular tactics for a certain strategic outcome is difficult in the best of times. It is made more difficult by the problems inherent in homeless organising (Rosenthal, 1996). Homeless activism, therefore, should be understood not merely as disruptive street protests, squatting or camping, but rather as a more complicated dance with city, county and state authorities involving both direct and non-direct action. Activists must develop cross-regional coalitions, utilise a variety of movement tactics, overcome their ideological differences, and move beyond the traditional single-issue focus if they want to propel urban development policy in a more progressive direction and ultimately eradicate economic and social inequality.

ACKNOWLEDGEMENTS

Both authors contributed equally to the conceptualisation, design, and implementation of the research project and to the writing of this chapter. The order in which the authors' names appears does not imply a difference in workload. We would like to acknowledge the assistance of Loyola University, Chicago and the Bay Area Homelessness Program at San Francisco State University for their support of this research.

NOTES

1 While Sonoma County remains relatively homogenous in racial ethnic make-up with whites comprising over 80 per cent of the population, those identifying themselves as Latino now make up 17.8 per cent of the county's population. African Americans and Asians constitute only 1.3 and 3.0 per cent of the population respectively (US Bureau of the Census, 2000).

2 All economic and demographic projections made by ABAG and the Northern California Council for the Community may turn out to be incorrect since Northern California entered into an unexpected recession in the last several months of 2001. This recession has hit the software and high-tech knowledge based industries particularly hard. Thousands of Bay Area

employees have been laid off, wages have been frozen, and numerous start-up companies have gone out of business. Further exacerbating this recession is the recent decline in regional tourism. After the September 11th attack on the World Trade Center many Americans and foreign visitors cancelled their travel plans because they were afraid to fly.

3 It is however expected that the number of foreign-born individuals living in San Francisco is higher than the 1990 Census due to under-counts of both documented and undocumented residents. For example, a post-1990 study found that San Francisco's Mission District, the neighbourhood with the largest percentage of Latino and Latin American residents, many of whom are undocumented, was undercounted by at least 19 per cent (Northern California Council for the Community, 1998).

4 The San Francisco County's population in 1990 was proportionally 46.6 per cent white, 28.4 per cent Asian–Pacific Islander, 13.9 per cent Latino, 10.6 per cent African American, and 0.36 per cent Native American. Currently it is estimated that San Francisco's population is 38.2 per cent white, 34.6 per cent Asian–Pacific Islander, 16.8 per cent Latino, 9.8 per cent African American, and 0.3 per cent Native American. Data indicate that between 1990 and 1998 the fastest growing groups were Asian–Pacific Islanders (25 per cent) and Latinos (24.1 per cent) respectively with a decrease in population among whites (–15.8 per cent), African Americans (–4.4 per cent), and Native Americans (–20.7 per cent).

5 Multi-Media Gulch is an area bounded on the north by Harrison Street, on the east by the Embarcadero, on the south by China Basin, and on the west by 7th Street.

6 In Sonoma County most of the homeless advocacy is organised and executed by service providers, many of whom are extremely concerned with losing state, local, and federal funding. In San Francisco, there are several homeless social movement organisations that do not rely on traditional sources of funding and are therefore more radical in their political activism. In all five counties we found that service provider advocates were systematically less likely to include the homeless in their activism than more radical direct action movement organisations. Subsequently, there is less external fragmentation between homeless advocates in Sonoma County since most of them don't include the homeless in their organisational hierarchies.

7 Similar to the discussion of including the homeless in movement activity, the discourse around effective organising is less divisive in Sonoma than in San Francisco County. In Sonoma, effectiveness is determined by whether or not services are delivered to homeless people and shelter beds are made available, not by the type of action utilised by movement advocates.

8 A few of the more radical direct action groups stopped applying for federal or private funding altogether. These advocates said they got tired of funders setting their political agenda and telling them what types of actions they could engage in. Despite the sentiment among most advocates in our study that funding agencies engaged in a pernicious form of social control, they nonetheless were forced to rely on them to stay afloat.

9 Jerry Brown is the Mayor of Oakland and is unrelated to Willie Brown, the Mayor of San Francisco.

11 GRAEME EVANS AND JO FOORD

Shaping the Cultural Landscape: Local Regeneration Effects

INTRODUCTION

Culture has a wide variety of meanings and values. To invoke 'culture' is to open up a Pandora's Box of different symbolic representations and complex personal, national and group identities (Eagleton, 2000). In the discourse that follows some groups and individuals have more ability to shape cultural representation and understanding than others. Cultural institutions, creative industries and different communities (of interest and/or of place) constantly struggle to construct and claim 'their' interpretation of culture. Increasingly the power to name, construct meaning and exert control over the dissemination of ideas about culture underpins divisions within society (Stevenson, 2001), and as Harvey observed: 'The ability to "name" things, acts and ideas – is a source of power' (1989: 388).

This chapter sets out to illustrate the current tensions over the meanings and values of culture as it emerges in local regeneration in terms of both the physical and symbolic landscape. Within the political economy of regeneration, 'culture' now has popular appeal and is pivotal; culture is seen as a new resource that, if correctly mobilised, could maximise the potential of local areas and neighbourhoods as well as whole cities (Smith, 1998a; Landry, 2000). This notion of culture as 'asset', however, obscures the contestation of cultural meanings and values that is played out within the regeneration process. Two sets of meanings are particularly prevalent in current urban policy and regeneration discourses. One set of meanings and values suggests culture is an essential element of everyday life and identity and is bound up in the processes of 'making, doing and enjoying' cultural activity – including the rites and rituals of everyday living, as well as the enjoyment of entertainment and the participation in cultural events. In this interpretation an active engagement in cultural activity improves or even represents a large element of the *quality of life*; encourages individuals and communities to be active and promotes engaged contributions to their local community and to society as a whole.

The second interpretation of culture is as an integral and substantive part of present day city economies. With the loss of other areas of economic activity, particularly manufacturing industry, various cultural industries have come to be seen as a saviour for many so-called post-industrial cities in Europe (Bianchini and Parkinson, 1993; Verwijnen and Lehtovuori, 1999), as well as in 'new world' cities. The interconnected, networked nature of this fast growing sector provides the antidote to obsolete rigid economies. Culture and cultural activity in this context contributes to local prosperity and economic growth, provides jobs with a high multiplier effect, and is identified with enhanced city status as place-based cultural capital. Underlying this economic development and exploitation of culture is not only the earning and jobs potential of the creative industries and their spin-offs (such as cultural tourism), but also a widespread assumption that future economies will be based on the trading of symbolic meanings underpinned by cultural knowledge (Leadbeater, 2000; Scott, 2001).

Through an exploration of a local case study in east London, this chapter explores the local consequences of this dualistic approach to culture within local regeneration. Furthermore it suggests that this dualism creates a lasting and sometimes detrimental impact on cultural development, indeed that in certain circumstances the power to define cultural meaning recreates social – and spatial – divisions in the name of regeneration. The chapter concludes by suggesting that this outcome of regeneration practice works against the current rhetoric of a positive and creative role of 'culture' as a catalyst for social cohesion, multiculturalism and sustainable regeneration.

CULTURAL LANDSCAPES

> Place and culture are persistently intertwined with one another, for any given place . . . is always a locus of dense human inter-relationships (out of which culture in part grows), and culture is a phenomenon that tends to have intensely local characteristics thereby helping to differentiate places from one another.
>
> (Scott, 2000: 30)

Particular cultural landscapes are created out of the complex interactions between cultural activity and communities, be they communities of identifiable social or interest groups and/or communities of place. Over time expressions of diverse cultures shape the senses of place attached to particular locations (Massey, 1994). In an increasingly transitory and globalised world, such senses of places are more likely to be cross-cutting and hybrid rather than singular, not least in the dense but atomised inner neighbourhoods of cities. Cultural landscapes can be read as a combination of, on the one hand, the role and value that cultural activity is given in communities and, on the other hand, the type of formal and informal spaces where cultural participation and production occurs. Thus, cultural landscapes express the identity of (different) communities, articulating their relational place in the world (spatially and culturally), and help define the vivacity of life in a particular shared location.

The 'regeneration' of urban communities that are assessed to be in physical, economic and social 'decline' has been the subject of public and private intervention and investment programmes for over twenty years in North American and European cities. Many of these programmes, faced with derelict land and buildings, poor environmental quality and lack of apparent 'social cohesion', have actively sought to re-shape the cultural landscape of central, inner city and waterfront sites. Yet, it is now evident that, in this process, the needs of those living within or alongside these areas have not been met by the type of flagship property development and cultural projects which were created as part of the first and second wave of regeneration strategies (Bianchini, 1990; Harvey, 1993; Robins, 1996; Evans, 2001) – a failure of the promised trickle-down effects: 'doubt has been cast as to what extent urban redevelopment policies have benefited the local community, which is what local culture policy should be about, rather than the construction industry. The urban high-fliers may have not gained as much as they may appear to have' (Kawashima, 1997: 38). Such projects, including the high-profile *Grands Projets Culturel,* may well have improved the central zones associated with cultural tourism and new infrastructure (transport, offices), but have done little to reflect and support the active production of local (diverse) cultural landscapes, or even enhanced cultural production and creativity overall. Indeed many of the flagship cultural projects implemented within these cities have drawn upon the model of the international style (Evans, 2001) – displaying, celebrating and elevating the art, architecture, artefacts and renaissance values of western Europe.

In contrast, the cultural landscapes of the locations deemed to be in need of regenerating are more accurately represented as ones carved out through ongoing non-European migration and in which cultural diversity 'dominates' (*sic*). As a result of successive diasporas these locations often exhibit rich, if largely unrecognised, histories of cultural expression. Cultural experiences and practices of different social and ethnic groups cross-cut one another in everyday life. The cultural landscape here is reflected in adaptations of domestic and commercial buildings and in the re-use of symbolic or sacred sites – from Huguenot church to Methodist hall to synagogue to mosque; from gin palace to music hall, cinema to bingo hall; from factory and hospital to studios, apartments and offices; and from pleasure garden, 'green lung', to millennial theme park. However, as British-Asian architect Rajan Gujral comments from Southall, west London: 'Ethnic communities are a permanent part of the society in the major cities of the country. There is no mistaking the areas favoured by the various ethnic groups; the writing on the shops, the rhythm in the streets, the faces, the dress. But somehow the communities live in spite of their environment rather than shaping it' (Gujral, 1994: 7).

These culturally rich, though impoverished and abandoned, urban areas have for several centuries attracted others who have left their particular mark on the cultural landscape. Cheaper land and rents, the availability of large spaces and, in the case of east London, the proximity of residential and workshop space, have attracted high concentrations of creative artists working in the visual arts, crafts and as designer-makers (Foord, 1999; Evans, 2001; Wedd, 2001). Such concentrations both reinforce traditional 'cultural workshop' locations within cities and provide a creative milieu that has proved attractive to the new media industries and associated leisure lifestyles. The positive aspects of the 'rich mix' cultural landscape of central city fringe locations has been subsequently identified and promoted as an essential backdrop to promoting new creative urban economies (Worpole and Greenhalgh, 1999). However, the resulting pressure on premises, forcing increases in rent and the loss of cheap residential and studio space, is giving cause for concern as these particular areas in close proximity to city centres become victims of their own success as creative hothouse locations (Vision in Art, 2000) and as places of commercial, retail and residential gentrification.

Major regeneration development projects, notably waterfronts, industrial quarters and entertainment centres, are now ubiquitous sights from Singapore (*Global City for the Arts*) (Chang, 2000) to Sheffield (*Cultural Industries Quarter*), and from Manchester (O'Connor and Wynne, 1996) to Montreal (*Cité Multimedia*). In contrast, much of the attention now paid to areas where such property-led regeneration failed to make an impact through sustained improvements in jobs, inward investment and new cultural development, is directed geographically at the 'socially excluded'.[1] Poor and so-called excluded neighbourhoods have become a target for intervention and their cultural landscapes are now being re-shaped. In the most recent culture-led regeneration strategies (cf. Department for Culture, Media and Sport, 1998, 2001; Cultural Strategy Partnership, 2001) the dualistic approach to culture has been mobilised through identification of the potential of the arts in neighbourhood renewal (Shaw, 1999) and the expectation that creative industries can kick start local economic development. Consequently regeneration policy and investment attempts to neatly reconcile the social and economic objectives of area renewal through a form of cultural democracy and cultural citizenship (Department for Culture, Media and Sport, 1999; Stevenson, 2001):

> Culture can offer a unique combination of benefits to urban regeneration. It contributes to the quality of the urban environment and to the local economy, bringing life and cultural opportunities to local town centres. It offers an effective means of exploring communities' and individuals' histories, experiences and needs. New uses are being found for historic buildings and neighbourhoods: churches and industrial buildings, for example, finding new roles as cultural centres and work and living space. Creative (and sporting) activities can play a critical role in developing community and individual confidence, skills, health and fitness.
>
> (Cultural Strategy Partnership, 2001)

This approach to reshaping cultural landscapes is also consonant with Bennett's argument that culture itself should be thought of as 'inherently governmental', so that 'culture is used to refer to a set of practices for social management deployed to constitute autonomous populations as self-governing' (Bennett, 1995: 884; and see Barnett, 1999: 371). Cultural projects funded through regeneration and 'social inclusion' programmes are increasingly harnessed in attempts to create new economically viable communities of self-reliant, creative citizens (Evans and Foord, 1999). Indeed a particular form of local governmentality is increasingly evident, in which the individual is encouraged to take up opportunities to participate in mainstream society – through education, training, volunteer work – but primarily through paid work. However, as Stuart Hall writes on the liberal nation-state: 'Its universality depends on its capacity to hold at bay, or indeed to abolish to the sphere of private life, all those cultural particularities which threaten to undermine its universality' (2000: 44). Such governmentality (Foucault, 1991b) is an expression of a dispersed form of power, and it relies upon the modes, rationales and techniques of governance that seek to 'shape, guide or affect the conduct of some person or persons' (Gordon, 1991: 2). Its purpose is to 'act on the conduct of conduct', influencing the actions of others through processes which encourage the self-regulation of behaviours. This form of governance does not rely on coercion or consent, rather it uses manipulation. Cultural landscapes become subject to this manipulation through regeneration programmes and the incentives they selectively employ.

Despite the current attraction of such an approach to governance and its managerial propensities, Bennett's confidence in its effectiveness when applied to reshaping cultural landscapes is somewhat naive, particularly if one accepts the 'Protean character of cities and their resistance to top-down planning or prediction' (Worpole and Greenhalgh, 1999: 38). Indeed, urban regeneration (and urban policy) as a whole 'has moved away from the old planning idea that populations were merely demographic statistics with easily identifiable needs for units of accommodation, employment, leisure, transport links, and social welfare needs' (ibid.). Over time policy has had to adjust to the realities of the demands and shifts of global capital, geopolitics and the supply and demand for urban resources (for example, office space; residential development; retail and leisure facilities; transport and information infrastructure). Strategic thinking now seems to be giving way to the realisation that communities are made up of individuals, subcultures, interest groups and coalitions all of whom have different needs, wants and 'life-strategies' – many of which may overlap and conflict with one another. For example, the application of *regime theory* to urban governance and growth coalitions has evolved from its neo-Marxist elite and pluralist 'schools', with recent acceptance, particularly in Europe, that things are more complex; power is not solely one of control of resources, but also of diverse interests, leadership and implied power, making up various (dynamic) partnership groups (Stoker, 1995). Nevertheless, attempts to intervene creatively in complex cultural landscapes are still in danger of being reduced to formulaic techniques of managerialism (cf. Landry, 2000) following an instrumental form of governmentality. This is manifest for instance in the appropriation of the terms 'stakeholder' and 'partnership' in social policy and urban regeneration programmes – as if all partners are equal, consultation means that views of community interests actually carry weight, and that the origin of 'stakeholder' does *not* in fact refer to the banker (i.e. state funder) in a game of cards played for money.

However, the now advanced stage of urban regeneration continues to look for opportunities to shape the cultural landscape by adding 'cultural value' and 'cultural quality' to major development projects through urban design, animation, public art installations, and improvements in the quality of the public realm. Popular interventions on the local scale include landscaping, temporary and permanent art, and in the grand manner, the 'hard branding' identified with art museums – for example, the Guggenheim in Bilbao, Berlin, Las Vegas and their aspirants from Lyon and Liverpool to Rio, and the Tate 'family' of galleries. Thus the provision and protection of landmark cultural buildings, strong symbols of shared cultural understanding and open spaces for collective cultural participation, becomes critical in any city plan or regeneration strategy. These aspects of the cultural landscape embody the sense of 'community' through which the city can create some sort of order. As Ellmeier and Rasky see it: 'today the tasks of city planning also include compensating for differences and creating necessary community in order to allow the city to function at all. If inhomogeneity becomes visible, if the idea of the homogenous national or city culture is no longer tenable, then the city, the urban space, becomes important' (1998: 80).

How far the multicultural, pluralism and 'identity' debates have seriously impacted on urban cultural development and regeneration processes is hard to identify. But as Robins warns, urban regeneration 'reflects a more acceptable face of rationalism, and fails to come to terms with the emotional dimensions of urban culture' (1996: 8, and see 1993). For instance, in the tradition of ethnic quarters (*Little Italy, Little Germany*), the promotion through regeneration programmes of cultural districts such as *Banglatown* (east London), *Balti Triangle* (Birmingham), *Currytown* (Bradford) and the ubiquitous *China Towns*, as attractions for cultural tourists and conspicuous consumption, can, however, create commercially driven pastiche replicas of what originated as authentic urban communities. Their effect locally creates displacement; economic and physical gentrification; spatial divides in amenity and investment; and the transformation of mixed-use and places of cultural production to mono-zones of consumption. The irony of this is that the proponents and formulation of cultural strategies can reinforce these distributive and exclusionary effects, where power is itself unevenly distributed and creativity is defined and valued narrowly and economically. Even the toolkits advocated for developing the *Creative City* acknowledge that whilst cultural planning is 'strategically principled and tactically flexible', the solutions 'will be infinitely diverse . . . . Some merely require imagination, *others require confronting deep power structures and entrenched attitudes of mind*' (Landry, 2000: 271, our emphasis).

INTERMEDIARIES

To an extent the patchwork of urban funding programmes converges, or at least purports to conflate, the social and economic rationales for culture within urban regimes. Perhaps key questions to be asked, therefore, given the meanings and values being attached to culture within urban regeneration, are about the role of *intermediaries* in the generation of policy objectives. Where cultural policy is 'located' in terms of professional, institutional and artistic interests therefore becomes important in framing, and funding, cultural landscapes. Intermediaries in this case are those professionals and semi-professionals working in the interface between cultural activity (creative production and arts organisations) and the regeneration system (local/regional authorities, regeneration companies, development agencies, housing providers), including what McGuigan identifies amongst the 'Professional-Managerial Classes' as cultural intermediaries: 'those particular sections that are directly employed in practices of cultural mediation and consumer management' (1996: 39). The critical issue raised by the role of intermediaries is how they relate to and operate within the local communities which are supposed to be the 'beneficiaries' of such regeneration.

The notion of 'community' itself has presented a shifting array of imposed definitions, not least in the archetypal working-class districts of the East End. Lash and Urry (1994), revisiting Bell and Newby's (1976) analysis of place-based community, argue that their nostalgic notion of East End life has re-emerged, not least in policy and local political circles, as a vehicle for rooting individuals and societies in a climate of social and economic uncertainty. As the restructuring of economies intensifies in pace and national political, cultural and social institutions begin to be questioned, so local communities have come to be seen as the essential building blocks in the new urban 'sociations'. Aldous (1992) and others have reclaimed the concept of the 'urban village', and although this re-visioning of local communities is somewhat over-sentimentalised, it has proved attractive as a way of imagining the new communities of regenerated neighbourhoods in a traditional form. As such it lies not too far below the surface of many town centre, creative quarter and neighbourhood renewal projects devised and promoted by regeneration and cultural intermediaries and by contemporary social reformers.

Furthermore, the East End has been the subject of well-meaning intervention since the nineteenth century. As Weiner demonstrates, during the late 1800s there was a prevailing 'vision of cultural transformation as a palliative ... offered up as a solution to the social question' (1989: 45). *People's Palaces* were central to this idea of transformative intervention. These secular institutions were supposed to bring 'cultural advantages' to 'culturally deprived' inner-city areas and in particular to alleviate an assumed air of 'boredom' and lack of 'artistic grace'. One such institution, Queen's Hall on Mile End Road, conceived by Walter Besant, was built to bring the cultural sensibilities, including the architecture of the West End's 'Palaces of Delight', to the 'joyless' and 'ugly' East End (ibid.). Reactions to this imposed transformation of the cultural landscape suggest that East End communities played unwilling 'hosts' to such recreation and social reforms as they both ignored and denigrated existing everyday cultural activity, including the long-standing music hall tradition. However, this paternalistic project was to be short-lived, it 'lacked a coherent objective understood and shared by the public' (ibid.: 48), a familiar sentiment today (Willis, 1991) and a risk which any imposed, top-down cultural solution runs.

In a less direct way current policy and interventions have shaped understanding of the 'potential' of east London. There are two very different, though related, interpretations. One sees areas of socially excluded populations as a threat. Inclusion policy is therefore driven by a desire to minimise this threat. The other interpretation presents the same areas as an opportunity. Here the goal is to make use of 'wasted' resources. Both interpretations, however, have led their intermediary protagonists to propose policies that fuse cultural initiatives with urban regeneration programmes. Both advocate the remaking of local cultural landscapes (Evans and Foord, 2000b). For those who construct socially excluded populations as a threat, spatialised urban inequality highlights the increase in social isolation of those who are outside the organisations and networks of the key social institutions (primarily those of work and education). They are therefore not in the environments in which normative social values are created, negotiated and absorbed. Such exclusion therefore does not simply pose moral questions of social injustice, it fuels more deep-seated fears.

The architect and masterplanner Lord Rogers – a highly influential intermediary – referred to the inner suburban ring around London's prosperous core as 'a powder keg', during a series of public lectures on the future of cities (Rogers and Gumuchdjian, 1997). He therefore suggested, in a manner somewhat reminiscent of the nineteenth-century 'middle-class ratepayers' and Progressives in London (Pennybacker, 1989), that these areas are unpalatable and liable to bouts of social unrest. Yet unlike the Victorian middle classes a number of today's new (predominantly white, though not exclusively so) urban professionals no longer wish to escape the urban core (at least those pre- or post- or never-to-have-children). Instead this influential, if numerically relatively small, urban elite wish to 'reclaim' the core city for themselves as workplace and pleasure zone.

This 'rediscovery' of the city has been widely explored, as has the role of 'culture' as lifestyle, animation and enterprise within the urban renaissance (Zukin, 1988, 1995, 1996; *Urban Studies*, 1998; *European Urban and Regional Studies*, 1999). The 'vision' of a culture/consumption/enterprise-led urban renaissance has come to underpin many of the detailed strategies proposed in the New Labour Government's *Urban White Paper* (Department of the Environment, Transport and the Regions, 2000) in which the objective is to *reconnect* potentially threatening areas to the core city by effectively *colonising* them. New regeneration strategies are increasingly altering inner city cultural landscapes in order to facilitate this colonisation. Criticisms of such an urban cultural revanchism (Smith, 1996a) focus on its potential to intensify both the social and spatial divisions within the core/inner areas of east London, and the displacement and alienating effects that result.

Other commentators and intermediaries identify perceived *opportunities*, rather than threats, within inner city areas such as east London where socially excluded populations are 'to be found'. The focus here is on the invaluable potential of new forms of cultural economic activity and innovative product development (Porter, 1995), ways of living creatively (Landry and Bianchini, 1995; Landry, 2000), and capitalising on the 'rich mix' of culturally diverse communities (Worpole and Greenhalgh, 1999). In this interpretation culture is a resource. Porter (1995), drawing on evidence from North American cities, argues that there are 'true advantages' of inner city locations (and see Gratz and Mintz, 1998). These locations provide cheap sites for rent or development close to the overheating city core. They have local niche 'ethnic' markets that can be tapped for 'new' cultural practices and products and a reliable, culturally literate, inexpensive and flexible labour force. Following this argument which has gained popular political support in the UK, areas such as the East End have cultural assets and resources which would add to the city's competitive advantage.

Landry (2000) takes another view of inner city opportunities. He outlines the key components of a *creative city* – whilst alert to the dangers of reducing creative strategies to flagship arts projects and property-led developments, he presents the economic potential of culture in cities as dynamic and inclusive. He assumes that creativity has become more important to city economies and that nurturing it has become critical to economic success. His definition of creativity is as a *social process*, rather than *individual* instances of original achievement. Mobilising creativity is seen as more likely in environments when past conventions and social practices have already been dismantled. The deindustrialisation and derelict urban landscapes found, for example, in east London are not presented as problems but are rethought as opportunities: they are re-cast as arenas in which innovative creative opportunities can emerge. The disruption of old shared rhythms of life based on the routines of the workshop, factory or the office which deindustrialisation produced, are not lamented. Instead their loss is seen as an opportunity to challenge inherited assumptions and ways of working. So the disruption (destruction) of regular working hours (and recreational hours, for example, licensing, Sunday trading) and regulated working conditions paves the way for the introduction (creation) of new working practices, attitudes, motivations and behaviours. These new work disciplines inspire self-regulated and networked work organisation in a flexible labour market where the (self-) employee takes on the risks and responsibilities of employment in a climate of economic uncertainty and flexible production (Foord, 1999). East London is a ready stage upon which this *creative city* can be acted out, and for which its histories provide both a symbolic and physical foundation – *viz* successive waves of migrants/refugees; located outside of the old city walls and crafts guilds/entertainment licence controls; popular culture (music hall/gin palace); cheap studios and artists' colonies, workshop/sweatshop culture, and so on.

In an extrapolation of this argument,

Worpole and Greenhalgh (1999) suggest that the creative city potential of east London lies in the very intensity and diversity that Rogers sees as a threat. Cities, and especially those like London, are now home to diverse populations with a multiplicity of languages and lifestyles, behaviours and aspirations. Difference – whether superficial or substantial – is often more familiar than sameness. As Hall points out: 'cultural diversity belongs to a relatively small area of society, to the cities and the places of high migratory settlement . . . (but) the term "multicultural" has to include the enormous pluralisation of forms of life amongst the majority population themselves' (2000: 46). This 'rich mix', they argue, is a breeding ground for creativity. Individuals and enterprises are more likely to succeed if they tap into cultural difference and make use of its opportunities – tastes, practices, interactions. They also, as some now argue (Sassen and Roost, 1999), tap into a global, eclectic market for skills and cross-cultural consumption – cosmopolitan culture as a form of comparative advantage.

How far the future of cities will be influenced by the aspirations and responses of these local diverse communities will in large part rest on the forms of governance, the role of policy and cultural intermediaries, and the trust placed in public engagement and governance (Sennett, 1986). Where there is power, there are also points of resistance and the power over memory and identity held by any dominant social group is rarely left unchallenged: 'These points of resistance are present everywhere in the power network. Hence there is no single locus of great Refusal . . . instead there is a plurality of resistances, each of them a special case . . . by definition they can only exist in the strategic field of power relations' (Foucault, 1990: 95–6).

SHAPING STEPNEY'S CULTURAL LANDSCAPE

> when I was roaming around, the day was so quiet and still, so cold, so undisturbed . . . that it had that mesmerising stillness and silence you get after snowfall. That light and openness worked on me and. . . . It was deception of course. Because the calm wasn't a real peace: it was just the absence of anything else, the absence of activity and life and business. Because that's all Stepney is, now: just the minimum of places to live, roofs and front doors and streets to walk down, the struggles of town planners and housing creators; but nothing, honestly, nothing else; the calm of no expectations.
>
> (Charles Jennings, 'Bleak houses?', *Guardian* 1999)

Jennings' Stepney is one of hopeless emptiness. In this powerful image nothing of any interest or importance seems to happen here. Worst of all, nothing is expected to happen. This perceived vacuum is reflected in local regeneration strategies in which Stepney has been constructed primarily in terms of its hopelessness – its poverty, ill-health, poor environment and fragility do not offer opportunities for (self-) development, particularly economic development. Where there is lively discussion it is around negative aspects of a community that is thought to be slowly imploding with problems of racism, youth crime, gangs and drugs. With this baggage Stepney is deemed, rightly so, to be 'in need'. The area has been the target of numerous rounds of funding from successive post-war urban programmes for over a quarter of a century – Urban Aid, Single Regeneration Budget (SRB), European Regional Development (ERDF), Estates Renewal, New Deal for Communities, to Action Zones for health and education. Culture, however, is apparently 'absent' in these programmes and plans. It is as if there is no cultural landscape into which the policy machinery can descend – no creative buzz of artists' colonies, no café society, no obvious formal or informal spaces where cultural activity and participation takes place. There are no visible signs of valued cultural or recreational pursuits – no posters of events, no signs to venues, no shops selling the artefacts of cultural consumption. What there is, significantly, lies on its outer edges (a re-opened cinema, a renovated but very part-time music hall, a local history museum, dance and visual artist studios and a theatre for young people). So there seems to be no colonising potential, no cultural resources, no 'rich mix'.

This perceived absence of culture, as we have seen, derives from both 'history' (above) and from several contemporary sources and these have colluded to perpetuate the myth of cultural absence in Stepney. First, the reconciliation of social and economic objectives, promised in national government guidance (Department for Culture, Media and Sport, 1999), was compromised at the local level. Heavily influenced by the ideas of creative city/milieu advocates and other local cultural intermediaries, the economic potential of creative and cultural activity came to override the social objectives of cultural planning and development. A commissioned report on the *Economic Importance of Cultural Industries in Tower Hamlets* (Landry *et al.*, 1997) identified clusters of creative industry potential almost exclusively located in the west of the borough (around Spitalfields and Brick Lane), and proposed strategies to support its development. This included the east London marketing initiative, *Hidden Art,* and a Cultural Industries Development Agency to provide enterprise support to small creative firms and artists, as has been created in other post-industrial cities (for example, Sheffield, Manchester, Cardiff). The implementation of the recommendations of this report became borough-wide cultural policy, resulting in an exaggerated emphasis on economic development and a geographical (and ethnic) polarisation in terms of public resource investment. Thus the entrenched east–west divide in London, as in other old-world industrial cities, has been mirrored locally.

Second, the recognition of cultural activity and the value attached to cultural participation suffered from a narrow interpretation of culture by local policy intermediaries. The assumed value attached to arts and culture was read from a menu of 'known' identifiers of cultural practice – theatres, dance studios, art galleries, festivals, etc. This elevated the public expression of professional, semi-professional and amateur participation in a European tradition of practice as well as social welfare provision. The effect of this was to leave unrecognised and 'hidden' forms of expression and cultural participation that did not have this kind of public face.

Within the resident population of Stepney there are some dominant groups and it is this social structure which goes a long way to explaining this lack of a public face to local cultural expression. The indigenous white European population is an ageing group with a particularly high concentration of very elderly residents. The mobility of this population is restricted and their forms of participation in community activity largely revolve around daytime clubs and residents' associations. The Bangladeshi population, representing the largest concentration of this ethnic group in Britain, originates from tight-knit village communities in the Sylhet region of Bangladesh, who are 'four times more disadvantaged on every measure you can think of than any other single population' (Hall, 2000: 47).

Strong ties between Stepney and Bangladesh are maintained and as Keith has claimed: 'Sylheti settlement in the East End is perhaps better understood as the first British case of post-Fordist labour migration' (1995: 17). There is regular consultation and communication with distant family members and considerable travel between the two locations. With reunification of families in the 1970s following the original male-dominated migration (ibid.), a baby boom has created a high proportion of young people within this group since the late-1980s/90s and the structures of authority, including the role of women, reflect traditional Muslim values. The everyday reproduction of cultural practices and values is played out in the domestic and private spaces of homes and community buildings and in the mosques. Whereas men and youths are much more visible in the public spaces of Stepney, women do also meet and engage in shared activity in their homes, in the community buildings, local junior schools and in a progressive local mosque.

Another smaller, but significant, group within Stepney is refugees from Northern

Somalia. This group is less established in terms of local representation in community forums and is more fragmented. However, as refugees, their expressed need to maintain their own cultural identity through active participation in Somalian arts, crafts and music has led to a wide-ranging programme of cultural activities which draws participation from the scattered Somalian community across London. However, once again this active cultural production and participation takes place behind the closed doors of non-arts venues including the meeting rooms of charitable organisations. At the policy level this non-recognition and invisibility of activity has reinforced the 'no-culture' label attached to Stepney. It was overlooked by the economic, legitimate and cultural values of both intermediaries and institutions.

A third and related factor shaping the cultural landscape of Stepney is an inward-looking attitude fostered by community leaders and by some key service providers. The social composition of residents, their lack of material resources and the poor infrastructure have also contributed. However, the boundaries drawn around specific regeneration and housing renewal projects does little to foster a collective or open response to problems. Individual estates compete for resources, development projects and community organisations set up similar support and cultural initiatives which compete with each other for the same active participants or to service the same client group, and there is a reluctance to learn from outside or to bring in external expertise.

STEPNEY'S HIDDEN CULTURAL LANDSCAPE

With resources increasingly targeted at the west of the borough, on cultural consumption and creative enterprise, Stepney Central Single Regeneration Budget commissioned an investigation to examine the assumed culture vacuum in Stepney (Evans *et al.*, 1999). This research contradicted the earlier economic study and found examples of professional arts organisations located in the area, individual creative artists and significant cultural activity within community and housing organisations and educational establishments. Once again the issue was one of visibility and diversity. The professional artists and arts organisations, though located within the Stepney area/catchment, had a low local profile for Stepney residents with most of their work directed at participants and audiences spread across east London, and beyond. Some arts organisations drew their workshop attendances exclusively from the white owner-occupied residents (who travelled by car to venues).

For example, ACME studios, a well-established non-profit organisation providing studio space and housing for visual artists, which specialises in reusing industrial buildings, have their offices and a block of studio facilities in Stepney. Artists renting the studios do not live in Stepney and the local multiplier effect is minimal. The economic and artistic networks created by the organisation are London-wide and international. Likewise Matt's Gallery, located next door, provides a space for international contemporary artists with no connection to, nor sense of obligation to communicate with, local communities. Both these organisations are well supported by the state arts funding system. In a less overtly separatist manner, a small number of enterprising designer-makers and artists had converted industrial and domestic spaces into workshops. They saw themselves as temporary, using the cheap space for as long as they needed to.

In contrast the Ragged School Museum, and the Half Moon Youth Theatre, did include local young people as far as their programmes allowed. The Ragged School, in the tradition of the *People's Palace* movement, used professional actors and volunteers to deliver holiday and weekend activities that aimed to encourage local children, and sometimes their parents, to make 'better' use of their leisure time. Most of the volunteers were older (white) former residents of Stepney who had campaigned vigorously to save the building and raise funds to refurbish the space as a museum. The link to

Stepney that they created was one step removed. Attempts to encourage local culturally diverse participation were beginning. The Half Moon is a nationally recognised professional youth theatre organisation which had benefited from earlier urban programme funding, and in so far as it is committed to working in the East London context of diversity, local schools were included in its programmes. However, specific outreach into the local community was limited by the wider aims of the theatre.

It is apparent that these professional arts and cultural organisations do exist in Stepney, as they do in other urban areas. However, though they tend to work in the location, they do not, by and large, work *with* or for the local area. This reflects a widening divide between the social 'local' arts (amenity) and professional arts (excellence, creative industry) within the state-funding regimes (Wilding, 1989; Hewison, 1995), despite compensatory programmes and cultural policy exhorting access and audience development (during the 1980s/90s audiences fell in the subsidised performing arts sector whilst entry prices rose in real terms – Evans, 1999).

LOCAL CULTURAL FLAGSHIPS

Mirroring the city centre/downtown and *Grands Projets* of world cities and self-styled cultural capitals, the creation of local flagships seeks to create identity – visually, economically and in terms of local pride (on the basis of the *Every Town Should Have One*, Lane, 1978 and Guggenheim effects). Their transformative potential is literal – *chesspieces, nodes, must-see attractions*, much as the anchor store is sought for the successful shopping mall. Stepney is no exception to this, and the two manifestations of the cultural flagship – the festival and regenerated facility (building/complex, landscape/ public art) – form part of the area's regeneration programme. In celebration of the Bangladeshi community and roots, an international festival was held with support from both the British and Balgladesh governments, and organised by a cultural intermediary organisation, Arts Worldwide. The programme included events and performances with visiting and UK-based artists at mainstream cultural centres (Barbican Arts Centre, Queen Elizabeth Hall, Canary Wharf, the Institute of Contemporary Arts, the Whitechapel Gallery) with smaller 'club' events (for example, poetry, music) in and around the west of the borough/city fringe in bars and restaurants. A festival *mela* (carnival) was held in Brick Lane, the well-known street market, a community event held in Toynbee Hall and the main 'community' project was a community musical with schoolchildren from the borough performing in an outdoor stage off Brick Lane (funded by the Single Regeneration Budget). None of these events was held in Stepney or targeted the Bangladeshi community concentrated there; local Bangladeshi bands and groups were not involved (80 artists from Bangladesh were brought over for the festival). The festival aimed to: 'develop a positive profile of the Bangladeshi community in Britain, to provide opportunities for community education and capacity building amongst young Bangladeshis living in the East End, to address issues of cultural diversity in London's cultural life, and to present a more positive image of Bangladesh than that currently portrayed in the media' (Arts Worldwide, 1999: 1). Key outputs, however, were distinctly economic – six jobs created, 480 schoolchildren with enhanced attainment, four Bangladeshis obtaining vocational qualifications (NVQs), twenty-two local residents gaining access to employment, 523 training weeks, two new business start-ups, thirty-five local businesses in the cultural industries receiving advice (ibid.) – a mere cultural festival was clearly not enough.

The flagship development project was in the form of the renovation of the Mile End Park, in the north-west corner of the Stepney regeneration area. Six kilometres long and only 100 metres wide in some places, Mile End Park is formed from World War II bomb sites and was one of Abercrombie's original 'green lungs' for London (1944). This linear park runs north to south, adjoined in the east side by the Grand

Union Canal, which new private housing developments overlook, and is crossed by three major roads and two railways. By the early 1990s the park had become a no-go area after dark, used for antisocial pastimes and dog walking (Hare, 2001). The idea of transforming it into a 'people's park fit for the twenty-first century' (Mile End Park, 2001) led to initial overtures to the Millennium Commission (Lottery-funded) and the formation of the Mile End Partnership with local council and regeneration agencies.

The original concept was to create 'miles of play, art, ecology, sport and fun', reflecting the view that inner-city parks should be full of life and activity, despite the fact that park users value their passive nature most highly and resist their 'development' (Evans and Bohrer, 2000). Buzzwords and catchphrases like 'sustainable', 'high quality', 'community involvement', 'regional regeneration effect' and 'completing the Abercrombie plan' were floated around. Initial consultation began in 1995, but had to be speeded up to satisfy the Millennium Commission's insistence that there should be a masterplan at bid stage. A quickly drawn plan subdivided the park into a number of activity areas, linked by a path for pedestrians and cyclists (Hare, 2001). A 'Green Bridge' designed by Piers Gough continues the park over the busy Mile End Road, whilst an Ecology Park and Arts Park, it is planned, will lie to the north of the bridge. A play and football area includes an arena for spectators and a jogging track. This benign facility takes on a different meaning when one considers that the Bangladeshi female population will not be able to use such a public facility (even if they wanted to). Their recreational needs are not met by existing facilities (for example, swimming pools), where their only option is to use a small school facility via minibus.

An *Xtreme Sport* area will also cater for rollerbladers, BMX bikes and skateboarders, and a franchised Electric Karts concession will, it is hoped, bring in revenue. This will be marketed to city workers in the nearby Canary Wharf Docklands office city for hire at lunch-times and after work. The Sports Park and Mile End Stadium is to be refurbished and a new swimming pool built, but as the Bangladeshi youth of Stepney have opined (below), they will be priced out of 'their' local facility, and as Hare comments: 'The idea that people (let alone wildlife) will use each space exactly as intended is naive. Vandalism suggests that the consultation work failed to reach some members of the community, despite the three to four permanently employed members on the community liaison team. An uncertain future lies ahead for Mile End Park – Britain's urban parks urgently need investment but there's a possibility that one of the first and most high-profile projects has not helped the cause' (Hare, 2001: 6).

In contrast to these cultural flagships, the informal 'cultural' activity that the closer investigation uncovered was, as would be expected, more representative of the local community and its needs for cultural expression. One example is the Youth Action Scheme (YAS), inspired and run by a local resident and supported by a charitable foundation. He responded to the increasing tension between groups of young men and the police, particularly around drugs, by setting up a youth programme of sports and leisure activity. A particularly popular aspect of this programme is participation in the Asian youth football league, with teams playing both London-wide and internationally. Participation in the league plays a very important role in positive self-determination. The scheme struggles to maintain this activity on poorly maintained pitches, as the cost of hiring the all-weather, floodlit pitches in the adjacent Mile End Park is prohibitive (above). However, by far the most important aspect of this project is its building, located in the centre of Stepney. This provides a safe and non-judgemental space in which young men can spend time together. Other examples of informal, everyday cultural participation can be found in the local community advice centre. This space is used by women's groups, meeting to study (the Koran) and to do crafts, music and children's activities.

A creative enterprise project using traditional skills to produce traditional funishings emerged out of a women's sewing group meeting informally in an estate-based community centre. Reaffirmation and development of cultural identity actively takes place within these settings. Self-help and support mixes traditional forms of consultation with respected members of the community, with more formal advice and advocacy sessions. The large-scale housing redevelopment in the neighbourhood has also thrown up opportunities for creative intervention and participation: video projects around safety, children's and youth arts projects, an artist-in-residence project, housing and public art, local estate fun days and festivals, all act to encourage participation by and in the local community. These, however, are very small-scale and though they drew favourable comment, there is no long-term investment in such basic cultural development, which therefore serves as a palliative, rather than what Raymond Williams saw as a continual opportunity for cultural development and change (1981).

CONCLUSION

The effect of local regeneration on the shape of the cultural landscape in this part of east London has been to overlook and undervalue the common and everyday culture of its residents. This has been as much a consequence of conceptualisation as implementation. The aspirational agendas of culture-led local regeneration, have proved difficult to implement. In this process the economic rationale (including skills and training) has remained the imperative. It is apparent that Stepney does have a cultural landscape, or rather cultural landscapes. Much of the everyday cultural practice is hidden in the informal spaces of the community. These activities are self-affirming but desperately in need of resources and support. This forms the 'other' cultural landscape which policy, limited both by vision and expertise, finds difficult to engage. The organisations which make up the professional cultural landscape are either isolated from the locality, being in the locality but not of it, or too stretched to provide specifically for local needs.

A future regeneration effect is emerging and will create further social and spatial division within Stepney itself, as in similar inner city and fringe zones. New canal-side housing development on the edge of Mile End Park and adjacent to ACME artists' studios, Matt's Gallery and the Ragged School Museum, is the basis of a putative 'cultural quarter'. The new facilities in the 'people's park' have been priced beyond the reach of many locals but will add to the infrastructure (and property values) supporting incoming residents. Such colonisation and reshaping of the cultural landscape is well within the remit of the current Urban White Paper and the re-use of brownfield sites for housing and mixed-use development. As such this part of Stepney may well turn out to be 'successfully' regenerated in these terms, and herein lies the paradox.

The cultural landscape – the amenity and policy Stepney deserves, but has not got because of this 'regeneration effect' – is one that needs to be rooted in an understanding of local needs and values, and is able to draw on the professional expertise of those already working to provide high-quality cultural resources and opportunity. Access to what exists needs to be matched by the resources – buildings and people – to provide a sustained cultural development programme which enhances the different aspirations for cultural expression, participation and production. However, where control over resources is absent, local governance becomes a hollow promise for local communities, as Borja and Castells point out, reflecting the pattern of urban regeneration and regional development programmes in Europe and North America:

> Lack of resources means that in practice higher layers of government replace local government through sectoral programmes or individual projects. In other cases action is taken by the private sector, without being integrated into a coherent urban programme. In yet other cases, a major area of the city and of inhabitants are simply left without any cultural facilities.
>
> (Borja and Castells, 1997: 113)

Cultural landscapes have been shaped by forms of power and control since the foundation of urban society and settlement. They represent the inherited urban morphology of cities: 'Land-use and culture are fundamental natural and human phenomena, but the combined notion and practice of culture and planning conjure up a tension between tradition, resistance and change; heritage and contemporary cultural expression, but also the ideals of cultural rights, equity and amenity' (Evans, 2001: 1). The place that cultural activity holds in a community and locale, the formal and informal spaces where participation and production occurs, on the one hand, defines that community and place (spatially and culturally), and on the other, helps define its quality of life. So as Worpole and Greenhalgh suggest: 'The city itself . . . may be regarded as a public good of a most enduring kind . . . public goods are the staples of a basic quality of life' (1999: 16). People looking for ways to improve community involvement and 'ownership' in their landscape and amenity look to models of community planning, public choice and regime theories and practice, including greater consultation and promotion of partnership-working in regeneration programmes and resource allocation. What such approaches tend to understate, however, is the attention to power, which Foucault refers to in terms of the 'conduct of conduct at the heart of governmental practice' (Keith, 1995: 369). As the Brick Lane Community Development Trust suggests more prosaically:

> Participation must continue beyond the planning stage of developments to their implementation and management. . . . We did not want simple results like planning permission granted or denied, or one-off planning gains, like money or a community centre . . . what we wanted was to join in designing strategy for the development of the area over the next 20 years.
>
> (Kate, 1994: 16)

As this chapter has revealed, the theory and practice of cultural development and regeneration within the multicultural city draws on a wide range of disciplines and subject areas – a reflection that none alone adequately deal with the complexities and challenges of contested social and cultural meanings and processes within either traditional structure and agency, or postmodernist critiques. The limitations of classical sociology, geography and cultural theories to explain these phenomena and the forces at play, and the failure of urban policy solutions and predictions (for example, deurbanisation) to reach those areas supposedly most in need, suggest a diverse and cross-disciplinary approach. Sources cited indicate a range of reading on these interlocking subjects, although it is too often case study or single-issue based.

For a detailed exposition of the relationship between culture, amenity and city planning, *Cultural Planning* (Evans, 2001) explores this both historically and as part of the urban renaissance. Community planning approaches are most detailed in Healey's work (1997, 2000). On the role of art and design in the public realm of cities, Malcolm Miles' *Art Space and the City* (1997) provides a comprehensive contemporary perspective. On the commodification of cultural space, Sharon Zukin's work on New York (1995) and Kearns and Philo's general text *The City as Cultural Capital* (1993) provide in-depth analysis, whilst particular analyses of globalisation include John Eade's *Living the Global City* (1997) and *Globalization and Culture* (Tomlinson, 1999).

Several collections, including chapters used here, provide a range of writing on the city, on city cultures and contemporary debate on urban society. Good examples include *Mapping the Subject* (Pile and Thrift, 1995; on the East End see Keith's 'Ethnic Entrepreneurs and Street Rebels'); *Culture, Globalization and the World-System* (1991) and *Re-Presenting the City* (1996), both edited by Anthony King; *The City Cultures Reader* (Miles *et al.*, 2001) and *The Urban Context* (Rogers and Vertovec, 1995, and therein Pickvance on 'Comparative Urban Studies'). Several of these writers and more feature in a major collection, *A Companion to the City* (Bridge and Watson, 2000). Finally, policy and planning studies and guidance (planning, practical, toolkits) exist in the form of government and agency reports, including those of the Urban Task Force, Cultural Ministry (Department for Culture, Media and Sport), and public policy think tanks (see Worpole, 1998 – namely DEMOS, Comedia, IPPR, Rowntree *et al.*).

NOTE

1 The term 'social exclusion' was first coined in France with reference to housing/homelessness in fringe Parisian estates, whilst in the USA the term 'ghetto' is still used, but usually with reference to inner urban 'black' neighbourhoods. Its wider politicised meaning has been adopted by the British New Labour (and other, for example Canadian) government and is overtly related to the current concept of social citizenship and therefore it is important to theorise and operationalise empirically – from what are individuals and groups being excluded and on what basis? An individual can only be regarded as being excluded from activities if they would like to participate. For example, in a five-year analysis of the British Household Survey (Le Grand, 1998) it was found from a sample of five types of activity, that it was not necessarily the *same* people who were excluded on each dimension and that no one was excluded on all five: 'social exclusion is a complex, messy phenomenon, as such it is not one that can be reduced to simplicities in terms of either questions or answers' (ibid.). Clearly in a diverse society the relationship between social exclusion, social integration and public choice is a complex one. In this respect we perhaps need to research the new modes of inclusion (which paradoxically may be individual and privatised), as a recent European Union study concluded, 'because social exclusion coexists with (or in fact be caused by) inclusion, research on exclusion cannot be disconnected from the analysis of change on the "included" side of society' (European Commission, 1998).

12 NIGEL CLARK

Turbulent Prospects: Sustaining Urbanism on a Dynamic Planet

> To Alsana's mind the real difference between people was not colour. Nor did it lie in gender, faith, their relative ability to dance to a syncopated rhythm or open their fists to reveal a handful of gold coins. The real difference was more fundamental. It was in the earth. You could divide the whole of humanity into two distinct camps, as far as she was concerned, simply by asking them to complete a very simple questionnaire, of the kind you find in *Woman's Own* on a Tuesday:
>
> (a) Are the skies you sleep under likely to open for weeks on end?
>
> (b) Is the ground you walk on likely to tremble and split?
>
> (c) Is there a chance (and please tick the box, no matter how small that chance seems) that the ominous mountain casting a midday shadow over your home might one day erupt with no rhyme or reason?
>
> (Zadie Smith, *White Teeth*, 2000)

UNRULY CITIES

Sociologist Ulrich Beck has likened living with environmentally hazardous technology to dwelling on a volcano (1992: 17). But many people quite literally live on shaky ground, as Bengali Londoner Alsana Iqbal reminds us in Zadie Smith's novel. Perhaps more of us than know it. Revisiting the earthquake-prone Los Angeles beloved of Hollywood scriptwriters, Mike Davis draws attention to the pervasive ordinariness of natural disasters. His *Ecology of Fear* (1998) depicts a sprawling metropolis whose patterns of land use places its citizens smack in the path of flood, tornado, wildfire, drought and even puma attack. Misleadingly marketed as a new world Garden of Eden, Davis suggests that Southern California should be more realistically viewed as a landscape configured by waves of dynamic instability. Far from aberrations, the 'catastrophes' he tracks are episodes of the very processes that have produced California's celebrated scenery. As well as reminding us that bursts of land-shaping activity remain a normal feature of urban environments, Davis points out that the periodic intensification of such episodes is an integral part of this normality. LA, he intimates, may be about to awake from a long climatic and seismic 'siesta' (1998: 35).

This is not just an issue for the citizens of Los Angeles. Their topography may be especially volatile, their land-use particularly extravagant, but many of the physical forces that impact on their region move in cycles that affect most of the globe. Our heavily urbanised planet, in other words, is also a chronically turbulent one. How these two dimensions of planetary ecology grate and grind together is an issue that should concern social scientists as much as our natural science counterparts across academia's great divide. Though they may have taken to ideas of cultural chaos and disorganisation, urban theorists have been slower to warm to the disturbances caused by 'inhuman' forces. And while social scientists have recently been obliged to take account of the human impact on the biophysical environment, this is not the same as engaging with ecological processes and events that impact on social worlds. Many natural scientists, on the other hand, now have a strong interest in the inherent dynamism of the earth's ecological systems, and are currently grappling with the issue of how anthropogenic changes interact with or overlay these other waves and bursts of transformation. Their theories and speculations could come as a shock to social scientists. Or perhaps as an inspiration.

In this chapter I look at some of the environmental implications of urban life: viewing the city as a set of physical processes whose 'footprint' impacts far beyond the fringes of urban space. But at the same time, I explore the way biological and geoclimatic forces stamp themselves on urban form – or patter quietly through the everyday spaces of city life. Around the themes of extreme weather, seismic unrest, plague and predation, I sketch something of the interface between the turbulent tendencies of our own species and those of the planet as a whole. This raises some serious questions – not only about the possible damage that current patterns of urban living might be inflicting on the environment, but about the more general sustainability or resilience of urbanism as we now conceive of it. But these are not necessarily the sort of questions that urban planners and administrators – even with the best scientific evidence at hand – are equipped to handle. While the connection may not be an obvious one, I raise the possibility that some of our 'wilder' and less organised experiences of city life might provide hints at how to live with the wildness that animates the rest of our planet.

URBAN SUSTAINABILITY

'The natural history of urbanization has not yet been written', Lewis Mumford observed in 1956, as he set about making a rather impressive stab at it (1972: 140). To begin to think of cities in terms of the biological or geophysical processes in which they are implicated is to bring into question the firm divide of densely built environments from their hinterlands: the duality which has underpinned so much of our experience and narration of urban life. When sociologist Louis Wirth, writing in the 1930s, famously portrayed the polarity of city and country, foremost in his mind was the dissociation of the urban from the natural realm. 'Nowhere has mankind been further removed from organic nature,' he claimed, 'than under the conditions of life characteristic of the great cities' (Wirth, 1996: 190). But Mumford suggested otherwise. Today, with the growing interest in the environmental impact of urbanism, his ideas about the material entanglement of country and city have come into their own. No longer viewed apart from nature, cities are being reconceived as physical forces in their own right. Contemplating the vast urbanised areas of the planet's surface, philosopher Michel Serres declares: 'the decisive actions are now, massively, those of enormous and dense tectonic plates . . . colossal banks of humanity as powerful as oceans, deserts or icecaps' (1995: 16–17).

It would be a gross simplification to see this concern with the ecological significance of the urban merely as an expression of a long-standing bias against city life. While there may be grounds for David Harvey's recent railing against the hostility of environmentalists towards cities (1996: 391–2), the radical-green defenders of the wilderness he has in his sights seem soft targets. Contemporary environmentalist critics of the city are much more likely to espouse a rigorous overhaul of existing patterns of urbanisation – backed by measurements of material-energetic flows – than they are to emote about the virtues of rustic simplicity. The discourses and practices of 'urban sustainability' aim at preserving the attractive aspects of city life, while mitigating its destructive side. And they locate their concern firmly with the interests and needs of all the human inhabitants of the city, as well as with the requirements of functioning ecosystems or a living nature. 'A "sustainable city,"' urban ecologist Herbert Girardet suggests, 'is organised so as to enable all its citizens to meet their own needs and to enhance their well-being without damaging the natural world or endangering the living conditions of other people, now or in the future' (1999: 13).

Outlined in this way, urban sustainability reveals its affiliation with sustainable development, a more encompassing concept that has been cementing itself as the dominant paradigm of environmental policy-making since the late 1980s (Baker *et al.*, 1997). Like sustainable development in general, the idea of sustainable cities foregrounds the need to close the gap between the poor and the privileged, both within and between societies. Because the biophysical systems vital to life appear already strained by existing demands, the improvement of standards of living across the social and geopolitical spectrum can only conceivably be met by radical transformations in the way resources are used. But whether further 'development' of the required scale is feasible remains contentious, with critics pointing to the unlikely prospect of coupling further economic growth with diminishing environmental impacts (Blowers and Leroy, 1996).

Most critical commentators concede, however, that discourses of sustainability do help focus debate on the realities of where and how we now live. Without advocating that cities should be wholly self-sufficient, proponents of sustainable cities draw attention both to the extent of the ecological demands of most existing patterns of urbanism and to the availability of strategies for lightening this burden (see Gordon, 1990; Badshah, 1996). Using the technique of 'ecological footprinting' – which maps the land area required to supply a city or country with the resources it consumes and to absorb the waste it excretes – environmental planners can offer a measure of changes required or achieved. London's footprint, for example, is around 125 times its surface area, which in abstract terms takes in the entire productive land area of Britain. By this logic it has been estimated that cities across the globe, with around half the human population of the planet, make demands on some 75 per cent of the world's resources (Girardet, 1999: 10, 27–8). On the positive side, urban ecologists point to the resource-saving or waste-diminishing potential of efficient recycling programmes, inner-city farms and gardens, renewable energy sources, urban forests and wetlands that function as living filtration systems, and well-integrated public transport schemes – each demonstrably successful in existing cities somewhere in the world (see Walter *et al.*, 1992).

Generalising, we can say that the drive for sustainable cities seeks some sort of equilibrium between urban needs and the resources available in the city and its hinterlands, and seeks to attain this through the participation of as many citizens as possible. While the aim seems admirable, we need to be aware of the tensions that course through the concept of sustainability. In most of its guises, the idea of sustainability evokes an ideal of ecological harmony, stability and order (Worster, 1993: 137). But how might such an image equate with the dynamism and vibrancy which is the appeal of city life for so many of us? As Amin, Massey and Thrift point out, in response to Richard Rogers' vision of a new balance between the built environment, nature and society: 'harmony in this sense is actually the last thing cities are about' (2000: 4). Moreover, many 'eco-city' exponents envisage a thorough greening of urban space, a transformation that would offer new opportunities for non-human life, both cultivated and wild. But can we really expect to make a variegated wilderness of the city, or even a garden? Or is there, perhaps, a more profound wildness already in our midst that threatens to shake up the very notion of sustainability as it now stands?

ENCOUNTERING TURBULENCE

While I suggested above that viewing urbanisation as a kind of ecological process began to destabilise the city/country dichotomy, we should be aware that thinking by way of binaries is not something that is easily relinquished. And nor should we necessarily wish to do so. In fact, there is a sense in which recent discourses on urban greening still draw much of their energy from a tension between city and country, or more broadly, between culture and nature.

Historical accounts suggest that whenever and wherever people live in cities, there is a tendency to fantasise about the freedom of nature from which civilisation has apparently delivered them (see Williams, 1973). In response to this tug-of-war between civilised present and wild untrammelled past, city dwellers tend to reach out for some sort of middle ground that embodies the best of both worlds. If the vision of 'Arcadia' is an ancient version of this dream, then the development of suburbia and garden cities can be seen as its modern actualisation (Eisenberg, 1998: chs. 13–14). Environmentally sustainable cities, as they are now being conjured, might be taken as the latest incarnation of this 'Arcadian' quest, extended in the context of ecological crises to embrace the city in its entirety.

So pervasive has 'the middle landscape' been in the western imagination, that even our visions of the natural world have tended to gravitate towards order and repose. In recent decades, however, evidence has been piling up that what we called 'nature' isn't, wasn't and never can be the peaceable kingdom of our reveries. Across a range of disciplines, natural scientists have come to new appreciation of the role of turbulence and convulsion in the shaping of the physical world. In the science of ecology, this shift is manifest in a turn away from the idea that ecosystems naturally succeed to stable 'climax' communities, towards a more 'discordant' vision in which repeated disturbance plays a vital part in the distribution and evolution of life (Botkin, 1990). Taking into account the impact of fire, storm, disease, landslide and other upheavals, it is now argued that most 'climactic' communities are actually a fairly heterogeneous mix of relatively mature and more recently perturbed patches. 'Equilibrium landscapes would therefore seem to be the exception, rather than the rule', as ecologists White and Pickett conclude (1985: 5).

This is not an easy lesson to absorb. 'Confined to the experiences of a human lifetime,' Stephen Budiansky writes, 'man (*sic*) has a hard time grasping the essential part that destructive forces play in nature' (1995: 75). For the majority of social scientists, who seem to have enough difficulty taking into consideration the apparently regular aspects of the biophysical world, the idea that upheaval in the atmosphere, lithosphere or biosphere might be anything less than extraordinary presents a serious challenge. There is, however, a growing interdisciplinary field of 'disaster studies' which explores the social interface with 'natural perturbations' (see Oliver-Smith, 1986). The general consensus of recent work is that while the translation of an episode of physical unrest into a 'disaster' is strongly dependent on a range of social variables, the impact works both ways. Which is to say, 'environmental extremes are powerful enough to exert a strong and consistent influence upon social and cultural systems' (Alexander, 2000: 55).

Where disaster studies converges with the scientific study of turbulence is in the view that extreme events do not simply punctuate 'normality', but upset the very notion of the normal or the average. Periodic upheavals shape the world precisely because of their potential to irreversibly set a physical system on a new course. 'The outcomes are not pre-ordained,' as Budiansky puts it; 'but for that hurricane or this fire, the world we see around us might well have turned out very different' (1995: 79). As some commentators have recognised, this presents serious problems for the concept of sustainability as it now stands (see Worster, 1993: 140–2; Luke, 1995: 28). For if it is in the 'nature' of physical systems to rather suddenly redirect their own trajectory from time to time, then the very idea of harmonising our demands with quantifiable ecosystemic capacities takes a heavy blow. And the sustainability paradigm is, after all, concerned with guaranteeing a level of well-being for current *and future* generations.

It might be assumed that such vagaries of nature are of greater concern beyond city limits, where livelihoods seem more closely hitched to 'the elements'. Disaster researchers suggest otherwise. 'In many, perhaps most, parts of the world, large cities constitute the greatest centre of hazard risk', David Alexander notes (2000: 94). While sheer density of population plays an important part, genealogical factors are also significant. Urban centres, like the rural clusters they so often grow out of, tend to form around certain features, taking advantage of topographic amenities that may be all but forgotten as development proceeds. Settlements have seeded themselves on alluvial deposits, at river fords, in rift valleys and passes, on rich volcanic soils, under mountain lees, or at coastal inlets. And in most of these cases, the elemental processes that shaped the landscape into a desirable site have not ceased their work. All too frequently, it appears, 'proximity to resources also involves proximity to hazard' (Oliver-Smith, 1999: 26).

What greatly exacerbates vulnerability to disaster in many urban areas around the world are the economic and political pressures that drive the less privileged to cluster in especially perilous zones (Burgess *et al.*, 1997: 67–9; Douglass, 1998: 119). Here, in what have been termed 'risk ghettos', the episodic hazards of flood, earthquake, landslide are likely to enter into dismal synergies with the everyday social problems characteristic of urban disadvantage (Alexander, 2000: 96, 188). Ironically, at the other end of the socio-economic scale, there is a mirroring of some the hazards of place that afflict the poor and marginalised. Citing a Southern Californian 'psychographic' survey showing that 98 per cent of potential home-buyers wanted to live near 'nature', Mike Davis proceeds to point out that actually attaining this goal can come at a price heavier than just the cost of real estate (1998: 90). For living by beaches, near woodlands or on spectacular incline, is at once to put one in the path of storm surge, wildfire and landslide.

HEAVY WEATHER

The need to negotiate between the physical advantages and drawbacks of particular places has played a constitutive role in the morphology of cities from earliest times. By channelling water, the farming communities of the Mesopotamian floodplains found a way of mediating between rainy and dry season, flood and drought. Their permanent settlements at once provided the labour for water control and consumers for the agricultural surplus that regulated flow made possible (Gordon Childe, 1996: 22; Eisenberg, 1998: ch. 9). But environmental variability could quickly undermine such arrangements. As the accounts of catastrophe that fill the Old Testament seem to imply, ancient cities were profoundly vulnerable to flood, drought, plague and quake (see Davis, 1998: 20). A wave of extreme weather in the medieval period could well have had impacts of similarly 'biblical' proportions. What may have been experienced as a welcome warm spell in Northern Europe, paleoclimatic evidence suggests, brought disaster to the hydraulic civilisations of drier regions as far apart as China and Central America (Davis, 1998: 20–5).

As Californian climatologist Bill Mork puts it, 'If you were to project [the medieval] conditions into today, it's obvious that we couldn't survive in the present infrastructure with the agriculture and the cities that we've got. We just wouldn't have enough water' (cited in Davis, 1998: 24). Far from moderating the vagaries of weather, modern urban form tends to exacerbate its effects. Deforestation and other changes in the ground cover of the hinterland – a perennial problem of urbanisation – decreases its ability to hold rainwater. Runoff is further accelerated by the canalising of river systems and by the general loss of porosity that comes with the paving of land. In the case of Los Angeles, with its high investment in the hard surfaces demanded by automobility, it has been calculated that there are some 250 tons of concrete blanketing the earth for every citizen (Davis, 1998: 80). But even then, the impact of modern motorised urbanism would be easier to account for, if it were not that its effects reached up to the atmosphere as well as down to the ground. Reliance on the car and the carbon-based economy it encapsulates adds a further complication to the interplay of urban form and climate. From low-grade coal to high-octane petrol, the tapping of fossil fuel has injected a surge of energy into the city-building process.

This ability to utilise reserves far in excess of the standard solar budget has enabled massive urban expansion into locations where heat, cold, dryness, infertility or the basic friction of distance would once have imposed severe limits. The current size, shape and distribution of cities, in other words, is unthinkable without the boost of salvaged carbon.

Over the latter third of the twentieth century, evidence began to mount of the climatic impact of releasing so much sequestered energy into the atmosphere. Initially couched in terms of a general warming or a destabilisation of current climatic conditions, more recent approaches emphasise that there is no simple base line against which we can plot anthropogenic impacts. The earth's climate, evidence now suggests, syncopates wildly of its own accord. Only recently have meteorologists begun to decipher the workings of the earth's principal 'weather engine' – the El Niño–Southern Oscillation (ENSO) – which cycles the excess of solar energy absorbed in the tropics in the poleward directions. ENSO, it now appears, pulses to a number of distinct rhythms that range from years to millennia: the complex interplay of these frequencies generating both small and epochal fluctuations in climate (Davis, 2001: ch. 7). Though they remain devilishly difficult to anticipate, charting such oscillations goes a long way towards explaining the swings between famine and plenty that punctuate human history, and they may well help us make some sense of our own current contribution to global weather conditions.

As one hypothesis has it, additional heat trapped by increased levels of carbon dioxide and other greenhouse gases could be joining the equatorial solar excess that drives El Niño. 'An enhanced ENSO cycle,' Davis informs us, '. . . may be the principal modality through which global warming turns into weather' (2001: 238). What such 'enhancement' seems to imply, at very least, is increased turbulence and extremity: more cyclones, worse desiccation in some places balanced by heavier deluges elsewhere. For cities already subject to seasonal water shortages or flood this has disturbing implications, as it does for settlements on hurricane-prone sites. In wealthier regions, impervious gentrifiers of the seafront migrate happily into storm-prone zones; the US Gulf and East Coasts having witnessed a 50 per cent increase in shore residents over the last two decades. But here, at least, citizens have the palliative of early warning systems and massive emergency services (Tenner, 1997: ch. 4). Those living on small tropical islands and atolls, or pressed by poverty onto cyclone-prone deltas and coasts like the Bay of Bengal, tend to lack such luxuries. Or simply have nowhere to retreat to (see Zaman, 1999).

It is not just that the extension of the urban fringe exposes more people to the prospect of disaster. The desire for intimacy with nature – amongst those urban populations affluent enough to act on the impulse – actually contributes to the likelihood of extreme environmental episodes. As Evan Eisenberg observes of the suburban manifestation of the middle landscape: 'Never before have so many people tried to live in Arcadia' (1998: 145). But the spread of low-density land-use eats into ecological buffer zones around cities and at the same time necessitates heavy expenditure of energy for maintenance and transportation. In this way, suburbia's middle ground adds considerably to the build-up of greenhouse gases, thereby exacerbating an already convulsive atmospheric–hydrological cycle. Paradoxically, the desire for a gentle, harmonious existence seems to be generating its own antithesis – though in many cases, the cost will be deferred to distant others who share few of the benefits.

WILD LIFE

The human addition to climatic fluctuation will have profound implication for other species, and our relationships with them. Disentangling the planet's rhythms of climate is enough of a challenge without bringing the vagaries of biological life into the equation. But unavoidably, our fellow creatures – animal, vegetable, algal, fungal and bacterial – interact profoundly with the planet's atmosphere and hydrosphere. If weather patterns change, with or without human influence, biodiversity will be seriously threatened – not least because human demands on the earth's surface can drastically narrow the options for other forms of life. Corralled on reserves, animal populations lose their normal ability to respond to perturbations, being unable to migrate in search of warmer, cooler, drier or wetter climes when the need arises (Grumbine, 1992: 44–60). But while many individual species may be vulnerable, life in general can be surprisingly resilient. Whether this is taken as reassuring or disconcerting depends on the life forms in question, and the circumstances of our encounters with them.

It is undeniable that urbanisation operates at the expense of other species. As Manuel De Landa points out, the urban ecosystem is one that is deliberately simplified by human beings in order to focus all flows of nutrient on ourselves (1997: 108). However, some of our more evasive competitors – rodents, insects, scavenging birds – have thrived in our midst, while others have made periodic comebacks when urban populations have been depleted or distracted. Up until the last few centuries, packs of wolves still preyed on humans throughout Europe, even breaching city walls when defences were down (De Landa, 1997: 109). But our era, rather surprisingly, has witnessed its own resurgence of wildlife, especially along the urban fringe. As Davis recounts, mountain lions are adding to the woes of Los Angeles by joining the urban food chain, having taken to a diet of garbage, household pets and even the occasional Californian citizen (1998: ch. 5). Like bears, coyotes, Australian dingoes and European foxes, cougars are proving more adaptable than expected, more 'polymorphous' in their predatory habits, often to the chagrin of suburbanites who prefer to take their nature without tooth and claw (see Gullo *et al.*, 1998).

The living nature so many urban dwellers love to commune with also has more surreptitious ways of biting back. Proponents of greener cities may have a point when they claim that '[a] city that has places for foxes, owls, geese . . . is more interesting and pleasant to live in' (Hough, 1990: 17). But as with the attractions of landform, the dynamic qualities that make life appealing also render it risky. Nearly all animals come with a complement of smaller life; parasites and pathogens whose survival prospects are greatly enhanced if they can hop from one species to another. A surprising number of suburban cohabitants or urban fringe dwellers are host to potentially deadly micro-organisms. Skunks carry rabies, deer ticks are linked with Lyme disease, opossum fleas can spread typhus, the deer-mouse harbours hantavirus, and the ground squirrel is a vector of bubonic plague (Davis, 1998: 249–51; Garrett, 1994: 553–5). As in the case of encounters with their larger hosts, it is the increasingly broad and convoluted interface between the urban and the (relatively) wild that encourages such intimacy. But the shiver induced by the spectre of plague is only a reminder that microbiological exchanges have played a long role in the shaping of urban life.

For by far the larger part of urban history, our species has willingly shared its urban spaces with the animals we prefer to feed on. In ancient and medieval cities, the proximity of pigs, fowl, goats, cattle and densely packed human bodies offered disease organisms the perfect conditions to make the leap between species – and from there to surge through urban populations, to spread from town to town and across whole continents. In this way, the urban populations of Europe and Asia eventually formed a unified 'disease pool', and settled into a relatively stable coexistence with the pathogenic micro-organisms their cities had spawned (De Landa, 1997: 109–26). When metropolitan Europeans and Asians travelled across oceans, the cocktail of diseases they inadvertently took with them wreaked havoc on the previously unexposed – or 'epidemiologically naive' – peoples they encountered. So much so that some historians argue that infectious diseases in many regions played at least as significant a role in European domination as any cultural or technological factor (McNeill, 1976: 208–20; De Landa, 1997: 132–3).

Most cities and their hinterlands, then, are palimpsests of plagues. The wild oscillations of population, the behaviour of survivors and the measures aimed at containing infection have each stamped their print on urban form and function (De Landa, 1997: 157–60). Even the love of countryside, it seems, is marked by the pox. 'After the experience of the great urban epidemics,' Klaus Eder observes, 'the country became sentimentally the locus of a better life, a life closer to nature' (1996: 177). But though we now realise that moving closer to nature itself brings us into contact with predators large and small, this may not turn out to be the most momentous way in which we encounter the ongoing exuberance of life. For under conditions of pronounced climatic fluctuation, new extremes impact on the rhythms and surges typical of coexisting life forms.

While changes in climate are generally hard on large, specialised organisms, they can be a boon to insects and microscopic organisms. Global warming, increased ultraviolet radiation, higher carbon dioxide levels and ecosystemic disturbance can all work in favour of microbial life, with its fast turnover and rapid adaptability (Garrett, 1994: 566–7). It is urban centres – the planet's 'microbe magnets' – that are most likely to reap the dubious benefits of this efflorescence, especially where poverty, civic unrest or disaster hinder basic sanitation. Moreover, relatively mild warming could extend the range of insect vectors, bringing the mosquito species which carry tropical diseases like yellow fever and dengue into the latitude of Tokyo, Rome, New York and other temperate conurbations (Garrett, 1994: 615, 567).

The intense linkages between cities around the globe, though sometimes presented as the recipe for microbial or viral apocalypse, in another sense serve to unify the planetary disease pool in a way that builds resistance through shared exposure. What rumours of pandemic should not detract from is the recognition that irruption and mobilisation of living beings – whether micro- or macrophysical – are ordinary phenomena; normal responses to changing conditions on a turbulent planet. Life, biologist Lynn Margulis reminds us, 'has fed on disaster and destruction from the beginning' (1998: 151). And human beings, a relatively versatile and adaptable species themselves, have capitalised on disturbance.

SHAKY GROUND

As I intimated above, it is not only the meshing of climate and life that sets the stage for our interventions, but also their interplay with another form of turbulence: that of the earth's crust. As is the case with other life forms, the grinding and rending of the planet's surface presents us with opportunities as well as danger. But perhaps even more so than the periodicities of weather and biological life, the dynamics of the lithosphere pose a challenge to the consciousness of urban peoples – even though we may be well-versed at shifting and shaping rock ourselves.

As Michel Serres' metaphor of 'dense tectonic plates' suggests, city-building itself can be viewed as a kind of geological process, one in which adept biological beings surround themselves with layers of stone or reworked minerals (see De Landa, 1997: 26–8). The trouble is that while urban dwellers may have grown used to the rhythms of demolition and rebuilding in the 'lithic landscape' of their own making, they tend to be less attuned to the ups and downs of the ground their cities cling to. But along with human earth-moving and construction, city morphology can also be shaped by siesmicity – the trembling and fracturing that results from movements of the plates comprising the earth's crust – and by volcanism – the release of pressure from the molten understorey of this crust. Apocalyptic though they may seem, such geotectonic events – no less than life and weather – are better viewed as ordinary agents of land-forming (Davis, 1998: 18). Emanating from the core of the earth, the energy that drives geological upheavals is independent of the solar flux that fuels the ocean–atmosphere interchange and biological life. But in practice, seismic and igneous activity interacts with the vagaries of weather and the living landscape, as when high rainfall disperses earthquake debris or when volcanic dust filters sunlight and inhibits photosynthesis. When earthquakes and ENSO-related extreme weather converge, as is often the case in the Andean regions of South America for example, the result can be complex, cascading synergies of catastrophe (Moseley, 1999: 59–64).

Relative to rural regions, cities amplify the risks of geological unrest by concentrating human beings and by stacking these populations in heavy, brittle structures. In the 1985 Mexico City earthquake, 200 high-rise buildings collapsed, while the Tangshan quake of 1976 levelled most of the Chinese city, with a death toll unofficially estimated at 800,000 (Davis, 1998: 40, 52). But unstable ground has its attractions. Volcanic soil offers exceptional fertility, a richness that may be topped up from time to time. Seismically hyperactive zones – like much of the Mediterranean basin or the Californian stretch of the 'Pacific Ring of Fire' – are often blessed with cultivation-friendly maritime climates and an inviting diversity of landforms. As Edward Tenner notes, 'The same features that make California earthquake-prone make it resource-rich as well as visually beautiful' (1997: 77). This could be said also of southern Italy, where the citizens of Naples cluster around the base of the volatile Vesuvius, with a density of urban population that is by far the highest in Europe (Alexander, 2000: 134).

What seems remarkable is the overall imperviousness of city building – from at least the early modern period onwards – to the frightening potentialities of geological turbulence. Accounts from around the world point to a widespread unwillingness to limit growth in unstable areas, coupled with persistent slowness to develop and implement tremor-resistant architecture (see Davis, 1998: 35–47). By contrast, a number of premodern civilisations demonstrate impressive earthquake avoidance and mitigation strategies. Pre-Columbian Incas, for example, eschewed urban development in zones of particularly high seismicity. Surviving structures reveal offset interlocking stonework and other construction techniques that minimised quake damage, and there is evidence of extensive storehousing of crop surpluses for disaster relief (Oliver-Smith, 1999: 78–9; Alexander, 2000: 66–7). As Davis argues, with regard to new world settler societies in particular, there is a tendency to work on timescales that are simply too compressed to take account of geological periodicities. By comparison with native Americans, who had some 140 lifetimes to come to terms with the physical dynamics of the Californian region, Anglo-Americans have had a mere two and a half lifetimes of experience. 'These spans,' Davis claims, 'are too short to serve as reliable proxies for ecological time or to sample the possibilities of future environmental stress' (1998: 36).

Such ecological myopia applies not simply to tectonic events, but to the broader oscillations of climate and the range of possible interactions between biological life, weather and geological transformation. Bringing all the variables together, Davis suggests that the urban development of Southern California may have taken place in a period of exceptional climatic and seismic quiescence: 'or put another way, twentieth-century Los Angeles has been capitalized on sheer gambler's luck' (1998: 37–8). But globally speaking, the extent of the gamble on urban living may operate on a far greater scale than this. Evidence from a range of disciplines has been converging for some time now into an acknowledgement of the unusual benignity of the last ten thousand years or so (Eisenberg, 1998: 4–6, 390; see also Davis, 2001: 237). That this phase is broadly congruent with the settling of our species into farm and village life may be far from coincidental. Over the course of several million years, humans weathered the advance and receding of dozens of ice ages without ever nestling into sedentary life. That a single interglacial – the one we still occupy – gave rise to farming and the civilisations it supports, it is now being suggested, may have a lot to do with its exceptional stability. By this logic, the entire human adventure of urbanism might be seen as coinciding with 'a fluke in earth history' (Eisenberg, 1998: 433).

FUTURE PROSPECTS

If there is some sense in which *Homo sapiens'* 'civilisational years' have been shaped by a kind of temporal bias or specificity, then this has surely been compounded over recent centuries by a particular geographical slant on the urban condition. The initial human experience of town and city formation and, indeed, the vast majority of our urbanised existence, has taken place in warmer regions – on floodplains and in fertile basins in the lee of mountain ranges. In other words, in the topographically complex 'basal zone' landscapes which are most likely to be subjected to high intensity events and correspondingly rapid ecological change (Davis, 1998: 18–20). Only in the last half millennia or so, a blink of the eye in anthropological time, has the epicentre of urbanism shifted northward to temperate, formerly forested lands where rainfall is comparatively regular and moderate, and geological forces relatively placid. It is the people of these regions, perhaps understandably, who have endowed us with visions of the natural world revolving around slow, incremental change and tendencies towards equilibrium or 'steady-states' (Davis, 1998: 14–15). And it is these same citizens, residing in the comparatively unshaky, flood-free, droughtless urban centres of north-western Europe and the American north-east Atlantic seaboard, who have crafted the now-dominant images of everyday urban life. And exported their urban template to the rest of the planet.

Along with its more official endorsement of social and spatial orderliness, the 'northern' metropolitan tradition deserves much of the credit for the current appreciation of the complex, contingent and disorganised aspects of urban life. But at the same time, we need to recognise that such attuning to the 'great teeming chaos of cities' (Sennett, 1996: 46) has been almost entirely restricted to the socio-cultural domain, with little corresponding sensitivity to the dynamic and generative role of biophysical forces in the turbulence of urban life. Though ancient 'basal zone' civilisations were by no means immune to dreams of a docile nature, the split between cultural dynamism and natural equilibrium as we now know it bears the hallmark of city living cradled in northern temperate ecologies. Whether we are dealing with suburbia's mass mobilisation of the Arcadian impulse or the recent drive for well-balanced eco-cities, the nature which 'northerners' have sought to embrace has been quietly shorn of its unruly elements.

It is this very image of nature – the compliant world of the middle landscape – that is looking increasingly unsustainable. As we learn more about the dynamics of interacting physical systems, about the exceptionality of steady states and the 'normality' of fluctuation, the idea of inhabiting a 'golden mean' begins to look decidedly shaky. Today, amidst gathering evidence that the planet is entering one of its spontaneously generated spells of intensified geoclimatic oscillation, we must also take account of the daunting signs that our species is adding its own surcharge to the rising peaks and deepening troughs of physical variability. As 'geophysiologist' James Lovelock puts it, 'Humans may have chosen a very inconvenient moment to add carbon dioxide to the air' (cited in Eisenberg, 1998: 420).

In this light, the sheer magnitude of some three billion people committed to urban lifestyles begins to appear as an immense gamble on global ecological stability. One which, because of its hunger for land and energy, stacks its own odds on the prospect of increased turbulence. Turbulent or extreme physical events, as disaster studies suggest, generally trigger fully fledged 'catastrophes' only when communities have failed – or are unable – to accommodate to the variability inherent in their environments (Hewitt, 1983: 24–8). Under conditions now known to be dynamically unstable at a global level, the question must be asked as to whether current patterns of vast and densely interconnected urbanism permit the sort of flexibility that may be required. And this is not simply a matter of extending the timescale on which we assess 'sustainability', for there is growing evidence that global change can be quite rapid once certain thresholds are reached (Alexander, 2000: 9).

Critical research in the fields of natural disasters and urban sustainability converges on the idea that improved grassroots participation in decision-making is a vital step towards better adaptation of city life to environmental conditions – as the UN-endorsed Local Agenda 21 acknowledges (Alexander, 2000: 27–8; Burgess *et al.*, 1997: 75, 86). Where communities have had a long period of coexistence with physical hazard, there is often an impressive accumulation of experience and strategy (Alexander, 2000: 34; Zaman, 1999: 208). But in many cases, from the Andean civilisations mentioned above to Qing state China and Mogul India, colonial influence has undermined local adaptations (Oliver-Smith, 1999: 80–4; Davis, 2001: ch. 9). Later urban migration has further distanced large sectors of hazard–prone populations from their customary means of crisis aversion (Alexander, 2000: 96).

There may be potential, however, to salvage traditional knowledge, as Oliver-Smith has suggested, for use in tandem with global technical and scientific expertise (1999: 87). There is also a sense in which the cultural fluidity and spatial mobility engendered by modern urbanism might have its own part to play in adapting to rapid ecological transition. But whatever socio-cultural resources might be tapped, the bottom line is that adaptation to environmental change on a broad regional or global front needs space. It requires fallow land, buffer zones, under-utilised resources in large enough supply to absorb massive displacements of human and non-human life (see Eisenberg, 1998: 398–409). Destructive episodes call for creative responses and, as Steven Jay Gould points out, 'the watchwords for creativity are sloppiness, poor fit, quirky design, and above all else, redundancy' (cited in diZerega, 1993: 39). 'Redundancy' is likewise the engineering term for surplus capacity, the term Mike Davis uses to sum up just what it is that the hazard-prone city of Los Angeles needs but will not countenance (1998: 53). In this sense, it becomes apparent that 'resilience' or 'flexibility' – with their connotation of 'uncommitted potentiality for change' (Bateson, 1973: 473) – are deeply at odds with the overtones of efficiency and economy – the drive to set all available resources to work – characteristic of the idea of development.

Is it possible that the 'ecologising' of the urban might be wrested from its enthralment to stability, continuity and, more alarmingly, development – in favour of concessions to chance, volatility and inbuilt redundancy? Might urbanism, under threat of catastrophes that are at once destructive and creative, be enticed 'to leave room within the programme, at the border of the programme, for the unprogrammable, for the incalculable' (Derrida, 2001: 259)? Or will the confrontation with environmental issues, as currently seems to be the case, lead us deeper into the terrain of scrutinisation, accountability and regulation? For these are precisely the aspects of urban existence that connoisseurs of city living have taken such pleasure in evading, as they have set about making the most of the unscripted, irruptive and exuberant dimensions of everyday life in the metropolis. But what if, in a roundabout way, it was this very warming to turbulence and surprise, nurtured in the heart of the city, that has helped pave the way for the scientific appreciation of the wider world's open-ended 'becoming'? And what might happen to environmental consciousness and action if the 'taste for the promiscuous and the unstable' that cosmopolitan city-dwellers bring to bear on cultural variability were to be extended to embrace the biophysical unruliness and instability that is no less a part of urban existence (Louis Aragon, cited in Clark, 2000: 15)? While we have every reason to be uneasy about the prospects of urban civilisation at a planetary scale, is there not, perhaps, a sense in which the turn towards a more resilient and flexible urbanism might hinge as much on fascination as on fear?

FURTHER READING

Alexander, David (2000) *Confronting Catastrophe*, Harpenden, Herts: Terra.

Clark, Nigel (2000) 'Botanizing on the asphalt? The complex life of cosmopolitan bodies', *Body & Society*, 6 (3–4), 12–33.

Davis, Mike (1998) *Ecology of Fear: Los Angeles and the Imagination of Disaster*, New York, NY: Vintage.

De Landa, Manuel (1997) *A Thousand Years of Nonlinear History*, New York, NY: Swerve.

Eisenberg, Evan (1998) *The Ecology of Eden: Humans, Nature and Human Nature*, London: Picador.

Girardet, Herbert (1999) *Creating Sustainable Cities*, Foxhole, Devon: Green Books.

CONCLUSION

In the preceding chapter Nigel Clark, citing Derrida (2001: 259), asks if urbanism can face the threat of catastrophe by admitting the unprogrammable at the border of its programme. Clark alludes to the enjoyment of unscripted encounters which he sees as characteristic of metropolitan life, and contrasts this with the heavy tread of regulation. Urbanism, it could be argued, is a discourse in relation to which various practices are situated, though it would be equally necessary to say that the discourse is situated in relation to the practices; but it would be difficult to argue that urbanism has a programme. Although the notion of a modern project has currency as an attempt to construct a society of the new, this is a figuration encapsulated in histories. Leaving that aside, the prospect of admitting the unprogrammable raises many questions. This conclusion examines three of them as a way to end the book without either summing up or unveiling a grand solution. If this is a diffusion rather than a bang, the grand finale is not the only way to end a piece of music. Sometimes the lingering sounds of a procession passing by give rise to unpredicted speculations. The questions are: first, how real and of what kind is the threat of catastrophe? second, what is the relation between the unprogrammable and claims for regulation in the public interest? and third, why is it relegated to the borders?

Clark adopts a concept of turbulence derived from the volatility of natural forces. He explains the urban desire for a countryside, either external to the city or in forms such as a landscaped park or suburban garden, as a quest for a safe middle ground between the wilderness from which city life gives freedom and its settled, static aspect. Disasters are most threatening to high-density populations, and Clark goes on to say that upheavals re-set patterns in ways beyond prediction. The implication is that the model of a city worked out in the relatively stable climatic, tectonic and social conditions of the northern hemisphere, with its middle grounds, may have limitations which will become increasingly evident. The brittleness of the concept then becomes the threat.

For others, outside the scope of this book, it is the rapidity of growth in cities in the non-affluent world which renders them unstable, and the pressures of global capital which wreck local economies: David Drakakis-Smith notes that the fastest expansion takes place in what are already the largest cities (Drakakis-Smith, 1990: 18), and Oswaldo de Rivera sees national economies in the non-affluent world as less viable when global production of commodities is relocated to centres of new technologies and raw material prices fall. He concludes that when 'the virus of scientific and technological poverty colludes with another non-viability virus, such as demographic explosion, non-development is virtually inevitable' (de Rivera, 2001: 118–19). A further threat is identified by Jennifer Elliott (1994: 34–5) in a rise in the rate of global water use over that of population growth, from which she predicts a water crisis. While Clark sees the city as failing to adapt to conditions (post-Darwin?), other voices see a need to adapt conditions (post-Marx?).

The threats perceived establish an axis from nature to culture, but also from the global to the local, the macro-scale of climate change and world economy to the micro-scale of impacts on and actions by specific groups. In chaos theory the points along the axis are connected; local incidents contribute to varied outcome paths (Byrne, 1997) while being produced within conditions which they re-produce. But perhaps the most immediate threat derives not from turbulence itself but from its uneven impact. Just as wealth is unevenly distributed, so some groups in societies are insulated against disturbance and others unduly liable to devastation. Indeed some devastations, economic and environmental, are produced by the operations of globalised capital. Money does its business while far-off places, such as the Ogoni lands of Nigeria, burn.

But this is more complex than disdain on the part of senior executives. Some transnational companies have links to repressive regimes, but while cocktails are served in the hospitality suite (if they are) the everyday use of products in suburban homes maintains the commercial viability of the producers. And while spectacular instances of destructiveness are distanced – an urban ghetto being as conceptually remote from centres of affluence as a rainforest – everyday demolition takes place on urban fringes. Jo Beall writes: 'The bulldozing of informal settlements to make way for the homes, shops and leisure facilities of more powerful groups is an all too common scandal' (Beall, 1997: 18). Among the leisure facilities may be golf courses consuming vast quantities of water. One outcome is a peripheralisation of the poor and abandonment of the migrant. In the Cathedral of São Paolo, Jeremy Seabrook (1993: 43–50) observes indigenous people huddled asleep whose flight from the degraded habitat of the Amazon has brought them to equally violent infringement in the city. The poor, women and members of groups regarded as minorities suffer most from uneven distributions of power, wealth and access to services. Can this polarisation continue without bringing some groups to a point at which they spontaneously, perhaps violently, assert a right to the city? Is the first threat that of riot – a destruction of the city not by flood but fire, as in Los Angeles in April 1992?

Probably the polarisation will continue: the dispossessed are members of a fragmented category not of their own making; their capacity for resistance is minimal compared with the legal, economic, or military force available against them. Poor countries can do little against the power block of the rich world where states resemble out-sourced governmental services providers for transnational corporations. But is there a possibility of rebellion from within when transnational capital exhibits itself as vicious? In Euripides' *The Trojan Women*, the Greek herald Talthybius sees his side's violent excess in the decision to kill Andromache's infant son. To the Greek commanders, as the play represents them to an audience remote from the scene, the necessity to extinguish the Trojan royal line is a consequence of defeat. For Talthybius, it stinks. He washes the infant's wounds himself and returns the body to the Trojan women for burial in restoration of a common rite. This is no help to Andromache (already on board ship, heading for slavery in Greece) and the comfort of his action is his own, yet it is an image of refusal, a possibility for redemption from barbarism antagonised upon a public stage.

Euripides' plot portrays desperation within the play's equivalent of a ruling class while twentieth-century narratives of social change locate revolt in an oppressed group, conventionally the working class. For Lenin (1975: 35–41) the problem is how to move from embryonic working-class consciousness in spontaneous acts to struggle as organised, or educated, consciousness. But if, as Marcuse argues, the working class no longer has the energy of change, it is to subversion within elites and institutions that attention turns. Marcuse writes:

> The integration of the largest part of the working class into the capitalist society is not a surface phenomenon; it has its roots in the infrastructure itself . . . [in] benefits accorded to the metropolitan working class thanks to surplus profits, neocolonial exploitation, the military budget, and gigantic government

> subventions. To say that this class has much more to lose than its chains may be a vulgar statement but it is also correct.
>
> (Marcuse, 1972: 6)

The question then is whether opposition among elites – which for Marcuse from the late 1960s onwards included the déclassé class, so to say, of students and academics in universities – remains spontaneous, follows Lenin's trajectory of organisation, or becomes self-organising.

To move, then, to the second question: what is the relation between the unprogrammable and claims for regulation in the public interest? Spontaneous resistance is not programmed and its means of going ahead may be formulated in the process of revolt; regulation relies on the institutional structures of a society, yet may be for the public good (if that can be identified). Taking the problems of urban mobility on which Hall writes, it would be possible (if unlikely given the strength of the oil lobby) to introduce a package of legislation and fiscal measures accompanied by infrastructural development to transfer most urban movement from roads to rail and tram. At the same time, the present situation in which market forces predominate is not a natural condition but made by deregulation, which is a form of regulation within a globalised political system. And, as Young (1990) argues, a regulatory framework of equality aimed at assimilation may do little to advance the interests of those categorised as minorities whose legal freedom is to be like rather than to be different. As Young also writes (2000: 43), the idea of a common good above individual interests tends to reflect the good of the group framing it, may be exclusionary, and may narrow the agenda for discussion of areas of negotiable discord. Young writes that 'the ideal of aiming to reach agreement normatively regulates meaningful dialogue' but that 'conflict and disagreement are the usual state of affairs' even in well-structured settings (Young, 2000: 44). Young's alternative is contingent cooperation between groups, perhaps a less abstract process than Habermas' communicative action.

Co-existent mobilisation may work locally, but how do groups address the limits of global power? Regulation currently seems part of the problem: trade rules favour the richest nations; in protocols such as Kyoto much is compromised without guarantee of delivery; organisations which regulate world trade and monetary policy – the World Trade Organization (WTO) and International Monetary Fund – affirm the power of capital. Bauman notes (1998: 85–9) that local impediments to profit can be appealed against in transnational fora, and Vandana Shiva writes that the WTO 'has earned itself names such as World Tyranny Organization because it enforces anti-people, anti-nature decisions to enable corporations to steal the world's harvests' (Shiva, 1999). Shiva claims that the WTO acts secretively and undemocratically, making the actions of the 50,000 demonstrators at its 1999 meeting in Seattle a legitimate resistance. So much for regulation? Not quite. Wilson (1991) calls for a non-patriarchal reinvestment of planning for a city for all, against abandonment of agency to market forces. This is not a choice between planning and laissez-faire: all economies are command economies, the question being whether the state or capital is in command. An alternative is a localised, non-money economy in which producers transact directly with consumers, as in a local economy trading scheme (LETS), or in the tradition of anarchism on a basis of 'from all according to ability and to all according to need'. LETS, though still a tiny part of the total economy in the US and UK, appear an interesting case of a self-regulating structure, perhaps with wider implications.

But can there be planning for all or does this conjure a communitarian illusion? James Donald proposes a 'pragmatic urbanity which can make the violence of living together manageable' (Donald, 1999: 167). He begins a chapter in *Imagining the Modern City* by citing Raymond Williams' adherence to a sense of home rooted in his background in rural Wales, and contrasts this with Doreen Massey's argument that communities are not defined only by place but also by interest, and that place-based communities have long since been rare (Massey, 1994: 153, cited in Donald, 1999: 148). Donald, with a nod to Lyotard, understands Williams' position as nostalgic and compensatory, and turns to Simmel for an image of ambivalence between closeness and remoteness in the stranger – a precursor of the unassimilable difference proposed as the urban condition by Young (1990). For Donald, efforts at a reconstruction of community founder because, when the tactics of communal politics are stated, they are as normative as before: 'They seldom quite capture the wiliness, the aggression, and the everyday paranoia which are inescapable features of sharing urban turf' (Donald, 1999: 157). And from this he moves to his pragmatic and collaborative image of damage limitation.

Planning, then, might be redefined as a process of managing the fora in which groups contest each other's claims for space while regulating for themselves the edges, however awkward, of their contact. This is a long way from Richard Rogers' vision of a latte-drinking piazza-sitting society, as nostalgic as Williams' attachment to his village but owing more to Tuscany than the Black Mountains and dependent on a social classlessness which is in effect the homogeneity of a bourgeois class (Rogers and Gumuchdjian, 1997); nor does it correspond to a pizza society lower down the social scale, though some of those edges might emerge when the piazza-goers tread in the litter of fast food containers and stuff dogs leave on streets, as they walk back to their apartments by a sparkling Thames. It raises many issues beyond the scope of this essay as to how the citizens in question – diverse according to multiple taxonomies – are described, perform themselves for others, or see themselves alone. If it is accepted that these selves are contingent, in formation in reciprocal relation to others' formations within a globalised culture and economy, as is evident in the work described by Rowe, then access to a public sphere of performance is only, as Donald (1999: 169) points out in a comment on Young, part of the problem.

A particular issue is where a public sphere – or *polis* – would be located when television brings publicity into domestic interiors and urban public space becomes commodified, as in Sony Plaza (Berlin, New York, wherever). Cities may allow escape from families but not from style, though consumers can be adept at playing games, as Steven Miles notes in his chapter's reference to youth in Australian malls. A further case of informally structured occupation is given by Jenny Ryan and Hilary Fitzpatrick (1996) in a description of the use by distinct groups of particular areas affording opportunities to gaze or be gazed at in the cafés of Manchester's gay district (used by gay and straight consumers of both genders). What emerges is that, without overt regulation but in highly specific environments, social groups constitute themselves and regulate their own multiple uses of space. At the same time, the case of Manchester illustrates also a depoliticisation of space: next to each other under the railway arches are two bars, one called Fat Cats, the other Revolution; further along by the canal side is another called Choice.

For Young it is civil society which limits the colonisation of the lifeworld by the state or capital. She sees civic associations as embodying participatory governance but sees a role, too, for the state in advancing social justice:

> Promoting social justice requires attending to issues of self-development as well as self-determination. Left to themselves, both the organization and consequences of capitalist market activity impede the self-development of many people .... Economic and infrastructure planning, redistributive policies, and the direct provision of goods and services by the state can minimize material deprivation.
>
> (Young, 2000: 189)

There is much to be said for Young's careful elaboration of a fluid relation between social groups and social organisation. Is there a possibility alongside this for direct action? Culturally, this may mean intervention in the cracks of the dominant society to reveal its contradictions. Artist Jane Trowell writes of art as a virus, the artist 'slipping a proposition into the blood-stream' (Trowell,

2000: 107) for unknown effect, which might be said also of the work of Mierle Ukeles on which Phillips writes.

A difficulty is that incidental or ephemeral acts tend to lack visibility. This may not matter if they nevertheless provoke a change of consciousness in individuals who experience them, but it leads into the third question: why does Derrida locate the unprogrammable at the border of the programme?

This could mean simply that the effects of impromptu acts nudge dominant images in new directions. Change, then, is incremental, accumulated gradually as encounters bring into question and slightly shift previously accepted norms. This has much in common with the idea of working in the crevices, slowly but irreversibly undermining the dominant society. But it is also possible to interpret the remark in terms of radical planning's avowal of a repositioning within mobilised but disempowered groups – the planner going AWOL as Sandercock puts it (1998a: 99–100). But is this self-marginalisation? Artists have been accused of that since they were thrust into the market economy, but planners might have been thought more reliable. Or is the margin merely acceptance of the centre's (non-legitimate) claim to power? The following report from an engineering magazine emphasises the need for a radical departure of some kind from the planning norm:

> A masterplan to rebuild the devastated centre of earthquake hit Indian city Bhuj has been delayed . . . [the planning firm] was told its five month study to produce a masterplan . . . was not detailed enough. [The firm] said the original brief did not demand the detail now requested by the development corporation. It has been asked to produce an extra plan of the walled city including the location of individual land plots, roads and open spaces.
>
> (*New Civil Engineer*, 2002: 12)

The information may be already known to dwellers in Bhuj – who will not be asked and for whom legal ownership might not be the most important factor. That the problem of restrictive models of knowledge which devalue that of dwellers, or of regeneration ignorant of local needs, is not confined to the non-affluent world is demonstrated by Evans and Foord in their chapter. Loftman and Nevin also show that much development is driven by economic objectives not easily compatible with an agenda of social inclusion. But perhaps it is a sense of local knowledges – diverse in kind and expression as well as content – which has most to offer urban sustainability. In one way the unprogrammable is at the borders; in another its position throws attention onto the limitations of the centre-margin model, as onto the restrictions of predictability.

Borden, in his chapter, states that radical architecture begins where architecture's death is proclaimed. This leads to new forms of education, production, and self-management, in permanent revolution. Collins and Goto, too, abandon the artworld, using skills of visualisation and organisation to initiate processes which are social and dialogic. If radical practices are marginal this again questions, or disrupts, centralism. Public monuments, for example, construct national histories as a means to convey national identities, but the cultivation of allotments is no less an articulation of a nation's way of life; what differs is the kind and extent of recognition. To assimilate one to the other tends to mean colonising the subjected, as in adding women to men's cultural history on men's exclusionary terms, and makes no difference. Neither do attempts to democratise the monument by taking away the plinth, or making statues of what are called ordinary people, detract from its historically acquired meaning and status, any more than allowing the lower classes to pay to look round royal palaces abolishes class divisions. If one story remains the norm, hegemonies thrive. The alternative is complexity: the intricacies of multiple but non-hierarchically ordered identities, contacts, and perceptions. At the end of *Transnational Urbanism*, Michael Peter Smith writes:

> To understand the future of urban change we must . . . focus our attention upon the communicative circuits, no matter how complex, by which people are connected to each other, make sense of their lives, and act upon the worlds that they see, in which they dwell, and through which they travel
>
> (Smith, 2001: 194)

If these circuits construct no centres there are no margins either, only borders, overlaps, and intersections – or spaces in between, thresholds of emancipatory impulses where Klee's (and Rendell's) angel stands.

BIBLIOGRAPHY

Abercrombie, P. (1944) *The Greater London Development Plan*, London: HMSO.

Adam, B., Beck, U. and van Loon, J. (2000) *The Risk Society and Beyond: Critical Issues for Social Theory*, London: Sage.

Adorno, T.W. (1991) *The Culture Industry*, ed. J.M. Bernstein, London: Routledge.

Adorno, T.W. (1997) *Aesthetic Theory*, London: Athlone.

Adorno, T.W. and Horkheimer, M. (1997) *Dialectic of Enlightenment*, London: Verso.

Ainley, R. (ed.) (1998) *New Frontiers of Space, Bodies and Gender*, London: Routledge.

Aitken, S. (2001) *Geographies of Young People: The Morally Contested Spaces of Identity*, London: Routledge.

Akbar, Shireen (ed.) (1999) *Shamiana: The Mughal Tent*, London: V&A Publications.

Aldous, T. (1992) *Urban Villages: A Concept for Creating Mixed-use Urban Developments on a Sustainable Scale*, London: The Urban Villages Group.

Alexander, David (2000) *Confronting Catastrophe*, Harpenden, Herts: Terra.

al-Khalil, S. (1991) *The Monument: Art, Vulgarity and Responsibility in Iraq*, Berkeley: University of California.

Amin, A., Masse, D. and Thrift, N. (2000) *Cities for the Many Not the Few*, Bristol: The Policy Press.

Appleyard, D. (1981) *Livable Streets*, Berkeley, CA: University of California Press.

Arendt, H. (1958) *The Human Condition*, Chicago, Ill.: The University of Chicago Press.

Arts Worldwide (1999) *Proposals for the Arts Worldwide Bangladesh Festival*, London.

Ashcroft, B., Griffiths, G. and Tiffin, H. (eds) (1995) *The Post-Colonial Studies Reader*, London: Routledge.

Association of Bay Area Governments (1999) *ABAG Projections 2000: City, County, and Census Tract Forecasts, 1999–2020*, Oakland, CA: Association of Bay Area Governments.

Association of Bay Area Governments (2001) *Regional Housing Needs: 1999–2006 Allocation*, Oakland, CA: Association of Bay Area Governments.

Attali, J. (1985) *Noise: The Political Economy of Music*, Minneapolis: University of Minnesota Press.

Auping, M. (1983) 'Earth Art: A Study in Ecological Politics' in A. Sonfist (ed.) *Art in the Land: A Critical Anthology of Environmental Art*, New York, NY: E.P. Dutton, Inc.

Bachelard, G. (1994) *The Poetics of Space: The Classical Look at How We Experience Intimate Places*, Boston: Beacon Press.

Badshah, Akhtar A. (1996) *Our Urban Future: New Paradigms for Equity and Sustainability*, London: Zed Books.

Baker, S., Kousis, M., Richardson, D. and Young, S. (1997) 'Introduction: The Theory and Practice of Sustainable Development in EU Perspective' in S. Baker, M. Kousis, D. Richardson and

S. Young (eds) *The Politics of Sustainable Development: Theory, Policy and Practice within the European Union*, London: Routledge.
Ballard, J.G. (1994) *Empire of the Sun*, London: Flamingo.
Banham, R. (1960) *Theory and Design in the First Machine Age*, London: Architectural Press.
Barber, S. (1995) *Fragments of the European City*, London: Reaktion.
Barnett, C. (1999) 'Culture, government and spatiality. Reassessing the "Foucault effect" in cultural policy studies', *International Journal of Cultural Studies*, 2 (3), 369–97.
Barth, U. (2000) 'Uta Barth in Conversation with Sheryl Conkelton, 1996' in Sheryl Conkelton and Uta Barth, *In Between Places*, University of Washington, Seattle: Henry Art Gallery.
Bateson, G. (1973) *Steps to an Ecology of Mind*, London: Paladin.
Bauman, Z. (1988) *The Consequences of Modernity*, Cambridge: Polity Press.
Bauman, Z. (1991) *Modernity and Ambivalence*, Cambridge: Polity Press.
Bauman, Z. (1993) *Postmodern Ethics*, Oxford: Blackwell.
Bauman, Z. (1998) *Globalization: The Human Consequences*, Cambridge: Polity Press.
Beall, J. (1997) 'Valuing Difference and Working with Diversity', and 'Participation in the City' in J. Beall (ed.) *A City for All*, London: Zed Books.
Beall, J. (ed.) (1997) *A City for All*, London: Zed Books.
Beardsley, J. (1977) *Probing the Earth: Contemporary Land Projects*, Washington, DC: Hirschorn Museum, Smithsonian Institution.
Beardsley, J. (1984) *Earthworks and Beyond*, New York, NY: Abbeville Press.
Beauchamp-Byrd, M.J. (1997) 'London Bridge: Late twentieth century British art and the routes of "national culture"', *Transforming the Crown: African, Asian and Caribbean Artists in Britain 1966–1996* (exhibition catalogue), New York: Franklin H. Williams Caribbean Cultural Centre/African Diaspora Institute.
Beck, U. (1992) *Risk Society: Towards a New Modernity*, London: Sage.
Begg, I. (1999) 'Cities and competitiveness', *Urban Studies*, 36 (5–6), 795–809.
Bell, C. and Newby, H. (1976) 'Communion, Communialism, Class and Community Action: The Sources of New Urban Politics' in D. Herbert and R. Johnson (eds) *Social Areas in Cities*, Vol. 2, Chichester: Wiley.
Bell, D. (1999) 'You can call me Al', *Birmingham Evening Mail*, 10 May, 7.
Bell, M.M. (1998) 'Culture as Dialogue' in M.M. Bell and M. Gardiner, *Bakhtin and the Human Sciences*, London: Sage, 49–62.
Bell, M.M. and Gardiner, M. (1998) *Bakhtin and the Human Sciences*, London: Sage.
Bellah, R., Madson, R., Sullivan, W.M., Swidler, A. and Tipton, S.M. (1985) *Habits of the Heart: Middle America Observed*, London: Hutchinson.
Benjamin, W. (1973) *Charles Baudelaire: A Lyric Poet in the Era of High Capitalism*, London: New Left Books.
Benjamin, W. (1979) *One-Way Street*, London: Verso.
Benjamin, W. (1985) *The Origins of German Tragic Drama*, London: Verso.
Benjamin, W. (1986) *Moscow Diary*, ed. G. Smith, Cambridge, Mass.: Harvard University Press.
Benjamin, W. (1992) *Illuminations*, London: Fontana Press.
Bennett, S. and Butler, J. (eds) (2000) *Locality, Regeneration and Diversity*, Bristol: Intellect Books.
Bennett T. (1995) 'The multiplication of culture's utility', *Critical Enquiry*, 21, 861–89.
Berman, M. (1983) *All That Is Solid Melts Into Air: The Experience of Modernity*, London: Verso.
Bhaba, Homi K. (ed.) (1990) *Nation and Narration*, London: Routledge.
Bhaba, Homi K. (1994) *The Location of Culture*, London: Routledge.

Bianchini, F. (1990) *Flagship Projects in Urban Regeneration*, Centre for Urban Studies, Liverpool: University of Liverpool.

Bianchini, F., Dawson, J. and Evans, R. (1992) 'Flagship Projects in Urban Regeneration' in P. Healey, S. Davsudi, M. O'Toole, S. Tavsanoglu and D. Usher (eds) *Rebuilding the City: Property-led Urban Regeneration*, London: E. and F.N. Spon, 245–55.

Bianchini, F. and Parkinson, M. (eds) (1993) *Cultural Policy and Urban Regeneration: The Western European Experience*, Manchester: Manchester University Press.

Birmingham City Council (n.d. a) *Eastside Story*, Birmingham: Birmingham City Council.

Birmingham City Council (n.d. b) *Millennium Point: The Midlands' Project for the Year 2000 and Beyond*. . . Birmingham: Birmingham City Council.

Birmingham City Council (1991) *Facing the Challenge in Birmingham: East Birmingham City Challenge*, Birmingham: Birmingham City Council.

Birmingham City Council (1992) *City Centre Strategy*, Birmingham: Birmingham City Council.

Birmingham City Council (1997) *City Living Birmingham Draft Strategy*, Birmingham: Birmingham City Council.

Birmingham City Council (1998) *Arena Central – Overview*, Report of the Director of Planning and Architecture to Planning Committee, 2 April 1998, Birmingham: Birmingham City Council.

Birmingham City Council (2000) '£400m development set for West End', *Birmingham Briefing*, March/April.

Birmingham City Council (2001) *Highbury 3: Report of Proceedings*, Birmingham: Birmingham City Council.

Birmingham City Council (2002) *Birmingham Residents' Survey 2001/2 Final Report*, Birmingham: Birmingham City Council.

Birmingham City Pride (1994) *Birmingham City Pride First Prospectus: Appendices*, Birmingham: Birmingham City Pride.

Birmingham Evening Mail (1999) 'Massive challenge for new leader', 10 May, 10.

Birmingham Marketing Partnership (1999a) 'Mailbox delivers', *Partnership Birmingham*, Autumn, 12.

Birmingham Marketing Partnership (1999b) 'Another first for Birmingham', *Partnership Birmingham*, Spring, 1.

Biswas, S. (2001a) 'The Autobahn' in A. Ghosh and J. Lamba (eds) *Beyond Frontiers: Contemporary British Art by Artists of South Asian Descent*, London: Saffron Books.

Biswas, S. (2001b) in *Chanting Heads 2001: A Glimpse of Eleven Visual Artists Working in Britain Today* (CD ROM), AAVA, NSEAD and the John Hope Franklin Centre in association with the University of East London, the Clore Duffield Foundation and the Arts Council of England. Essays by Rohini Malik Okon; teacher's notes by Paul Dash.

Bloch, E. (1986) *The Principle of Hope*, Cambridge, Mass.: MIT Press.

Bloch, E. (1991) *Heritage of Our Times*, Cambridge: Polity Press.

Blowers, A. (ed.) (1993) *Planning for a Sustainable Environment*, London: Earthscan.

Blowers, A. and Leroy, P. (1996) 'Environment and Society: Shaping the Future' in Andrew Blowers and Pieter Glasbergen (eds) *Environmental Policy in an International Context: Prospects for Environmental Change*, London: Arnold.

Blowers, A. and Pain, R. (1999) 'The Unsustainable City' in S. Pile, C. Brook and G. Mooney (eds) *Unruly Cities? Order/Disorder*, London: Routledge/Open University.

Bookchin, M. (1974) *Our Synthetic Environment*, New York, NY: Colophon.

Borden, I. (2001) *Skateboarding, Space and the City*, Oxford: Berg.

Borja, J. and Castells, M. (1997) *Local and Global: Management of Cities in the Information Age*, London: Earthscan.

Botkin, D. (1990) *Discordant Harmonies: A New Ecology for the Twenty-First Century*, New York, NY: Oxford University Press.

Boyce, S. (1992) 'Interview with Zarina Bhimji', *I Will Always Be Here* exhibition, African and Asian Visual Artists Archive, University of East London.

Bracewell, M. (1997) 'London – A Modern Subject' in Rut Blees Luxemburg, *Rut Blees Luxemburg, 'London – A Modern Project'*, London: Black Dog Press.

Bradshaw, A.D. (1995) 'Goals of Restoration' in J.M. Gunn (ed.) *Restoration and Recovery of an Industrial Region*, New York, NY: Springer-Verlag.

Breheny, M. (1995) 'The compact city and transport energy consumption', *Transactions of the Institute of British Geographers*, 20 (1), 81–101.

Breheny, M. and Hall, P. (1996) 'Four million households – where will they go?' *Town and Country Planning*, February, 39–41.

Bridge, G. and Watson, S. (eds) (2000) *A Companion to the City*, Oxford: Blackwell.

Brooke, P. (1994) 'A strategy for supporting millennium projects', *AMA Leisure and Tourism Circular* 28/94, Association of Metropolitan Authorities.

Brotchie, J. (1992) 'The changing structure of cities', *Urban Futures*, February, 13–23.

Brown, B., Green, N. and Harper, R. (eds) (2002) *Wireless World: Social and Interactional Aspects of the Mobile Age*, London: Springer Press.

Buck-Morss, S. (1991) *The Dialectics of Seeing: Walter Benjamin and the Arcades Project*, Cambridge, Mass.: MIT Press.

Buck-Morss, S. (2001) 'A global public sphere?' *Radical Philosophy*, 111, 2–10.

Budiansky, S. (1995) *Nature's Keepers: The New Science of Nature Management*, London: Weidenfeld & Nicolson.

Bull, M. (2000) *Sounding out the City: Personal Stereos and the Management of Everyday Life*, Oxford: Berg.

Burgess, E.W. (1925) 'The Growth of the City: An Introduction to a Research Project' in R.T. LeGates and F. Stout (eds) *The City Reader*, London: Routledge.

Burgess, R., Carmona, M. and Kolstee, T. (1997) *The Challenge of Sustainable Cities: Neoliberalism and Urban Strategies in Developing Countries*, London: Zed Books.

Burgin, V. (1986) *The End of Art Theory: Criticism and Postmodernity*, New York: Humanities Press International.

Burgin, V. (1996) *In/Different Spaces: Place and Memory in Visual Culture*, Berkeley, CA: University of California Press.

Burgin, V. (2000) *Shadowed*, London: AA Publications.

Burgin, V. (2001) *Barcelona*, Madrid: Fundacio Antoni Tapies.

Burnham, J. (1974) *Great Western Salt Works: Essays on the Meaning of Post-Formalist Art*, New York, NY: George Braziller, Inc.

Byrne, D. (1997) 'Chaotic Places or Complex Places? Cities in a Post-industrial Era' in Sallie Westwood and John Williams (eds) *Imagining Cities: Scripts, Signs, Memory*, London: Routledge.

Cao, P. (1999) *Alquile un cuerpo/Rent-a-Body*, Oviedo: Maguncia Artes Graficas (with Creative Time, Inc.).

Carey, J. (1999) *The Faber Book of Utopias*, London: Faber & Faber.

Carroll, W.K. and Ratner, R.S. (1996) 'Master framing and cross-movement networking in contemporary social movements', *The Sociological Quarterly*, 37 (4), 601–25.

Casey, B., Dunlop, R. and Selwood, S. (1996) *Culture as Commodity: The Economics of the Arts and Built Heritage in the UK*, London: Policy Studies Institute.

Cashmore, E.E. (1984) *No Future: Youth and Society*, London: Heinemann.

Caygill, H. (1998) *Walter Benjamin: The Colour of Experience*, London: Routledge.

Cervero, R. (1995) 'Changing Live–Work Spatial Relationships: Implications for Metropolitan Structure and Mobility' in J. Brotchie, M. Batty, E. Blakely, P. Hall and P. Newton (eds) *Cities in Competition: Productive and Competitive Cities for the Twenty-first Century*, Melbourne: Longman.

Chambers, I. (1994) *Migrancy, Culture, Identity*, London: Routledge.

Chambers, I. and Curti, L. (eds) (1996) *The Post-Colonial Question: Common Skies, Divide Horizons*, London: Routledge.

Champion, A.G. and Townsend, A.R. (1990) *Contemporary Britain – A Geographical Perspective,* London: Edward Arnold.

Chaney, D. (1990) 'Subtopia in Gateshead: The MetroCentre as cultural form', *Theory, Culture and Society*, 7 (4), 49–68.

Chang, T.C. (2000) 'Renaissance Revisited: Singapore as a "Global City for the Arts"', *International Journal of Urban and Regional Research*, 24 (4), 818–31.

Cherry, D. (2000) *Beyond the Frame: Feminism and Visual Culture in Britain 1850–1900*, London: Routledge.

Chorneau, T. (2000) 'Housing hard times', *The Press Democrat*, 29 October, 1–7.

Clark, N. (2000) 'Botanizing on the asphalt? The complex life of cosmopolitan bodies', *Body & Society*, 6 (3–4), 12–33.

Clarke, D. (ed.) (1997) *The Cinematic City*, London: Routledge.

Clarke, J. (1975) 'The Skinheads and the Magical Recovery of Community' in S. Hall and T. Jefferson (eds) *Resistance Through Rituals: Youth Sub-cultures in Post-war Britain*, London: Hutchinson.

Coalition on Homelessness (1999) *Homelessness in San Francisco*, San Francisco, CA: Coalition on Homelessness.

Colls, R. and Dodd, P. (1986) *Englishness: Politics and Culture 1880–1920*, London: Croom Helm.

Commission of the European Communities (1990) *Green Paper on the Urban Environment*, Brussels: European Commission.

Conkelton, S. (2000) *Uta Barth: In Between Places*, University of Washington, Seattle: Henry Art Gallery.

Connor, S. (1997) 'The Modern Auditory I' in R. Porter (ed.) *Rewriting the Self: Histories from the Renaissance to the Present*, London: Routledge.

Cooper, D. (ed.) (1968) *The Dialectics of Liberation*, Harmondsworth: Penguin.

Copjec, J. and Sorkin, M. (eds) (1999) *Giving Ground: The Politics of Propinquity*, London: Verso.

Corrin, L. (1999) *Louise and Jane Wilson's Stasi City*, London: Serpentine Gallery, and Hannover: Kunstverein Hanover, Deutscheer Akadamischer Austauschdienst Berliner Künstlerprogramm.

Crary, J. (1992) *Techniques of the Observer*, Cambridge, Mass.: MIT Press.

Crawford, J.H. (2000) *Carfree Cities*, Utrecht: International Books.

Cress, D.M. and Snow, D.A. (1988) 'Mobilization at the Margins: Organizing by the Homeless' in Anne N. Costain and Andrew S. McFarland (eds) *Social Movements and American Political Institutions: People, Passions, and Power*, New York, NY: Rowman & Littlefield.

Cress, D.M. and Snow, D.A. (2000) 'The outcomes of homeless mobilization: The influence of organization, disruption, political mediation and framing', *American Journal of Sociology*, 105 (4), January, 1063–104.

Crilley, D. (1993) 'Architecture as Advertising: Constructing the Image of Redevelopment' in G. Kearns and C. Philo (eds) *Selling Places: The City as Cultural Capital Past and Present*, Oxford: Pergamon Press.

Cultural Strategy Partnership (2001) *Culture and the City*, London: Cultural Strategy Partnership for London.

Curtis, K. (1999) *Our Sense of the Real: Aesthetic Experience and Arendtian Politics*, Ithaca, NY: Cornell University Press.

Davies, J. (1995) 'Less Mickey Mouse, more Dirty Harry: Property, policing and the postmodern metropolis', *Polemic*, 5 (2), 63–9.

Davis, M. (1990) *City of Quartz*, London: Verso.

Davis, M. (1998) *Ecology of Fear: Los Angeles and the Imagination of Disaster*, New York, NY: Vintage.

Davis, M. (2001) *Late Victorian Holocausts: El Nino Famines and the Making of the Third World*, London and New York: Verso.

De Certeau, M. (1984) *The Practice of Everyday Life*, Berkeley, CA: University of California Press.

De Landa, M. (1997) *A Thousand Years of Nonlinear History*, New York, NY: Swerve.

de Rivera, O. (2001) *The Myth of Development: The Non-Viable Economies of the 21st Century*, London: Zed Books.

Dean, T. (2001) *Tacita Dean*, London: Tate Britain.

Debord, G. (1977) *Society of the Spectacle*, Detroit: Black and Red.

DeNora, T. (2000) *Music in Everyday Life*, Cambridge: Cambridge University Press.

Denzin, N.K. (1995) *The Cinematic Society*, London: Sage.

Department for Culture, Media and Sport (1998) *Creative Industries Mapping Document*, London: Department for Culture, Media and Sport.

Department for Culture, Media and Sport (1999) *Draft Guidance for Local Cultural Strategies*, London: Department for Culture, Media and Sport.

Department for Culture, Media and Sport (2001) *Creative Industries Mapping Document 2001*, London: Department for Culture, Media and Sport.

Department of the Environment (1993) *Sustainable Development: The UK Strategy*, Cmnd 2426, London: HMSO.

Department of the Environment, Transport and the Regions (1998) *New Deal for Transport: Better for Everyone*, London: HMSO.

Department of the Environment, Transport and the Regions (1999) *An Urban Renaissance, Final Report of the Urban Task Force*, London: HMSO.

Department of the Environment, Transport and the Regions (2000) *Our Towns and Cities: The Future – Delivering an Urban Renaissance*, London: HMSO.

Derrida, J. (2001) 'A Roundtable Discussion with Jacques Derrida' in L. Simmons and H. Worth (eds) *Derrida Downunder*, Palmerston North: Dunmore Press.

Deutsche, R. (1991) 'Uneven Development: Public Art in New York City' in D. Ghirardo (ed.) *Out of Site: A Social Criticism of Architecture*, Seattle: Bay Press.

Diller, E. and Scofidio, R. (eds) (1994) *Back to the Front: Tourisms of War*, New York, NY: FRAC Basse-Normandie/Princeton Architectural Press.

diZerega, G. (1993) 'Unexpected harmonies: Self-organization in liberal modernity and ecology', *Trumpeter*, 10 (1), Winter, 25–32.

Donald, J. (1999) *Imagining the Modern City*, London: Athlone.
Douglass, M. (1998) 'World City Formation on the Asia Pacific Rim: Poverty, "Everyday" Forms of Civil Society and Environmental Management' in J. Friedman and M. Douglass (eds) *Cities for Citizens*, Chichester: Wiley.
Douglass, M. and Friedman, J. (eds) (1998) *Cities for Citizens*, Chichester: Wiley.
Doy, G. (1999) *Black Visual Culture: Modernity and Postmodernity*, London: I.B. Tauris.
Drakakis-Smith, D. (1990) *Third World Cities*, London: Routledge.
du Gay, P. and Hall, S. (eds) (1997) *Doing Cultural Studies: The Story of the Sony Walkman*, London: Sage.
Duncan, J.S. (1996) 'Me(trope)olis: Or Hayden White Among the Urbanists' in A. King (ed.) *Re-Presenting the City: Ethnicity, Capital and Culture in the 21st Century Metropolis*, Basingstoke: Macmillan.
Duncan, N. (ed.) (1996) *Body Space: Destabilising Geographies of Gender and Sexuality*, London: Routledge.
Durning, A.T. (1996) *The Car and the City: 24 Steps to Safe Streets and Healthy Communities*, Seattle WA: Northwest Environmental Watch.
Eade, J. (ed.) (1997) *Living the Global City: Globalization as Local Process*, London: Routledge.
Eagleton, T. (2000) *The Idea of Culture*, Oxford: Blackwell.
Eaton, M.M. (1997) 'The Beauty that Requires Health' in J.I. Nassau (ed.) *Placing Nature: Culture and Landscape Ecology*, Washington, DC: Island Press.
Edensor, T. (1998) 'The Culture of the Indian Street' in N. Fyfe (ed.) *Images of the Street: Planning, Identity and Control in Public Space*, London: Routledge.
Eder, K. (1996) *The Social Construction of Nature*, London: Sage.
Ehrenfeld, J.G. (2000) 'Defining the limits of restoration: The need for realistic goals', *Restoration Ecology*, 8 (1), 2–9.
Eisenberg, E. (1998) *The Ecology of Eden: Humans, Nature and Human Nature*, London: Picador.
Elliott, J. (1994) *An Introduction to Sustainable Development*, London: Routledge.
Ellmeier, A. and Rasky, B. (1998) *Cultural Policy in Europe – European Cultural Policy? Nation-state and Transnational Concepts*, Vienna: International Archive for Cultural Analysis.
Engwicht, D. (1999) *Street Reclaiming: Creating Livable Streets and Vibrant Communities*, Gabriola Island, BC: New Society Publishers.
Escobar, A. (1996) 'Constructing Nature' in R. Peet and M. Watts (eds) *Liberation Ecologies*, London: Routledge.
Ettorre, E. and Miles, S. (2002) 'Young People, Drug Use and the Consumption of Health' in S. Henderson and A. Petersen (eds) *The Commodification of Health*, London: Routledge.
European Commission (1998) *Social Indicators: Problematic Issues*, Brussels: European Commission.
European Urban and Regional Studies (1999) Special Issue: 'Cities, Culture, Political Economy', *European Urban and Regional Studies*, 6 (4).
Evans, G.L. (1999) 'The economics of the national performing arts – exploiting consumer surplus and willingness-to-pay: A case of cultural policy failure?' *Leisure Studies*, 18 (2), 97–118.
Evans, G.L. (2001) *Cultural Planning: An Urban Renaissance?* London: Routledge.
Evans, G.L. and Bohrer, J. (2000) 'Urban Parks and Green Space in the Design and Planning of Cities' in J. Benson and M. Roe (eds) *Urban Lifestyles: Spaces, Places, People*, Rotterdam: A.K. Balkema, 147–54.

Evans, G. and Foord, J. (1999) 'Cultural Policy and Urban Regeneration in East London: World City, whose City?' in *Proceedings of the International Conference on Cultural Policy Research*, Bergen: University of Bergen, 457–94.
Evans, G.L. and Foord, J. (2000a) 'European funding of culture: Promoting common culture or regional growth?' *Cultural Trends*, 36, 53–87.
Evans, G.L. and Foord, J. (2000b) 'Landscapes of Cultural Production and Regeneration' in J. Benson and M. Rose (eds) *Urban Lifestyles: Spaces, Places, People*, Rotterdam: A.T. Balkema, 249–56.
Evans, G.L., Foord, J. and White, J. (1999) *Putting Stepney Back on the Cultural Map: An Investigation into the Potential for Local Cultural Activity*, London: London Borough of Tower Hamlets.
Evans, J. and Hall, S. (eds) (1999) *Visual Culture: A Reader*, London: Routledge.
Exley, R., 'Sites of absence', *Contemporary Visual Arts*, 20, 63–8.
Fainstein, S.S. (1994) *The City Builders: Property and Planning in London and New York*, Oxford: Blackwell.
Fischer, E. (1973) *Marx in his Own Words*, Harmondsworth: Penguin.
Fisher, S. and Holder, S. (1981) *Too Much Too Young?* London: Pan.
Fiske, J. (1989) *Reading the Popular*, London: Routledge.
Foord, J. (1999) 'Creative Hackney: Reflections on hidden art', *Rising East,* 3 (2), 38–66.
Fortunati, V. (1993) 'The Metamorphosis of the Apocalypse Myth: From Utopia to Science Fiction' in K. Kumar and S. Bann (eds) *Utopias and the Millennium*, London: Reaktion.
Foucault, M. (1986) 'Heterotopias', *Diacritics*, Spring, 22–7.
Foucault, M. (1990) *The History of Sexuality, Volume 1*, London: Penguin.
Foucault, M. (1991a) *Discipline and Punish: The Birth of the Prison*, Harmondsworth: Penguin.
Foucault, M. (1991b) 'Governmentality' in G. Burchell, C. Gordon and P. Miller (eds) *The Foucault Effect: Studies in Governmentality*, London: Harvester Wheatsheaf.
Frantz, D. and Collins, C. (2000) *Celebration, USA: Living in Disney's Brave New Town*, New York: Henry Holt.
Freidberg, A. (1993) *Window Shopping*, Berkeley, CA: California University Press.
Freire, P. (1972) *Pedagogy of the Oppressed*, Harmondsworth: Penguin.
Frisby, D. (1985) *Fragments of Modernity: Theories of Modernity in the Work of Simmel, Kracauer and Benjamin*, Cambridge: Polity Press.
Frisby, D. (1992*) Simmel and Since: Essays on Georg Simmel's Social Theory*, London: Routledge.
Frisby, D. and Featherstone, M. (eds) (1997) *Simmel on Culture*, London: Sage.
Furlong, A. and Cartmel, F. (1997) *Young People and Social Change: Individualization and Risk in Late Modernity*, Buckingham: Open University Press.
Gallaccio, A. (1999) *Chasing Rainbows*, London: Tramway/Locus +.
Garreau, J. (1991) *Edge City: Life on the New Urban Frontier*, New York, NY: Doubleday.
Garrett, L. (1994) *The Coming Plague: Newly Emerging Diseases in a World Out of Balance*, New York: Penguin.
Garrett, M. and Taylor, B. (1999) 'Reconsidering social equity in public transport', *Berkeley Planning Journal*, 13 (1), 6–27.
Geoghegan, V. (1987) *Utopianism and Marxism*, London: Methuen.
Ghirardo, D. (ed.) (1991) *Out of Site: A Social Criticism of Architecture*, Seattle: Bay Press.
Ghosh, A. and Lamba, J. (eds) (2001) *Beyond Frontiers: Contemporary British Art by Artists of South Asian Descent*, London: Saffron Books.
Gilloch, G. (1996) *Myth and Metropolis: Walter Benjamin and the City*, Cambridge: Polity Press.

Gilloch, G. (2002) *Walter Benjamin: Critical Constellations*, Cambridge: Polity Press.
Girardet, H. (1999) *Creating Sustainable Cities*, Foxhole, Devon: Green Books.
Goffman, E. (1969) *The Presentation of Self in Everyday Life*, London: Penguin.
Goffman, E. (1971) *Relations in Public: Microstudies of Public Order*, London: Penguin.
Gordon, C. (1991) 'Governmental Rationality: An Introduction' in G. Burchell, C. Gordon and P. Miller (eds) *The Foucault Effect: Studies in Governmentality*, London: Harvester Wheatsheaf.
Gordon, D. (ed.) (1990) *Green Cities: Ecologically-Sound Approaches to Urban Space*, Montreal: Black Rose Books.
Gordon, P., Richardson, H.W. and Jun, M.J. (1991) 'The commuting paradox: Evidence from the top twenty', *Journal of the American Planning Association*, 57 (4), 416–20.
Gordon Childe, V. (1996, first published 1951) 'The Urban Revolution' in R.T. LeGates and F. Stout (eds) *The City Reader*, London: Routledge.
Graham, S. and Marvin, S. (1996) *Telecommunications and the City: Electronic Spaces, Urban Places*, London: Routledge.
Gratz, R.B. and Mintz, N. (1998) *Cities Back from the Edge: New Life for Downtown*, New York, NY: Wiley.
Grieco, M. (1995) 'Time pressures and low income families: The implications for "social" transport policy in Europe', *Community Development Journal*, 30 (4), 347–63.
Grieco, M., Turner, J. and Hine, J. (2000) 'Transport, employment and social exclusion: Changing the contours through information technology', *Local Work*, 26 (np).
Griffin, C. (1993) *Representations of Youth: The Study of Youth and Adolescence in Britain and America*, Cambridge: Polity Press.
Griffin, C. (1997) 'Troubled teens: Managing disorders of transition and consumption', *Feminist Review*, 55, Spring, 4–21.
Grimley, T. (1988) 'Focus on the city's future', *Birmingham Post*, 17 February, 7.
Groundwork Trust, (1999–2001) <http://www.groundwork.org.uk/public/reports/index.html>.
Grumbine, R.E. (1992) *Ghost Bears: Exploring the Biodiversity Crisis*, Washington, DC: Island Press.
Guha, R. and Martinez-Alier, J. (1997) *Varieties of Environmentalism*, London: Earthscan.
Gujral, R. (1994) 'Opinion', *Architecture Today*, 50, 7–8.
Gullo, A., Lassiter, U. and Wolch, J. (1998) 'The Cougar's Tale' in Jennifer Wolch and Jody Emel (eds) *Animal Geographies: Place, Politics and Identity in the Nature–Culture Borderlands*, London and New York: Verso.
Gupta, S. (ed.) (1993) *Disrupted Borders: An Intervention in Definitions of Boundaries*, London: Rivers Oram Press.
Gussin, G. and Carpenter, E. (eds) (2001) *Nothing*, London: August.
Habermas, J. (1989) *The Structural Transformation of the Public Sphere: An Inquiry into a Category of Bourgeois Society*, Cambridge: Polity Press.
Hall, S. (1990) 'Cultural Identity and Diaspora' in J. Rutherford (ed.) *Identity*, London: Lawrence & Wishart.
Hall, S. (2000) 'Multicultural Citizens, Monocultural Citizenship?' in N. Pearce and Hallgarten (eds), *Tomorrow's Citizens: Critical Debates in Citizenship and Education*, London: Institute for Public Policy Research, 43–51.
Hall, S. and Sealy, M. (eds) (2001) *Different*, London: Phaidon.
Hamilton, A. (1994) *Mneme*, Liverpool: Tate Gallery.
Hamilton, A. (1999) *Whitecloth*, Connecticut: The Aldrich Museum of Contemporary Art.

Hamilton, K., Ryley Hoyle, S. and Jenkins, L. (1999) *The Public Transport Gender Audit: The Research Report*, London: Transport Studies, University of East London.
Hannigan, J. (1998) *Fantasy City: Pleasure and Profit in the Postmodern Metropolis*, New York, NY: Routledge.
Hare, C. (2001) 'Mile End Park', in *Landscape London: A Guide to Recent Gardens, Parks and Spaces*, London: Ellipsis, 6, 6.
Hart, T. (2001) 'Transport and the City' in R. Padison (ed.) *The Urban Studies Handbook*, London: Sage.
Hartshorn, T. and Muller, P.O. (1989) 'Suburban downtowns and the transformation of metropolitan Atlanta's business landscape', *Urban Geography*, 10, 375–95.
Harvey, D. (1989) *The Condition of Postmodernity: An Enquiry into the Origins of Cultural Change,* Oxford: Blackwell.
Harvey, D. (1993) 'Goodbye to all that? Thoughts on the social and intellectual condition of contemporary Britain', *Regenerating Cities*, 5, 11–16.
Harvey, D. (1996) *Justice, Nature and the Geography of Difference*, Cambridge, Mass.: Blackwell.
Hathaway, W. and Meyer, D.S. (1997) 'Competition and Cooperation in Movement Coalitions: Lobbying for Peace in the 1980s' in Thomas R. Rochon and David S. Meyer (eds) *Coalitions and Political Movements: The Lessons of the Nuclear Freeze*, Boulder, CO: Lynne Rienner Publishers, 61–80.
Haughton, G. and Hunter, C. (1994) *Sustainable Cities*, London: Regional Studies Association.
Hawkins, R. (1986) 'A road not taken: Sociology and the neglect of the automobile', *California Sociologist*, 9, 61–79.
Hay, A. and Trinder, E. (1991) 'Concepts of equity, fairness and justice expressed by local transport policy makers', *Environment and Planning C: Government and Policy*, 9 (4), 453–65.
Haymann, N. (1994) *Resumed in Protest: The Human Cost of Roads*, Sydney: Bungoona Books.
Hays, S. (1959) *Conservation and the Gospel of Efficiency*, Boston, Mass.: Harvard University Press.
Healey, P. (1997) *Collaborative Planning: Shaping Places in Fragmented Societies*, London: Macmillan.
Healey, P. (2000) 'Planning in Relational Space and Time: Responding to New Urban Realities' in G. Bridge and S. Watson (eds) *A Companion to the City*, Oxford: Blackwell.
Hertmans, S. (2001) *Intercities*, London: Reaktion.
Hess, T.B. and Ashbery, J. (eds) (1968) *Avant-Garde Art*, London: Collier-Macmillan.
Hewison, R. (1995) *Culture and Consensus: England, Art and Politics since 1940*, London: Methuen.
Hewitt, K. (1983) 'The Idea of Calamity in a Technocratic Age' in K. Hewitt (ed.) *Interpretations of Calamity*, Winchester, Mass.: Allen & Unwin.
Hobbs, R. (1981) *Robert Smithson: Sculpture*, London: Cornell University Press.
Hodge, D.C. (1995) 'My Fair Share: Equity Issues in Urban Transportation' in S. Hanson (ed.) *The Geography of Urban Transportation*, New York, NY: Guilford Press.
Home Base (1997) *Homelessness at a Glance*, San Francisco, CA: Home Base.
Honneth, A. (1995) *The Fragmented World of the Social: Essays in Social and Political Philosophy*, New York, NY: SUNY Press.
Horkheimer, M. (1993) *Between Philosophy and Social Science: Selected Early Writings*, Cambridge, MA: MIT Press.

Hosokawa, S. (1984) 'The Walkman Effect', *Popular Music*, 4, 165–80.
Hough, M. (1990) 'Formed by Natural Processes – A Definition of the Green City' in David Gordon (ed.) *Green Cities: Ecologically-Sound Approaches to Urban Space*, Montreal: Black Rose Books.
Houtart, F. and Polet, F. (eds) (2001) *The Other Davos: The Globalization of Resistance to the World Economic System*, London: Zed Books.
Huby, M. and Burkitt, N. (2000) 'Is the "New Deal for Transport" really "Better for Everyone"? The social policy implications of the UK's 1998 Transport White Paper', *Environment and Planning C: Government and Policy*, 18 (4), 379–92.
Hughes, M. (1992) 'Regional Economics and Edge Cities' in Federal Highway Administration (ed.) *Edge City and ISTEA – Examining the Transport Implications of Suburban Development Patterns*, Washington, DC: Federal Highway Administration.
Hyde, L. (1998) *Trickster Makes This World: Mischief, Myth, and Art*, New York, NY: Farrar, Straus & Giroux.
Irigaray, L. (1994) *Thinking the Difference: For a Peaceful Revolution*, London: Athlone.
Jacobs, J. (1961) *The Death and Life of Great American Cities*, New York NY: Random House.
Jarvis, H., Pratt, A.C. and Cheng-Chong, P. (2001) *The Secret Life of Cities: The Social Reproduction of Everyday Life*, Harlow: Prentice Hall.
Jay, M. (1993) *Downcast Eyes: The Denigration of Vision in Twentieth Century French Thought*, Berkeley, CA: University of California Press.
Jones, G. and Wallace, C. (1992) *Youth, Family and Citizenship*, Buckingham: Open University Press.
Jordan, W.R. III (1984) 'A Perspective of the Arboretum at 50' in *Our First Fifty Years: The University of Wisconsin, Madison Arboretum, 1934–1984*. <http:// libtext.library.wisc.edu/ Arboretum/>.
Jordan, W.R. III (ed.) (1987) *Restoration Ecology: A Synthetic Approach to Ecological Research*, <http://libtext/library.wisc.edu/Arboretum/>.
Jordan, W.R. III (1994) 'Sunflower Forest: Ecological Restoration as the Basis for a New Environmental Paradigm' in A.D. Baldwin, J. De Luce and C. Pletsch (eds) (2000) *Beyond Preservation, Restoring and Inventing Landscapes*, Minneapolis, Minnesota: University of Minneapolis Press. Subsequently published in W. Throop (ed.) (2000) *Environmental Restoration: Ethics, Theory and Practice*, Amherst, NY: Humanity Books, an imprint of Prometheus Books.
Jordan, W.R. III, Gilpin, M.E. and Aber, J.D. (1987) *Restoration Ecology: A Synthetic Approach to Ecological Research*, Cambridge: Cambridge University Press
Kahn, D. and Whitehead, G. (eds) (1992) *Wireless Imagination: Sound, Radio and the Avant Garde*, Cambridge, Mass.: MIT Press.
Kastner, J. and Wallis, B. (Eds) (1998) *Land and Environmental Art*, London: Phaidon Press.
Kate ten K. (1994) 'Claws for thought: People power is fashionable with developers and planners once more. But does it work?' *Guardian*, 10 June, 16.
Katz, J. (1999) *Connections: Social and Cultural Studies of the Telephone in American Life*, London: Transaction Publishers.
Kawashima, N. (1997) 'Local authorities and cultural policy: Dynamics of recent developments', *Local Government Policy Making*, 23 (5), 31–9.
Kear, A. (2001) 'Parasites', *Parallax*, 7 (2), 31–46.
Kearns, G. and Philo, C. (eds) (1993) *Selling Places: The City as Cultural Capital; Past and Present*, Oxford: Pergamon.

Keith, M. (1995) 'Ethnic Entrepreneurs and Street Rebels: Looking Inside the Inner City' in S. Pile and N. Thrift (eds) *Mapping the Subject: Geographies of Cultural Transition*, London: Routledge, 335–70.

King, A.D. (1991) 'The Global, the Urban and the World' in A.D. King (ed.) *Culture, Globalization and the World System*, Basingstoke: Macmillan, 149–54.

King, A. (ed.) (1996) *Re-Presenting the City: Ethnicity, Capital and Culture in the 21st-Century Metropolis*, Basingstoke: Macmillan.

King, A. (2000) 'Postcolonialism, Representation and the City' in G. Bridge and S. Watson (eds) *A Companion to the City*, Oxford: Blackwell, 261–9.

King, C. (ed.) (1999) *Views of Difference: Different Views of Art*, New Haven and London: Yale University Press.

Knox, P. (1991) 'The restless urban landscape: Economic and socio-cultural change and the transformation of metropolitan Washington, DC', *Annals: Association of American Geographers*, 81 (2), 181–209.

Knox, P. and Pinch, S. (2000) *Urban Social Geography: An Introduction*, Harlow: Prentice Hall.

Kotler, P., Haider, D.H. and Rein, I. (1993) *Marketing Places*, New York: Free Press.

Kubricki, K. (2000) 'Review of Rut Blees Luxemburg, *Leibeslied II*', Laurent Delaye Gallery, London, *Make*, 91.

Kumar, K. and Bann, S. (1993) *Utopias and the Millennium*, London: Reaktion.

Kureshi, H. (1986) 'Bradford', *Granta*, 20 (Winter).

Laclau, E. (1996) *Emancipation(s)*, London: Verso.

Landry, C. (2000) *The Creative City: A Toolkit for Urban Innovators*, London: Earthscan.

Landry, C. *et al.* (1997) *Cultural Industries Strategy for Tower Hamlets*, Stroud: Comedia.

Landry, C. and Bianchini, F. (1995) *The Creative City*, London: Demos.

Lane, J. (1978) *Arts Centres – Every Town Should Have One*, London: Paul Elek.

Lang, P. (ed.) (1995) *Mortal City*, New York, NY: Princeton Architectural Press.

Langman, L. (1992) 'Neon Cages: Shopping for Subjectivity' in R. Shields (ed.) *Lifestyle Shopping: The Subject of Consumption*, London: Routledge.

Lash, S. and Urry, J. (1994) *Economies of Signs and Space*, London: Sage.

Law, C. (1996) *Tourism in Major Cities*, London: International Thomson Business Press.

Le Grand (1998) *Social Exclusion in Britain Today*, ESRC/CASE Seminar, London: LSE.

Leadbeater, C. (2000) *Living on Thin Air: The New Economy*, Harmondsworth: Penguin.

Lefebvre, H. (1991a) *A Critique of Everyday Life*, London: Verso.

Lefebvre, H. (1991b) *The Production of Space*, Oxford: Blackwell.

LeGates, R.T. and Stout, F. (eds) (1996) *The City Reader*, London: Routledge.

Lenin, V.I. (1975, first translation 1952) *What is to be Done?* Peking: Foreign Languages Press.

Leslie, E. (2000) *Walter Benjamin: Overpowering Conformism*, London: Pluto.

Lewis, P. (1983) 'The Galactic Metropolis' in R.H. Platt and G. Macinko (eds) *Beyond the Urban Fringe*, Minneapolis: University of Minnesota Press.

Ley, D. (1983) *A Social Geography of the City*, New York, NY: Harper & Row.

Light, A. (2000) 'Restoration, the Value of Participation and the Risks of Professionalization' in P.H. Gobster and B.R. Hull (eds) *Restoring Nature: Perspectives from the Social Sciences and Humanities*, Washington, DC: Island Press.

Lippard, L.R. (1983) *Overlay: Contemporary Art and the Art of Prehistory*, New York, NY: The New Press.

Litman, T. (1995) *Transportation Cost Analysis: Techniques, Estimates and Implications*, Victoria: Victoria Transport Policy Institute.

Locke, Hew (2001) *Chanting Heads 2001: A Glimpse of Eleven Visual Artists Working Today* (CD ROM), AAVA, NSEAD and the John Hope Franklin Centre in association with the University of East London, the Clore Duffield Foundation and the Arts Council of England.

Loftman, P. and Middleton, A. (2001) 'Emasculating public debate and eroding local accountability: City promotion of urban development projects in Birmingham', *Geographische Zeitschrift*, 89 (2–3), 85–103.

Loftman, P. and Nevin, B. (1992) *Urban Regeneration and Social Equity: A Case Study of Birmingham 1986–1992*, Faculty of the Built Environment Research Paper No. 8, Birmingham: University of Central England in Birmingham.

Loftman, P. and Nevin, B. (1994) 'Prestige project development: Economic renaissance or economic myth: A case study of Birmingham', *Local Economy*, 8 (4), 307–25.

Loftman, P. and Nevin, B. (1995) 'Prestige projects and urban regeneration in the 1980s and 1990s: A review of benefits and limitations', *Planning Practice and Research*, 3 (1), 31–9.

Loftman, P. and Nevin, B. (1996) 'Going for growth: Prestige projects in three British cities', *Urban Studies*, 33 (6), 991–1019.

Loftman, P. and Nevin, B. (1998) 'Pro-growth Local Economic Development Strategies: Civic Promotion and Local Needs in Britain's Second City, 1981–1996' in T. Hall and P. Hubbard (eds) *The Entrepreneurial City: Geographies of Politics, Regime and Representation*, Chichester: John Wiley & Sons.

Logan, J., Taylor-Gooby, P. and Reuter, M. (1992) 'Poverty and Income Inequality' in Susan S. Fainstein, Ian Gordon and Michael Harloe (eds) *Divided Cities: New York and London in the Contemporary World*, Cambridge, MA: Blackwell Press.

Lucas, K., Grosvenor, T. and Simpson, R. (2001) *Transport, the Environment and Social Exclusion*, York: Joseph Rowntree Foundation.

Luke, T.W. (1995) 'Sustainable Development as a Power/Knowledge System' in F. Fischer and M. Black (eds) *Greening Environmental Policy: The Politics of a Sustainable Future*, London: Paul Chapman Publishing.

Lull, J. (1990) *Inside Family Viewing*, London: Routledge.

Luxemburg, R.B. (1997) *Rut Blees Luxemburg, 'London – A Modern Project'*, London: Black Dog Publishing.

Luxemburg, R.B. (2000) *Rut Blees Luxemburg, Liebeslied*/Alex Garcia Duttman, *My Suicides*, London: Black Dog Publishing.

Lyon-Callo, V. (1998) 'Constraining responses to homelessness: An ethnographic exploration of the impact of funding concerns on resistance', *Human Organization*, 57 (1), 1–6.

Maat, K. and Louw, E. (1999) 'Mind the gap: Pitfalls of travel reduction measures', *Built Environment*, 25 (2), 151–61.

MacCannell, D. (1999) 'New Urbanism and its Discontents' in J. Copjec and M. Sorkin (eds) *Giving Ground: The Politics of Propinquity*, London: Verso.

McDowell, L. and Sharp, J. (eds) (1997) *Space, Gender, Knowledge: Feminist Readings*, London: Arnold.

McGuigan, J. (1992) *Cultural Populism*, London: Routledge.

McGuigan, J. (1996) *Culture and the Public Sphere*, London: Routledge.

Macintyre, S., Kearns, A., Ellaway, A. and Hiscock, R. (2000) *The Thaw Report*, Glasgow: Medical Research Council/University of Glasgow.

McNeill, W.H. (1976) *Plagues and Peoples*, Garden City, NY: Anchor Press.

Malbon, B. (1998) 'Consumption, Identity and the Spatial Practices of Everynight Life' in T. Skelton and G. Valentine (eds) *Cool Places: Geographies of Youth Cultures*, London: Routledge.

Malik Okon, R. (2001) 'Supta Biswas essay', *Chanting Heads 2001: A Glimpse of Eleven Visual Artists Working In Britain Today* (CD ROM), AAVA, NSEAD and the John Hope Franklin Centre in association with the University of East London, the Clore Duffield Foundation and the Arts Council of England.

Malitsky, B.C. (1992) *Fragile Ecologies: Contemporary Artists' Interpretations and Solutions*, New York, NY: Rizzoli.

Marcuse, H. (1968) 'Liberation from the Affluent Society' in D. Cooper (ed.) *The Dialectics of Liberation*, Harmondsworth: Penguin.

Marcuse, H. (1969) *Essay on Liberation*, Harmondsworth: Penguin.

Marcuse, H. (1972) *Counter-Revolution and Revolt*, Boston, Mass.: Beacon Press.

Marcuse, H. (1978) *The Aesthetic Dimension*, Boston, Mass.: Beacon.

Marcuse, P. (1996) 'Space and Race in the Post-Fordist City: The Outcast Ghetto and Advanced Homelessness in the United States Today' in Enzo Mingione (ed.) *Urban Poverty and the Underclass: A Reader*, Cambridge, MA: Blackwell Publishers.

Margulis, L. (1998) *The Symbiotic Planet*, New York: Phoenix.

Marshall, S. (1999a) 'Introduction: Travel reduction – means and ends', *Built Environment*, 25 (2), 89–93.

Marshall, S. (1999b) 'Restraining mobility while maintaining accessibility: An impression of the "city of sustainable growth"', *Built Environment*, 25 (2), 168–79.

Marshall, S., Banister, D. and McLellan, A. (1997) 'A strategic assessment of travel trends and travel reduction strategies', *Innovation*, 10 (3), 289–304.

Marturana, H.R. and Varela, F.J. (1987) *The Tree of Knowledge: The Biological Roots of Human Understanding*, Boston, Mass.: Shambala Publications, Inc.

Marx, K. (1973, first published 1947) *The German Ideology*, R. Pascal (ed.) New York, NY: International Publishers.

Marx, K. and Engels, F. (1967) *The Communist Manifesto*, Harmondsworth: Penguin.

Massey, D. (1993) 'Power-geometry and a Progressive Sense of Place' in J. Bird, B. Curtis, G. Robertson and L. Tickner (eds) *Mapping the Futures: Local Cultures, Global Changes*, London: Routledge, 59–69.

Massey, D. (1994) *Space, Place and Gender*, Cambridge: Polity Press.

Matilsky, B. (1992) *Fragile Ecologies*, New York: Rizzoli.

Mayor's Office of Community Development (1995) *1995 Consolidated Plan: City and County of San Francisco*, San Francisco: San Francisco Mayor's Office of Community Development.

Melucci, A. (1988) 'Getting involved: Identity and mobilization in social movements', *International Social Movement Research*, 1, 329–48.

Melucci, A. (1989) *Nomads of the Present: Social Movements and Individual Needs in Contemporary Society*, Philadelphia, PA: Temple University Press.

Meskimmon, M. and West, S. (eds) (1995) *Visions of the Neue Frau: Women and the Visual Arts in Weimar Germany*, Aldershot: Ashgate.

Middleton, A., Chapman, D., Loftman, P., Mann, R., Nevin, B. and Watson, P. (1990) *The Proposed Redevelopment of the Bull Ring, Birmingham*, Birmingham: Built Environment Development Centre, Birmingham Polytechnic.

Mile End Park (2001) *Mile End Park – A Celebration – Summer*, London: London Borough of Tower Hamlets.

Miles, M. (1997) *Art Space and the City: Public Art and Urban Futures*, London: Routledge.
Miles, M. (2000) *The Uses of Decoration: Essays in the Architectural Everyday*, Chichester: J. Wiley & Sons.
Miles, M., Hall, T. and Borden, I. (eds) (2000) *The City Cultures Reader*, London: Routledge.
Miles, S. (1998) *Consumerism as a Way of Life*, London: Sage.
Miles, S. (2000) *Youth Lifestyles in a Changing World*, London: Sage.
Miles, S. and Paddison, R. (1998) 'Urban consumption: An historiographical note', *Urban Studies,* 35 (5–6), 815–24.
Miller, D., Jackson, P., Thrift, N., Holbrook, B. and Rowlands, M. (1998) *Shopping, Place and Identity*, London: Routledge.
Mitchell, W.J.T. (1994) *Picture Theory: Essays on Verbal and Visual Representation*, Chicago, Ill.: University of Chicago Press.
Mohan, J. (1999) *A United Kingdom? Economic, Social and Political Geographies*, London: Arnold.
Mokhtarian, P.L. (1990) 'A typology of relationships between telecommunications and transportation', *Transportation Research A*, 24A (3), 231–42.
Monbiot, G. (2000) 'Complainers who are facing ruin', *Guardian*, 10 February, 22.
Morley, D. (1992) *Television, Audiences and Cultural Studies*, London: Routledge.
Morris, R. (1979) 'Robert Morris Keynote Address' in *Earthworks: Land Reclamation as Sculpture*, Seattle, WA: Seattle Art Museum.
Morris, R. (1993) *Notes on Art as/and Land Reclamation*, in *Continuous Project Altered Daily: The Writings of Robert Morris*, Boston, Mass.: Massachusetts Institute of Technology.
Moseley, M.E. (1999) 'Convergent Catastrophe: Past Patterns and Future Implications of Collateral Natural Disasters in the Andes' in A. Oliver-Smith and S.M. Hoffman (eds) *The Angry Earth: Disaster in Anthropological Perspective*, London and New York: Routledge.
Muller, P.O. (1995) 'Transportation and Urban Form: Stages in the Spatial Evolution of the American Metropolis' in S. Hanson (ed.) *The Geography of Urban Transportation*, New York, NY: Guilford Press.
Mumford, L. (1972, first published 1956) 'The Natural History of Urbanization' in R.L. Smith (ed.) *The Ecology of Man: An Ecosystem Approach*, New York: Harper & Row.
Naess, A. (1989) *Ecology, Community and Lifestyle,* ed./trans. D. Rothenberg, New York, NY: Cambridge University Press.
Nead, L. (1995) *Chila Kumari Burman: Beyond Two Cultures*, London: Kala Press.
Neri, L. (ed.) (2000), *Looking Up: Rachel Whiteread's Water Tower*, New York City: Public Art Fund.
New Civil Engineer (2002) News item, *New Civil Engineer*, 17 January, 12.
Newman, P. (1999) 'Transport: Reducing Automobile Dependence' in D. Shatterthwaite (ed.) *The Earthscan Reader in Sustainable Cities*, London: Earthscan.
Nochlin, L. (1968) 'The Invention of the Avant-Garde in France, 1830–80' in T.B. Hess and J. Ashbery (eds) *Avant-Garde Art*, London: Collier-Macmillan.
Northern California Council for the Community (1998) *Building a Healthier San Francisco: A Citywide Collaborative Community Assessment*, San Francisco: Northern California Council for the Community.
O'Connor, J. and Wynne, D. (1996) *From the Margins to the Centre: Cultural Production and Consumption in the Post-Industrial City*, Aldershot: Arena.

O'Toole, M. and Usher, D. (1992) 'Editorial' in P. Healey, S. Davoudi, M. O'Toole, S. Tavsanoglu and D. Usher (eds) *Rebuilding the City: Property-led Urban Regeneration*, London: E. and F.N. Spon.

Oakes, B. (1995) *Sculpting with the Environment*, New York, NY: Van Nostrand Reinhold.

Office for National Statistics (2000) *Family Spending Survey*, London: Office for National Statistics.

Office of Federal Housing Enterprise Oversight (2001) *House Price Index: Second Quarter 2001*, Washington, DC: United States Government, Office of Federal Housing Oversight.

Oliver-Smith, A. (1986) 'Introduction – Disaster Context and Causation: An Overview of Changing Perspectives in Disaster Research' in A. Oliver-Smith (ed.) *Natural Disasters and Cultural Responses: Studies in Third World Societies No. 36*, Williamsburg, Virginia: Department of Anthropology, College of William and Mary.

Oliver-Smith, A. (1999) 'What is a Disaster? Anthropological Perspectives on a Persitent Question' in A. Oliver-Smith and S.M. Hoffman (eds) *The Angry Earth: Disaster in Anthropological Perspective*, London: Routledge.

Oliver-Smith, A. (1999) 'Peru's Five-Hundred-Year Earthquake: Vulnerability in Historical Context' in A. Oliver-Smith and S.M. Hoffman (eds) *The Angry Earth: Disaster in Anthropological Perspective*, London: Routledge.

Osgerby, B. (1998) *Youth in Britain Since 1945*, Oxford: Blackwell.

Palladino, G. (1996) *Teenagers: An American History*, New York: Basic Books.

Peet, R. and Watts, M. (eds) (1996) *Liberation Ecologies*, London: Routledge.

Pennybacker, S. (1989) '"The Millennium by Return of Post": Reconsidering London Progressivism, 1889–1907' in D. Feldman and G. Stedman Jones (eds) *Metropolis London*, London: Routledge, 129–62.

Pile, S. (1996) *The Body and the City: Psychoanalysis, Space and Subjectivity*, London: Routledge.

Pile, S. and Thrift, N. (1995) *Mapping the Subject: Geographies of Cultural Transformation*, London: Routledge.

Piven, F.F. and Cloward, R.C. (1992) 'Normalizing Collective Protest' in A.D. Morris and C. McClurg Mueller, (eds) *Frontiers in Social Movement Theory*, New Haven, CT: Yale University Press.

Plumwood, V. (1993) *Feminism and the Mastery of Nature*, London: Routledge.

Polet, F. (2001) 'Some Key Statistics' in F. Houtart and F. Polet (eds) *The Other Davos: The Globalization of Resistance to the World Economic System*, London: Zed Books.

Pollock, Griselda (1999) *Differencing the Canon: Feminist Desire and the Writing of Arts Histories*, London: Routledge.

Porter, M.E. (1995) 'The competitive advantage of the inner city', *Harvard Business Review*, May/June, 55–71.

Porter, M.E. (1996) 'Competitive advantage: Agglomeration economies and regional policies', *International Regional Science Review*, 19, 85–90.

Poster, M. (1995) *The Second Media Age*, Cambridge: Polity Press.

Pradervand, P. (1989) *Listening to Africa*, New York: Praeger.

Presdee, M. (1986) *Agony or Ecstasy: Broken Transitions and the New Social State of Working-Class Youth in Australia*, Occasional Papers, Magill, S. Australia: Australian Centre for Youth Studies, SA College of AE.

Rattray, F. (1999) 'The art of discovery', *Blueprint*, June, 50–2.

Reade, E.J. (1997) 'Planning in the Future or Planning of the Future?' in A. Blowers and B. Evans (eds) *Town Planning into the Twenty-First Century*, London: Routledge.

Ree, J. (1999) *I See a Voice: Language, Deafness and the Senses, A Philosophical Enquiry*, London: HarperCollins.

Roberts, J. (2001) 'Art, politics and provincialism', *Radical Philosophy*, 106, 2–6.

Roberts, K. (1995) *Youth Employment in Modern Britain*, Oxford: Oxford University Press.

Robins, K. (1993) 'Prisoners of the City' in E. Carter, J. Donald and J. Squires (eds) *Space and Place*, London: Lawrence & Wishart.

Robins, K. (1996) 'Collective Emotion and Urban Culture' in B. Brandner, S. Mattl and V. Ratzenbock (eds) *Kulturpolitik und Restrukturierung der Stadt*, Vienna: 73–96.

Rogers, A. and Vertovec, S. (1995) *The Urban Context: Ethnicity, Social Networks and Situational Analysis*, Oxford: Berg.

Rogers, R. and Gumuchdjian, P. (1997) *Cities for a Small Planet*, London: Faber & Faber.

Rosenthal, Rob (1996) 'Dilemmas of Local Anti-homelessness Movements' in J. Baumohl (ed.) *Homelessness in America*, Phoenix, AZ: Oryx Press.

Rosler, M. (1991) 'Fragments of a Metropolitan Viewpoint' in B. Wallis (ed.) *If You Lived Here*, Seattle: Bay Press.

Rowe, D. (1995) 'Georg Simmel and the 1896 Berlin Trade Exhibition', *Journal of Urban History*, 22 (2), 216–28.

Rowe, D. (2003) *Representing Berlin: Sexuality and the City in Imperial and Weimar Germany* Aldershot: Ashgate, Scolar Press.

Ruddick, S. (1997) 'Modernism and Resistance: How "Homeless" Subcultures make a Difference' in T. Skelton and G. Valentine (eds) *Cool Places: Geographies of Youth Cultures*, London: Routledge.

Ryan, J.C. (1994) *State of the Northwest*, Seattle, WA: Northwest Environmental Watch.

Ryan, J.C. (1995) 'Greenhouse gasses on the rise', *New Indicator*, August.

Ryan, J. and Fitzpatrick, H. (1996) 'The Space that Difference Makes: Negotiation and Urban Identities Through Consumption Practices' in J. O'Connor and D. Wynne (eds) *From the Margins to the Centre: Cultural Production and Consumption in the Post-Industrial City*, Aldershot: Ashgate.

Sachs, W. (1992) *For Love of the Automobile: Looking Back into the History of our Desires*, Berkeley: University of California Press.

Saloman, I. (1986) 'Telecommunications and travel relationships: A review', *Transportation Research – A*, 20a (3), 223–38.

Sandercock, L. (1998a) *Towards Cosmopolis*, Chichester: Wiley.

Sandercock, L. (1998b) 'The Death of Planning: Radical Praxis for a Postmodern Age' in M. Douglass and J. Friedman (eds) *Cities for Citizens*, Chichester: Wiley.

Sandercock, L. (ed.) (1998c) *Making the Visible Invisible: A Multicultural Planning History*, Berkeley: University of California Press.

Sassen, S. (1991) *The Global City: New York, London, Tokyo*, Princeton, NJ: Princeton University Press.

Sassen, S. and Roost, F. (1999) 'The City: Strategic Site for the Global Entertainment Industry' in D.R. Judd and S.S. Fainstein (eds) *The Tourist City*, New Haven and London: Yale University Press, 143–54.

Savage, M. and Warde, A. (1993) *Urban Sociology, Capitalism, and Modernity*, Basingstoke: Macmillan.

Sawicki, D.S. and Moody, M. (2000) 'Developing transportation alternatives for welfare recipients moving to work', *APA Journal*, 66 (3), 306–18.

Schafer, R.M. (1977) *The Tuning of the World*, New York: Knopf.

Scott, A. (1988) *Metropolis: From the Division of Labour to Urban Form*, Berkeley, CA: University of California Press.

Scott, A. (2000) *The Cultural Economy of Cities*, London: Sage.

Scott, A. (2001) 'Capitalism, cities and the production of symbolic forms', *Transactions of the Institute of British Geographers*, NS 26 (1), 11–23.

Scottish Executive (2000) 'The role of transport in social exclusion in urban Scotland'. <http://www.scotland.gov.uk/cru/kd01/blue/rtseuc_03.htm> (accessed 5 September 2001).

Seabrook, J. (1993) *Victims of Development*, London: Verso.

Sekhon, J. (1997) 'Putting Sociology in its Place, in Ballard' in C.J. Gubbay and C. Middleton (eds) *The Student's Companion to Sociology*, Oxford: Blackwell.

Sennett, R. (1986) *The Fall of Public Man*, New York: Norton.

Sennett, R. (1990) *The Conscience of the Eye*, London: Faber & Faber.

Sennett, R. (1996, first published 1970) *The Uses of Disorder*, London: Faber & Faber.

Sennett, R. (1998) *The Corrosion of Character: The Personal Consequences of Work in the New Capitalism*, New York: Norton.

Serres, M. (1982) *The Parasite*, translated by Lawrence Schehr, Baltimore: Johns Hopkins University Press.

Serres, M. (1995) *The Natural Contract*, Ann Arbor: University of Michigan Press.

Sessions, G. (1995) *Deep Ecology for the 21st Century*, Boston, Mass.: Shambala Publications.

Shaw, P. (1999) *The Arts and Neighbourhood Renewal: A Research Report by Policy Action Team 10*, London: Department for Culture, Media and Sport.

Shearman, C. (1999) *Local Connections: Making the Net Work for Neighbourhood Renewal*, London: Department of Trade and Industry.

Sheller, M. and Urry, J. (2000) 'The city and the car', *International Journal of Urban and Regional Research*, 24 (4), 737–57.

Shields, R. (1999) *Lefebvre, Love and Struggle*, London: Routledge.

Shiva, V. (1999) 'This round to the citizens', *Guardian*, 8 December.

Short, B. (ed.) (1992) *The English Rural Community: Image and Analysis*, Cambridge: Cambridge University Press.

Short, J.R. (1991) *Imagined Country: Society, Culture and Environment*, London: Routledge.

Sibley, D. (1995) *Geographies of Exclusion: Society and Difference in the West*, London: Routledge.

Silverstone, R. (1994) *Television and Everyday Life*, London: Routledge.

Silverstone, R. and Hirsch, E. (eds) (1992) *Consuming Technologies: Media and Information in Domestic Spaces*, London: Routledge.

Simmel, G. (1997a, first published 1903) 'The Metropolis and Mental Life' in D. Frisby and M. Featherstone (eds) *Simmel on Culture*, London: Sage, 174–85.

Simmel, G. (1997b) 'Sociology of the Senses' in D. Frisby and M. Featherstone (eds) *Simmel on Culture*, London: Sage.

Skelton, T. and Valentine, G. (eds) (1998) *Cool Places: Geographies of Youth Cultures*, London: Routledge.

Smith, C. (1998a) *Creative Britain*, London: Faber & Faber

Smith, D. (1998b) 'Bakhtin and the Dialogic of Society' in M.M. Bell and M. Gardiner (eds) *Bakhtin and the Human Sciences*, London: Sage.

Smith, M.P. (2001) *Transnational Urbanism*, Oxford, Blackwell.

Smith, N. (1996a) *The New Urban Frontier: Gentrification and the Revanchist City*, London: Routledge.
Smith, N. (1996b) 'After Tompkins Square Park: Degentrification and the Revanchist City' in Anthony D. King (ed.) *Re-Presenting the City: Ethnicity, Capital and Culture in the 21st-Century Metropolis*, New York, NY: New York University Press.
Smith, Z. (2000) *White Teeth*, Harmondsworth: Penguin Books.
Social Exclusion Unit (2000) *Closing The Digital Divide: Information and Communication Technologies in Deprived Areas, A Report by Policy Action Team 15*, London: Social Exclusion Unit.
Soja, E.W. (1989) *Postmodern Geographies: The Reassertion of Space in Critical Social Theory*, London: Verso.
Soja, E. (2000) *Postmetropolis*, Oxford: Blackwell.
Sonfist, A. (ed.) (1983) *Art in the Land: A Critical Anthology of Environmental Art*, New York, NY: E.P. Dutton, Inc.
Spivak, G.C. (1995) 'Can the Subaltern Speak?' in B. Ashcroft, G. Griffiths and H. Tiffin (eds) *The Post-Colonial Studies Reader*, London: Routledge.
Starkey, M. (1989) *Born To Shop*, Eastbourne: Monarch.
Stevenson, N. (2001) 'Culture and Citizenship: An Introduction' in N. Stevenson (ed.) *Culture and Citizenship*, London: Sage.
Stewart, F. (1992) 'The Adolescent as Consumer' in J.C. Coleman and C. Warren-Anderson (eds) *Youth Policy in the 1990s: The Way Forward*, London: Routledge.
Stewart, J. (2000) *Fashioning Vienna: Adolf Loos and Cultural Criticism*, London: Routledge.
Stoker, G. (1995) 'Regime Theory and Urban Politics' in D. Judge, G. Stoker and H. Wolman (eds) *Theories of Urban Politics*, London: Sage.
Stone, C. (1972) *Should Trees Have Legal Standing? Toward Legal Rights for Natural Objects*, Palo Alto, CA: Tioga Publishing.
Strelow, H. (1999) *Natural Reality: Artistic Positions Between Nature and Culture*, Daco Verlag, Stuttgart: Ludwig Forum for Internationale Kunst.
Stutz, F.P. (1995) 'Environmental Impacts' in S. Hanson (ed.) *The Geography of Urban Transportation*, New York, NY: Guilford Press.
Suleri, S. (1992) 'Women skin deep: Feminism and the postcolonial condition', *Critical Inquiry*, 18 (4), 756–69.
Swartz, S. (1999) 'Wanted: Housing; more people can find a job, but not a place to live', *The Press Democrat*, 18 January, B1.
Tafuri, M. (1976) *Architecture and Utopia: Design and Capitalist Development*, Cambridge, Mass.: MIT Press.
Tafuri, M. (1987) *The Sphere and the Labyrinth: Avant-Gardes and Architecture from Piranesi to the 1970s*, Cambridge, Mass.: MIT Press.
Tenner, E. (1997) *Why Things Bite Back*, London: Fourth Estate.
Throop, W. (2000) *Environmental Restoration: Ethics, Theory and Practice*, Amherst, NY: Humanity Books, an imprint of Prometheus Books.
Tillyard, E.M.W. (1943) *The Elizabethan World Picture*, London: Chatto & Windus.
Tomlinson, J. (1999) *Globalization and Culture*, Oxford: Blackwell.
Toon, I. (2000) 'Finding a Space in the Street: CCTV Surveillance and Young People's Use of Urban Public Space' in D. Bell and A. Haddour (eds) *City Visions*, Harlow: Longman.
TraC (2000) *Social Exclusion and the Provision and Availability of Public Transport*, London: Department of the Environment, Transport and the Regions.

Trowell, J. (2000) 'The Snowflake in Hell and the Baked Alaska: Improbability, Intimacy, and Change in the Public Realm' in S. Bennett and J. Butler (eds) *Locality, Regeneration and Diversity*, Bristol: Intellect Books.

Turkle, S. (1995) *Life on the Screen: Identity in the Age of the Internet*, New York: Simon & Schuster.

Turner, F. (2000, first published 1998) 'A Field Guide to the Synthetic Landscape' in W. Throop (ed.) *Environmental Restoration: Ethics, Theory and Practice*, Amherst, NY: Humanity Books, an imprint of Prometheus Books.

Twombly, J.G., Crowley, S., Ferris, N. and Dolbeare, C.N. (2001) *Out of Reach 2001: America's Growing Wage–Rent Disparity*, Washington, DC: National Low Income Housing Coalition.

UNCHS (Habitat) (1994) *Urban Public Transport in Developing Countries*, Nairobi: UNCHS (Habitat).

United States Department of Housing and Urban Development (1998) *HUD News Internet Site*, <http://www.hud.gov/news.html>.

Urban Studies (1998) Special Issue: 'Urban consumption and urban lifestyles', 35(5).

Urban Taskforce (2000) *Towards an Urban Renaissance*, London: DETR.

Urry, J. (1995) *Consuming Places*, London: Routledge.

US Bureau of the Census (2000) *Census 2000*, Washington, DC: US Department of Commerce.

Valentine, G. (2001) *Social Geographies: Space and Society*, Harlow: Prentice Hall.

van den Berg, L. and Braun, E. (1999) 'Urban competitiveness, marketing and the need for organising capacity', *Urban Studies*, 36 (5–6), 987–99.

Vanderbeck, R. and Johnson, J. (2000) '"That's the only place where you can hang out": Urban young people and the space of the mall', *Urban Geography*, 21 (1), 5–25.

Veash, N. and O'Sullivan, J. (1997) 'Forget about Swampy. All they want is their own car', *Independent*, 19 November, 5.

Verwijnen, J. and Lehtovuori, P. (eds) (1999) *Creative Cities: Cultural Industries, Urban Development and the Information Society*, Helsinki: UIAH.

Vision in Art (2000) *Regeneration and the Premises Crises*, Queen Mary and Westfield College, July, Conference Papers, <http://www.raimes.com/conf1.htm>.

Vulliamy, E. (2001) 'Americans warned of erosions of freedom', *Observer*, 11 November.

Wallis, B. (ed.) (1991) *If You Lived Here*, Seattle: Bay Press.

Wallis, C. (2001) 'Introduction' in Tacita Dean, *Tacita Dean*, London: Tate Britain.

Walter, B., Arkin, L. and Crenshaw, R. (1992) *Sustainable Cities: Concepts and Strategies for Eco-City Development*, Los Angeles: Eco-Home Media.

Walters, J. (2001) 'War on the car sparks driver rage', *Observer*, 26 August, 1.

Washton Long, R.-C. (1975) 'Kandinsky's abstract style: The veiling of apocalyptic folk imagery', *Art Journal*, XXXIV, 217–28.

Watt, P. and Stenson, K. (1998) '"It's a bit dodgy around there": Safety, danger, ethnicity and young people's use of public space' in T. Skelton and G. Valentine (eds) *Cool Places: Geographies of Youth Cultures*, London: Routledge, 249–65.

Wedd, K. (2001) *Creative Quarters: The Art World in London*, London: Museum of London.

Weiner, D. (1989) 'The People's Palace: An Image for East London in the 1880s' in D. Feldman and G. Stedman Jones (eds) *Metropolis London Histories and Representations since 1800*, London: Routledge.

Wells, M. (2001) 'Repeat showings of towers pornographic', report of Newsworld Conference, Barcelona, *Guardian*, 15 November.

White, J. *et al.* (1995) 'Newcomer children in San Francisco: Their health and well being', *A Report by the Child Health Initiative for Immigrant/Refugee Newcomers Project*, San Francisco: Department of Public Health.

White, P.S. and Pickett, S.T.A. (1985) 'Natural Disturbance and Patch Dynamics: An Introduction' in S.T.A. Pickett and P.S. White (eds) *The Ecology of Natural Disturbance and Patch Dynamics*, Orlando, FL: Academic Press.

White, R. (1992) 'Young people, community space and social control', *Juvenile Justice National Conference AIC Conference Proceedings*, Adelaide, 189–202.

White, R. (1993) 'Youth and the conflict over urban space', *Children's Environments*, 10 (1), 85–93.

White, R. (1996) 'No-go in the fortress city: Young people, inequality and space', *Urban Policy and Research*, 12 (1), 37–50.

Whitelegg, J. (1993) *Transport for a Sustainable Future: The Case of Europe*, London: John Wiley.

Whiteread, R. (1996) *Shedding Life*, Liverpool: Tate Gallery.

Whiteread, R. (1997) Venice: British Pavilion, XLVII Venice Bienalle.

Wilding, R. (1989) *Supporting the Arts – Review of the Structure of Arts Funding*, London: Office of Arts and Libraries.

Williams, R. (1973) *The Country and the City*, London: Chatto & Windus.

Williams, R. (1981) *Culture*, London: Fontana.

Williams, R. (1988) *Keywords: A Vocabulary of Culture and Society*, London: Fontana.

Willis, P. (1991) *Towards a New Cultural Map, Discussion Document No.18, National Arts and Media Strategy*, London: Arts Council.

Wilson, E. (1991) *Sphinx in the City*, Berkeley: University of California Press.

Wilson, E. (2001) *The Contradictions of Culture: Cities, Culture, Women*, London: Sage.

Wirth, Louis (1996, first published 1938) 'Urbanism as a Way of Life' in R.T. LeGates and F. Stout (eds) *The City Reader*, London: Routledge.

Wolin, R. (1994) *Walter Benjamin: An Aesthetic of Redemption*, Berkeley, CA: University of California Press.

Woods, L. (1995) 'Everyday War' in P. Lang (ed.) *Mortal City*, New York: Princeton Architectural Press.

Worpole, K. (1998) 'Think-tanks, consultancies and urban policy in the UK', *International Journal of Urban and Regional Research*, 22 (1), 147–55.

Worpole, K. and Greenhalgh, L. (1999) *The Richness of Cities: Urban Policy in a New Landscape – Final Report*, Stroud: Comedia/Demos.

Worster, D. (1993) 'The Shaky Ground of Sustainability' in W. Sachs (ed.) *Global Ecology: A New Arena of Political Conflict*, Halifax, Nova Scotia: Fernwood.

Wright, G.H. von (1997) 'The crisis of social science and the withering away of the nation state', *Associations*, 1, 49–52.

Wright, T. (1997) *Out of Place: Homeless Mobilizations, Sub-Cities, and Contested Landscapes*, Albany, NY: SUNY Press.

Wyn, J. and White, R. (1997) *Rethinking Youth*, London: Sage.

Yass, C. (2000) *Works 1995–2000*, London: Asprey Jacques.

Young, I.M. (1990) *Justice and the Politics of Difference*, Princeton: Princeton University Press.

Young, I.M. (2000) *Inclusion and Democracy*, Oxford: Oxford University Press.

Zaman, M. Q. (1999) 'Vulnerability, Disaster, and Survival in Bangladesh' in A. Oliver-Smith and S.M. Hoffman (eds) *The Angry Earth: Disaster in Anthropological Perspective*, London: Routledge.

Zukin, S. (1982) *Loft Living: Culture and Capital in Urban Change*, Baltimore: Johns Hopkins University Press. (1988, Radius Books, London).
Zukin, S. (1992) 'Postmodern urban landscapes: Mapping culture and power' in S. Lash and J. Friedman (eds) *Modernity and Identity*, Oxford: Blackwell.
Zukin, S. (1995) *The Cultures of Cities*, Oxford: Blackwell.
Zukin, S. (1996) 'Space and symbols in an age of decline' in A.D. King (ed.) *Re-Presenting the City: Ethnicity, Capital and Culture in the 21st-Century Metropolis*, Basingstoke: Macmillan, 43–59.

INDEX

Page references for plates, figures and tables are in *italics*; those for notes are followed by n

Adorno, T.W. 47–8, 62
air pollution 96–7
Aitken, S. 74
Akbar, Shireen 11, 30–1
Alexander, David 185–6
. . . *and of time* (Barth) 18
Angelus Novus (Klee) 10–11, 13, 26, 62
Apocalypse 44–5, 46–7
Appleyard, Don 100
architecture: and politics 111–15, *see also* radical architecture
Architecture Centres 108
Arena Central, Birmingham 77, 83
Arendt, Hannah 55, 56, 109, 123, 131
art: earth-art 138–9; landscapes 138; public 109, 122–33; and radical ecology 140–3
As Above, So Below (Driscoll) 124
audience 132
automobility *see* cars

Baker, Mickey 36
Bakhtin, Mikhail 55
Ballard, J.G. 45, 47
Bangladeshis 147, 175, 177
Bankside (Yass) 17
Barber, Stephen 47
Barcelona Pavilion 21–2
Barth, Uta 14, 17–18
Bauman, Zygmunt 49–50, 51, 56
Beall, Jo 196
Beardsley, John 138, 139
Beauchamp-Byrd, Mora J. 41n
Beck, Ulrich 69, 70, 182
Bell, D. 81
Bell, Michael Mayerfeld 55
Benhabib, Seyla 56, 57
Benjamin, Walter 21, 22, 26; *Angelus Novus* 10–11, 12, 13; film 47–8, 54
Bennett 170
Berlin 19–20, 62
Berman, Marshall 52, 96, 115, 116
Bhaba, Homi 27–8, 29, 31, 35
Bhimji, Zarina 11, 29, 31
Bibliotheque Nationale de France, Paris 21, 22
Birmingham 63, 64, 77–8, 82; Bore era 81–2; Eastside Regeneration Initiative 84–6; Knowles era 78–9, *80*; social equity impact 88–90; Stewart era 79–81; Westside 82–4
Biswas, Sutapa 11, 35–6, 39
Blade Runner 51
Bloch, Ernst 46
Blowers, A. 103–4
body 126
Bookchin, Murray 140
Bore, Albert 78, 79, 81–2, 83–4
Borja, J. 179–80
boundaries 11, 28; public art 109, 122, 130, 133
Boyce, Sonia 11, 28–9, 31, 33
Bracewell, Michael 14
Bradshaw, Anthony 136
Breheny, Michael 102–3
Bridge, Gary 28
Brindleyplace, Birmingham 77, 78, 83, 87
Broad Street Redevelopment Area 77, 82–4
Brooke, Peter 77
Buck-Morss, Susan 48, 55, 58
Budiansky, Stephen 185
Bull Ring, Birmingham 77, 84, 85, 90
Burgess, E.W. 5, 6, 52
Burgess, Jacquelin 42–3n
Burgin, Victor 14, 21–2, 120
Burman, Chila Kumari 28, 33, 41n
Byrne, David 6

Caliban Towers I and II (Luxembourg) 14–15, *15*
Cao, Paco 126, 131
cars 63–4, 92–3, 96, 105; economic impacts 97; environmental impacts 96–7; health impacts and road safety 97–8; social impacts 98–100; travel reduction strategies 100–5; and urban form 94, *95*
Cashmore, E.E. 66
Castells, M. 179–80
CCTV 68–9
Celebration, Florida 5, 51
Centenary Square, Birmingham 83
Cézanne, Paul 46
Chaney, D. 71
Changing Places (Groundwork Trust) 136–7
Chauda, Meera 31

Chicago School 2, 4, 5
Chowdhary, Lubna 33, 35, *38*, 39–40, *39*
City Living, Birmingham 82, 86–8, 90
City Park, Birmingham 77, 84, 85–6
civil society 198
civilisation 114
Clarke, John 68
climate change 96, 186–7, 189
Collins, Catherine 5
Commonwealth Games 80
community 171–2, 197–8
compact city model 102–4, 105
congestion 97
congestion fees 105
conservation 110, 134–5
consumption 65, 67, 70–4
Convention Centre Quarter 82–4
Corridor (Yass) 17
creative cities 171, 173–4
Crilley, D. 76
crisis 45
critical theory 5–6, 26
Cross-Bronx Expressway 96
Crowhurst, Donald 20
cultural ecologies 138–40
cultural landscapes 168–71, 179–80; Stepney 174–7, 179
Cultural Strategy Partnership 169
culture 1, 3–4, 167–8, 180–1; prestige projects 76, 77, 177–9; and urban regeneration 147; young people 67–9, *see also* art
Curtis, Kimberly 56

Davis, Mike 11, 44, 48–9, 54, 56, 146, 148, 182, 186, 188, 190
Dawn (Kolbe) 21
De Certeau, Michel 39, 74
De Landa, Manuel 188
de Rivera, Oswaldo 195
De Voorst, Hollans 20
Dean, Tacita 11, 14, 20–1
Debord, G. 59
deep ecology 142
Delft Hydraulics (Dean) 20
demolition 47
Department of the Environment 102
Derrida, Jacques 192, 195, 199
Deutsche, Rosalyn 53
development studies 55
difference 6, 11, 12, 28, 55, 56–8, 108
differential space 114–15
Diller, Elizabeth 133
disaster scenario 11–12, 44, 185–6; Los Angeles 48–54; September 11 attacks 47–8
discourse 55, 120
Discovery Centre, Birmingham 85, 90
Disney Corporation 5, 51, 143
Donald, James 4–5, 197–8
Doors, The 58–9
Douglas, William O. 143
D'Oyly Carte Opera 79
Drakakis-Smith, David 195
Dresden 47
Driscoll, Ellen 124
Duncan, James 5, 50
Durning, Alan Thein 94, 96, 97–8
Duttman, Alexander 15–16
dystopia 44, 45–6, 54

earth-art 138–9
earthquakes 189–90
Eastside Regeneration Initiative, Birmingham 77, 81, 82, 84–6
eco-art 140, 143–4; dualities 140–1; and paradigmatic change 142–3; philosophies 141–2; theory and interdiscipline 140
eco-feminism 141–2
ecological footprinting 147, 183, 184
ecology 134, 147–8; cultural 138–40, *see also* restoration ecology
ecosystem approaches 139–40
Edensor, T. 69
education 84–5, 90
Ehrenfeld, Joan G. 137, *138*
El Nino-Southern Oscillation 187
Elective Affinities (Burgin) 21
electronic technology and communication 104–5
Elliott, Jennifer 195
Empire of the Sun (Ballard) 45, 47
employment 94
English National Sports Stadium 81
Escobar, Arturo 51–2
Euripides 196
European Capital of Culture 81
European Commission 102
Eurovision Song Contest 81
Evans, Stephen 47

Falkirk Asian Women's Group 31, *32*
fiction, disaster scenario 44
Field (Barth) 17–18
film 44, 47–8
Fisher, S. 70–1
Fiske, J. 67
flagship projects *see* prestige projects
Flow City (Ukeles) 109, 128
Foley Artist (Dean) 21
footprinting 147, 183, 184
form, radical architecture 119
Fortunati, Vita 44, 45, 50, 51
Foucault, Michel 45, 49, 174, 180
Frankfurt Institute for Social Research 5–6
Frantz, Douglas 5
Freire, Paolo 55
Fresh Kills Landfill, Staten Island 109, 128
frontier 53
Future Systems 117
futurism 108, 111

G8 Summit 81
Gallaccio, Anya 14, 24–5
Gamma (Wilson and Wilson) 19
Genette, Gérard 29
gentrification 146–7, 153–4
geological turbulence 189–90
Ginzel, Andrew 124–6
Girardet, Herbert 183
Glaschu (Gallaccio) 24
Gordon, P. 104
Grand Central Terminal, New York 109, 123–4
Greenham Common 19
greenhouse gases 96

Griffin, C. 72–3
Grimley, T. 79
Ground (Barth) 17, 18
Ground Zero 47–8, 58, 109, 128
Groundwork Trust 136–7
Gujral, Rajan 169
Gumuchdjian, P. 92, 100

Habermas, Jürgen 55
Half Moon Youth Theatre, Stepney 176–7
Hall, Stuart 27, 170, 174
Hamilton, Ann 14, 25
Hart, T. 94, *95*
Hartford, Connecticut 126–7
Harvey, David 183
Hays, Samuel 134
Heidegger, Martin 29
Hemmed in Two (Locke) 33, 35, *36*, *37*
Highbury 3 Symposium 81–2
Highbury Initiative Symposia 79, *80*
historical materialism 10, 21, 26
Hobbs, Thomas 138, 139
Holder, S. 70–1
Holocaust Memorial (Whiteread) 23, *23*
Holt, Nancy 139
home 11, 31, 33
homelessness 146, 149–50; advocacy and activism 157–62, 164–5; policing actions 154–7; regional activist networks 162–4; San Francisco and Sonoma County 152–4
Hope (Hopscotch Asian Women's Group) 33, *34*
Hopscotch Asian Women's Group 33, *34*
Horkheimer, Max 5–6, 47–8
House (Whiteread) 22
housing 108; Birmingham *88*, 89–90; San Francisco and Sonoma County 152–4
Hoxton 146–7
Hub, Birmingham 85, 90
Hull 24–5
Hyatt Hotel, Birmingham 78, 83
Hyde, Lewis 122, 129

I Will Always Be Here (Bhimji) 29, 31
imminence 45–6
. . . in passing (Barth) 18
information technology 104–5
installations: Bhimji 29; Gallaccio 24–5; Hamilton 25
intermediaries 171–4, 175
International Convention Centre (ICC), Birmingham 77, 78, 80, 82, 83, 89
Islam Now (Falkirk Asian Women's Group) 31, *32*

Jacobs, Jane 4, 108
Jarvis, H. 92
Jennings, Charles 174
Johnson, J. 74
Jones, Kristin 124–6
Jordan, William, III 110, 135–6, 137, 142

Kate ten K. 180
Kear, Adrian 28
Keep off the Grass (Gallaccio) 24
Kermode, Frank 129
keys 18–19
Khan, Tasneem 33
King, Anthony 11, 40, 41n
King, Catherine 29
Klee, Paul 10–11, 13, 26, 62
Knowles, Richard 78–9
Knox, P. 71
Kolbe, George 21
Kureshi, Hanif 31, 33

Laclau, Ernesto 6
Landry, C. 171, 173, 175
landscapes 138
Langman, L. 72
Lay Back, Keep Quiet and Thin About what Made Britain So Great (Boyce) 30
Le Corbusier 111
Leavitt, Jacqueline 57–8
Lefebvre, Henri 7, 54, 114–15, 117
Lenin, V.I. 196–7
Leopold, Aldo 135
Leslie, Esther 10–11, 47, 48
Liebeslied (Luxembourg) 15–16, 18
Light, Andrew 136
Lippard, Lucy 138, 139
Lisbon *145*
local economy trading schemes (LETS) 197
local regeneration *see* regeneration
Locke, Hew 33, 35, *36*, *37*
London: Biswas 39; Boyce 29–30; footprint 147, 184; Luxembourg photographs 14–15, *15*; regeneration 146–7, 172–4, *see also* Stepney
London: a modern City (Luxembourg) 14–15, *15*
Lorrain, Claude 138
Los Angeles 48–52, *56*, 146, 182, 186, 188, 196
Louw, E. 101–2
Lucas, K. 98, 99
lustre 14, 16–17, 18
Luxembourg, Rut Blees 11, 14–16, *15*, 18

Maat, K. 101–2
McCallum, Bradley 126–8, 131–2
McGuigan, J. 72
Macintyre, S. 98
Madison Arboretum 135–6
Mailbox, Birmingham 77, 83, 88
malls 62, 67, 72, 73, *see also* Bull Ring
Manchester *9*, *61*, 198
Manhole Cover Project, The (McCallum) 126–7
Marcuse, Herbert 46, 59, 196–7
Marshall, S. 100, 101, 102
Martineau Galleries, Birmingham 84, 85
Marturana, H.R. 144
Marx, Karl 58, 116
masculinity 109, 110
Massey, Doreen 18, 44, 51, 197–8
materials 118–19
Matilsky, Barbara 139, 140
May, Ernst 111
Meimsuchung/Affliction (Luxembourg) 16
Metronome (Jones and Ginzel) 124–5

Metropolis (Chowdhary) *38*, 39–40, *39*
Mies van der Rohe, Ludwig 21
Mile End Park, Stepney 177–8, 179
Millennium Dome 77
Millennium Point, Birmingham 77, 81, 84–5, 90
Miller, D. 73
Miss Pollard's Party (Pollard) 40
Mitchell, W.J.T. 122–3
mixed-use zoning 108, 113
Mnemonics (Jones and Ginzel) 125–6
mobility 62, 63–4, 92–3; and urban form 94, *95*, 96, *see also* cars
modernism 108, 111–12, 113, 115, 123
Mohan, J. 103
Monbiot, G. 92
Mork, Bill 186
Morris, Robert 139
Morris, William 30
motor races 79
motor vehicles *see* cars
movement 62
Muller, Peter 94, *95*
MultipliCity (Blowers) 104
Mumford, Lewis 147, 183

Naess, Arne 142
narratives 11
natality 55, 56
National Exhibition Centre, Birmingham 80
National Indoor Arena (NIA), Birmingham 77, 78, 80, 82, 83, 89
nature 184–5, *see also* turbulence
new urbanism 5
New York: gentrification 52–4; Ground Zero 47–8; public art 109, 123–6, 127–8
Newman, Peter 97
Nietzsche, Friedrich 22
Nine Mile Run Greenway 109–10
nostalgia 18
nowhere near (Barth) 18

Oakes, Bylai 139
Oculus (Jones and Ginzel) 125
Okon, Rohini Malik 36
Olmsted, Frederick Law 110, 135
Olympic Games 79, 82
One-Way Street (Benjamin) 13
oppression 68

Pain, R. 103–4
Palladino, G. 71
Paris 22
Pastoral Interludes (Pollard) 40
Peet, Richard 55
People's Palaces 172
Perrault, Dominique 21, 22
photography 14, 18, 47–8; Barth 17–18; Luxembourg 14–16, *15*; Pollard 40; Yass 16–17
piazzas 108, 114
pigeon lofts *107*
Pittsburgh 109–10
politics, and architecture 111–15
Pollard, Ingrid 40
Pollock, Griselda 11, 28
Postmetropolis (Soja) 50
postmodernism 108, 112–13
poststructuralism 55
Presdee, M. 67, 72
preservation 134–5
Preserve Beauty (Gallaccio) 24
Prestige (Gallaccio) 24
prestige projects 76–7, 147; Birmingham 77, 78–9, 81, 82–90; Stepney 177–9
private cars *see* cars
public art 109, 122–33
public space 129
public sphere 55, 198
public transport 64, 93

Queen's Hall, Mile End Road 172

radical architecture 108–9, 116–18, 120–1, 199; discourse 120; form and its experience 119; materiality and tectonics 118–19; relations of production 120; social relations and culture 119
radical ecology 140–3
Ragged School, Stepney 176–7
Rana, Samena 42n
Reade, E.J. 92
rebellion 71, 73, 75, 196–7
Red on Green (Gallaccio) 24
Rée, Paul 22
regeneration 76–7; and culture 167–71; intermediaries 171–4; London 146–7, 172–4, 176–9, *see also* Birmingham
regime theory 170
regulation 197
Rent-a-Body (Cao) 126
resistance 68–9, 174
restoration ecology 109–10, 134; concepts and practices 136–8, *138*; and eco-art 142–3; origins 134–6
risk 69–70
risk ghettos 186
road safety 97–8
Roberts, John 54
Rogers, Richard 108, 113–14, 147, 172, 198; cars 92, 100
Rosler, Martha 53

Saddam Hussein 45
Sadiq, Nadira 31
Sadlers Wells Royal Ballet 79
Salomé, Lou André 22
San Francisco 146, 150; anti-homeless policies 154–7; homelessness advocacy and activism 158–62, 164–5; political economic context 151–4; regional activist networks 162–4
Sandercock, Leonie 6, 44, 54, 56, 57–8
Sant'Elia, Antonio 108, 111
Savage, Mike 5
scientific rationalism 2–3
Scofidio, Ricardo 133
Scott, A. 168
Scottish Executive 98–9
sculpture: Gallaccio 24–5; Whiteread 22–4, *23*
Sea, with a ship, The (Dean) 21
Seabrook, Jeremy 196
self-reflection 18
September 11 attacks 44, 55, 58, 125, 128
Serres, Michel 28, 183, 189

Seurat, Georges 46
Shah, Fahmida 30, 31, 33
Shamania 11, 30–1, *32*, 33, *34*, 42n
Sheller, M. 93, 98, 99–100
shopping malls *see* malls
Sibley, David 5, 11, 28
Simmel, Georg 3, 4–5, 44, 62, 198
skateboarding 108–9
skinheads 68
Skinningrove *107*
Smith, Dorothy 55
Smith, Michael Peter 7, 199–200
Smith, Neil 52–4
Smith, Zadie 11, 27, 147, 182
Smithson, Robert 20, 139
social ecology 141
social exclusion 169, 181n; and cars 98–100; strategies 100–1, 104, 105
social relations, radical architecture 119
Society for Ecological Restoration 136
Soja, Edward 11, 50
Somalis 1765–6
Sonfist, Alan 139
Sonoma County 146, 150; anti-homeless policies 155, 156–7; homelessness advocacy and activism 158–62, 164–5; political economic context 150–1, 152, 154; regional activist networks 162–4
Sound Mirrors (Dean) 20
Stasi City (Wilson and Wilson) 19
Staten Island 109, 128
Stepney 147, 179; cultural landscape 174–7; local cultural flagships 177–9
Stewart, Janet 71
Stewart, Theresa 78, 79–81
Strange Days (The Doors) 58–9
Strelow, Heike 139–40
Stuyvesant High School, New York 125–6
Suleri, Sara 41n
Surfaces and Intensities (Gallaccio) 24
sustainability 93, 148, 183–4, 185, 191–2; and cars 100, 103–4
Swallow, Deborah 30–1
Symphony Court, Birmingham 87–8
systems theory, and ecology 139–40

Tafuri, Manfredo 6, 108, 117–18
tailor's chalk 22, 26
Talking Presence (Boyce) 28–30, 33
Tarry, Jacqueline 127–8, 131–2
Tate Modern 17
technology 118–19
Technology Innovation Centre, Birmingham 84–5
Teignmouth Electron (Dean) 20
Three Rivers, Second Nature 109–10
time 17–18, 130
Toon, I. 68–9, 74
Touch Sanitation (Ukeles) 109, 128
Trafalgar Square, London 108
traffic congestion 97
transgression 28, 58
travel reduction strategies 100–5
Trojan Women, The (Euripides) 196
Tropos (Hamilton) 25
Trowell, Jane 198
Trying to find the Spiral Jetty (Crowhurst) 20
turbulence 148, 182–3, 184–6, 195–6; climate change 186–7; future prospects 191–2; geological 189–90; wildlife 187–9
Turner, Frederick 137
Two Sisters (Gallaccio) 24–5

Ukeles, Mierle Laderman 109, 128, 133, 199
Union Square, New York 109, 124–5
United States: mobility and urban form 94, *95*, *see also* Los Angeles; New York; San Francisco; Sonoma County
University of the First Age, Birmingham 84
Untitled (The Trials and Tribulations of Mickey Baker) (Biswas) 36
urban regeneration *see* regeneration
urban sustainability *see* sustainability
urbanism 4, 6–7, 113, 192, 195
Urry, J. 71–2, 93, 98, 99–100
USSR 112
utopia 45–6

van Eyck, Aldo 108
Vanderbeck, R. 74
Varela, F.J. 144
Victoria and Albert Museum 11, 30–1, *32*, 33, *34*, 42n
video 14, 22; Burgin 21–2; Dean 20–1; Wilson and Wilson 19–20
Vitruvius 116–17
volcanos 189–90

Wallis, Clarrie 20
Wallpaper 115
Warde, Alan 5
Water Tower (Whiteread) 23–4
Watson, Sophie 28
Watts, Michael 55
weather 186–7, 189
Wesside, Knowles era 78
Westside, Birmingham 77, 82–4
Where is Home? (Chauda) 31
White, R. 68, 70, 73
White Teeth (Smith) 11, 27, 147, 182
Whitecloth (Hamilton) 25
Whitelegg, J. 104–5
Whiteread, Rachel 14, 17, 22–4, *23*
Wide 14–15
wildlife 187–9
Williams, Raymond 3–4, 197–8
Wilson, Elizabeth 4, 6, 52–3, 57, 63
Wilson, Jane 14, 19–20
Wilson, Louise 14, 19–20

Wirth, Louis 147, 183
Witness (McCallum and Tarry) 127–8, 131–2
Wolin, Richard 10
Woods, Lebbeus 44, 52
World Trade Center 44, 58, 125, 128
World Trade Organization 197
Wyn, J. 68, 70

Yass, Catherine 11, 14, 16–17, 18
Young, Iris Marion 6, 11, 44, 56, 57, 58, 197, 198
young people 62–3, 65–6, 74–5; rebellion 66–7; risk and consumption 69–71; stress of consumption 71–4; subcultures 67–9
Youth Action Scheme, Stepney 178

Zietzsche's Paris (Burgin) 22
Zukin, S. 147